AF588568

PRINCIPLES OF GEODYNAMICS

BY

ADRIAN E. SCHEIDEGGER

PH. D. (TORONTO); DIPL. PHYS. ETH. (ZÜRICH)
PROFESSOR OF PETROPHYSICS
UNIVERSITY OF ILLINOIS, URBANA, ILL. (U.S.A.)

SECOND EDITION

WITH 142 FIGURES

SPRINGER-VERLAG BERLIN HEIDELBERG GMBH
1963

Ursprünglich erschienen bei Springer-Verlag OHG., Berlin · Göttingen · Heidelberg 1963
MyCopy version of the original edition 1963

DOI 10.1007/978-3-662-12781-0

TO

MY WIFE

Preface

Geodynamics is an old science. Most of the basic theories have been conceived in principle during the 19th century and not many fundamental ideas have been added since. Some progress has been made in the following-up of these concepts and, in some instances, in the determination of some important facts about the Earth. Nevertheless, geodynamics has been a highly speculative subject for about a hundred years and it is not likely that this situation will change during the next hundred. It is also unlikely that many *basic* new ideas will be added in that time interval. The reason for this lies in the extreme difficulty of obtaining really relevant data about the mechanics of the Earth, partly due to the impossibility of probing into the depths of the Earth by direct means to any considerable extent and partly due to the fact that the time intervals in which "something happens" are of the order of millions of years, which is much too long for any human being to wait and experiment with.

The situation in geodynamics is therefore much akin to that which existed when the ancient Greek philosophers were speculating about the possibly atomic structure of matter: there was, at that time, absolutely no hope to either confirm or to reject the hypothesis. The subsequent historical developments proved indeed that two thousand years of technological advances were required before the question could be settled. Geodynamics is much in the same position now as the physics of matter was two millennia ago: the basic ideas that one can think of have all been thought of, but there seems to be no chance of settling the fundamental questions for a long time to come.

It seems, therefore, that the time is ripe for an evaluation of the existing ideas in the light of presently available facts. This, in spite of the early recognition of the subject, has never been done. All existing books, monographs and papers (of which there is legion) have been written to advance one or the other of the hypotheses as the "true" one. This led, in consequence, to much wishful thinking, to the inadvertant ignoring of unpleasant facts, and to the straining of others to fit preconceived ideas. The writer admits that he has been guilty of the same offense, falling in with the general trend and type of geodynamic speculations. It was only after much thought and disappointment that he arrived, so to speak, at being an "agnostic" on the subject. It is, however, his conviction that any real advances can only be expected if one starts with such a frame of mind. Otherwise, too much energy is needlessly wasted in the zealous promotion of concepts for which there can realistically be no hope of "proof" in the foreseeable future.

The present book represents, therefore, the writer's notes and ideas on the principles of geodynamics. It is not a comprehensive literature survey, but rather a compilation of the most competent presentations of each one of the,—usually very old—, basically possible hypotheses. Much of the material has been taken from the writer's own earlier attempts at struggling for the "proof" of one or the other of the ideas in which he was inclined to believe at the time, some of it from similar attempts of other geophysicists. He has particularly heavily drawn from those of his earlier articles which appeared in the Canadian Journal of Physics, in the Transactions of the American Geophysical Union, in Geofisica Pura e Applicata, in the Journal of the Alberta Society of Petroleum Geologists, in the New Scientist, in the Journal of Geology (published by the University of Chicago Press), in Canadian Oil and Gas Industries and in the Bulletin of the Geological Society of America. Permission to do this has kindly been granted by the editors of the journals in question and this is here gratefully acknowledged.

The first two chapters of the book give a brief summary of the physical facts about the Earth as far as they are known, the third puts together the principles of the theory of deformation of continuous matter which is the basic mechanical background of geodynamics, and the rest represents a synoptic view of the subject, much in the same way as one might present a synoptic view of the world's philosophies, without taking sides for one or the other.

The present edition is already the second edition of the book; in it, numerous revisions were made. Naturally, the writer is most pleased by the ready acceptance which quantitative methods have found in the Earth Sciences. Between the first and second editions of this book, the writer had an opportunity to write his "Theoretical Geomorphology" and it is hoped that the two books together form a reasonably coherent presentation of theoretical quantitative geology.

The writer is indebted to many colleagues and friends for helpful discussions and stimulating criticism. It was Dr. J. Tuzo Wilson in Toronto who started him off on the subject of theoretical geophysics in the first place. Many of the ideas presented here, especially those on the physiography of the Earth, can therefore be backtraced to Dr. Wilson's stimulating influence. The writer owes much to Dr. J. A. Jacobs, of the University of British Columbia in Vancouver, for the discussion of mathematical matters, and especially to Dr. Egon Orowan of the Massachusetts Institute of Technology whose influence on the writer's conception of fracture and failure has been great. As usual, the Springer-Verlag has been most efficient in producing the present book and has always been willing to accede to the writer's numerous requests.

Urbana, Illinois, U.S.A.,
November 3, 1962. A. E. Scheidegger

Table of Contents

I. Physiographic and Geological Data Regarding the Earth

1.1 Introduction

The science of geodynamics aims at an explanation of those of the present-day surface features of the Earth which are presumably caused by internal (endogenetic) forces. Geodesy, geology and geophysics have accumulated a wealth of information about our globe. Since the Earth is a physical object, it would be quite inconceivable that its present-day physiography would not be the result of well-defined physical processes. Since it is one of the most fundamental postulates of modern science that the laws of physics be universally valid, such changes of the surface features of the Earth as may have occurred, must have taken place in strict conformity with these laws.

The Earth is part of the solar system, the solar system is part of the universe. Ultimately, therefore, the surface features of the Earth are conditioned by the manner in which the evolution of the universe took place. Whether there is a mountain in a particular spot on the Earth's surface depends in the end on how the universe was created. However, it is quite obvious that a mountain on the Earth's surface is only a minute detail in the universe as a whole. It stands to reason, therefore, that processes can be defined which are not too intimately tied up with the universe as a whole, but which would be sufficient to explain the Earth's surface features. That these processes do have *some* connection with the evolution of the universe, is just one more instance demonstrating that there is in reality only one single nature.

Geodynamics confines itself to the study of the Earth's crust. Because of the unity of all nature, reference will have to be made occasionally to conditions above or below the crust, i.e. to conditions obtaining in the universe or in the interior of the Earth. However, we indulge in such diversions only if it is necessary for the understanding of the mechanics of the Earth's crust.

A serious handicap in the study of geodynamics is connected with the fact that it is extremely difficult to encompass geological data in quantitative terms. Traditionally, physical laws can be most easily applied to such phenomena which can be expressed by numbers. On the other hand, geology traditionally has been a descriptive science whose findings cannot easily be encompassed in numbers. Much space in the present book is therefore devoted to the discussion of this difficulty and to the task of abstracting numbers or simple geometrical shapes from the wealth of physiographic facts.

The principal physical processes governing the evolution of the Earth's crust are not yet definitely known. The approach is therefore one of trying out various theories and checking their consequences with regard to features accessible to observation. Sometimes, much mathematics is needed to follow up a particular hypothesis to its ultimate conclusions, particularly if reference has to be made to the mechanics of deformation of continuous matter. It is therefore expected that the reader is familiar with infinitesimal calculus, and in some sections, also with tensor calculus. However, pains have been taken to supply all the necessary *physical* background in sufficient detail to make the book, in this regard, self-contained. In general, the writer aimed at presenting the material in such a fashion that the reader who is interested in a particular topic can seek out the corresponding chapter, read it and understand it if he follows up the cross-references. It will be found that many topics can be understood without the necessity of referring to *all* that has been said on previous pages. Some of the theories have reached only a descriptive stage and can therefore be understood without any reference to mathematical analysis at all.

Although books on geodynamics are not lacking (a partial list is given in References[1–15] below), most of these deal with special physical[1–6] or special geological[7–15] aspects of the problem, and no comprehensive

[1] Aslanyan, A. T.: Исследование по теории тектонической деформации земли. Erevan: Iz-vo Akad. Nauk Armyansk. SSR 1955.

[2] Gutenberg, B.: Physics of the Earth's Interior. New York: Academic Press Inc. 1959.

[3] Haalck, H.: Physik des Erdinnern. Leipzig: Akademische Verlagsgesellschaft 1959.

[4] Jacobs, J. A., R. D. Russell and J. T. Wilson: Physics and Geology. New York: McGraw-Hill Book Co. 1959.

[5] Jeffreys, H.: The Earth, 4th ed. London: Cambridge Univ. Press 1959.

[6] Love, A. E. H.: Some Problems of Geodynamics. London: Cambridge Univ. Press 1912.

[7] Belousov, V. V.: Основые вопросы геотектоники. Moscow: Gos. Iz-vo Nauch. Tekh. Lit. 1954.

[8] Bucher, W. H.: The Deformation of the Earth's Crust, 2nd ed. New York: Hafner 1957.

[9] Hapgood, C. H.: The Earth's Shifting Crust. New York: Pantheon 1958.

[10] Kirsch, G.: Geomechanik. Leipzig: Johann Ambrosius Barth 1938.

[11] Roubault, M.: La génèse des montagnes. Paris: Pr. Univ. France 1949.

[12] Sitter, L. U. de: Structural Geology. New York: McGraw-Hill 1956.

[13] Termier, H., et G. Termier: Formation des continents et progression de la vie. Paris: Masson & Cie. 1954.

[14] Termier, H., et G. Termier: L'évolution de la lithosphère. II. Orogénèse. Paris: Masson & Cie. 1957.

[15] Umbgrove, J. H. F.: The Pulse of the Earth. The Hague: M. Nijhoff 1942.

synthesis of the *whole* problem seems ever to have been attempted. It is hoped, therefore, that the present book will represent a useful work of reference for all those who are interested in the broad aspects of geodynamics.

1.2. Geological Evolution

1.21. The Basic Rock Types. A study of geodynamics of necessity has to start with a review of some basic observational facts about the Earth that have been established by field investigations. The collection and classification of such facts, i.e. the taxonomic part of the Earth sciences, is primarily the domain of geology. Through the incessant efforts of generations of geologists, many facts have been learned about the constitution of the rocks in many parts of the world. For a detailed description of these facts, the reader is referred to any one of the many excellent textbooks on physical geology that are in existence; in the present context, our review must of necessity be held brief.

The appearance of rocks is the result of their geological past. Amongst the great wealth of rock types, however, a broad classification can be made. The two main rock types are *sedimentary* rocks and *igneous* rocks. Sedimentary rocks are separated into more or less distinguishable parallel layers, whereas no such structure is evident in igneous rocks.

Amongst *igneous* rocks, we encounter lava, which may be thought to have been exuded from the deeper parts of the Earth during volcanic activity. Other types of igneous rocks, such as the granites and granodiorites, were at one time thought[1] to have a similar history as lavas, with the difference that the cooling process had a much longer duration and took place at great depth. Hence the name "batholiths" (from Greek *βάθος*, depth and *λίθος*, stone) for masses of such granites found in the interior of mountain ranges. However, the present-day[2] view inclines toward assuming that the batholiths were formed *in situ* by a process called *metamorphose*. In the case of batholiths, this process must have been very complete as it must have involved melting of the present rocks in order to give them the igneous appearance. In other metamorphic rocks, it has been less complete.

The rocks on the surface of the Earth are continuously subject to detrition by the action of wind and water. Ground down by atmospheric influences, the débris is carried in rivers to larger bodies of water where deposition takes place. The accumulation of such débris, under further

[1] NEUMAYR, M.: Erdgeschichte, 3rd ed. by F. E. SÜESS. Leipzig: Bibliographisches Institut 1920.

[2] HOLMES, A.: Principles of Physical Geology. New York: The Ronald Press Co. 1945.

consolidation, gives rise to the sedimentary rocks mentioned above. The process of accumulation itself is called *sedimentation*. Sedimentary rocks, in accordance with their mode of formation, are "stratified". Corresponding types of strata can often be traced to various parts of the world.

One thus arrives at a *cycle* of evolution of rocks. Sedimentary rocks become gradually metamorphosed, possibly even entirely molten, until they have the appearance of igneous rocks. Then the process of detrition starts, the débris is deposited somewhere and eventually, new sedimentary rocks are formed.

The Earth is generally assumed to have begun as a hot, molten body. (For a more detailed discussion of this point, see Sec. 5.12.) If this be true, all "first" rocks must have been igneous. However, no such "first" rocks can be found. It appears that even the oldest known igneous rocks are not "first" rocks, but show signs of having been metamorphosed from even earlier sedimentary rocks (cf. HOLMES[1]). The beginning of the evolution of rocks (except lavas) is therefore not known.

1.22. Geological Time Scale. The fact that sedimentary rocks have been formed by deposition of débris yields a powerful means of dating them, at least relative to each other. During the process of deposition, it is inevitable that living and dead organisms become entrapped which are then preserved as fossils. It is thus possible not only to obtain an idea of the age of a stratum in which a fossil is found, but also to obtain a picture of the evolution of life. A drawback of this method of dating is that it is naturally confined to such times from which traces of life have been preserved to the present day. The traditional geological time scale, therefore, begins with that epoch from which the oldest fossils were found.

Detailed descriptions of the methods of setting up geological time scales have been given, for instance, by HIERSEMANN[2] and by HEDBERG[3]. A recently published traditional time scale is shown in Table 1. The absolute ages shown there are after LONGWELL[4], who made use of all presently known means, including radioactive age determinations (cf. Sec. 2.5).

1.23. Paleoclimatic Data. From a geological investigation of the various sedimentary strata it becomes evident that various parts of the Earth must have undergone large climatological changes. Thus, it is

[1] HOLMES, A.: Principles of Physical Geology. New York: The Ronald Press Co. 1945.

[2] HIERSEMANN, L.: Bergakademie **11**, 370 (1959).

[3] HEDBERG, H. H.: Bull. Geol. Soc. Amer. **72**, 499 (1961).

[4] LONGWELL, C. R.: Geo Times **2**, No. 9, 13 (1958).

Table 1. *Traditional Geological Time Scale.* (After LONGWELL[1]; reprinted by permission from GeoTimes)

Era	System and Period	Series and Epoch	Stage and Age: North America	Stage and Age: Europe	Absolute Age
Cenozoic	Quaternary	Recent			(Duration in years) Approximately the last 10000 years
		Pleistocene	In glaciated regions (Glacial stages in italics)		
			Wisconsin Sangamon *Illinoisan* Yarmouth *Kansan* Aftonian *Nebraskan*	*Würm* Würm-Riss *Riss* Riss-Mindel *Mindel* Mindel-Günz *Günz*	10000 ± to > 35000 years ago
	Tertiary	Pliocene	(Atlantic and Gulf Coast) Upper	(Europe) Astian	(Millions of years ago)
			Lower	Plaisancian	
		Miocene	Upper	Sahelian: Pontian Sarmatian	21
			Middle	Tortonian Helvetian	
			Lower	Burdigalian Aquitanian	
		Oligocene	Upper Middle Lower	Chattian Rupelian Tongrian	
		Eocene	Jackson	Ludian Bartonian	39
			Claiborne	Auversian Lutetian	
			Wilcox	Cuisian Ypresian	
		Paleocene	Midway	Thanetian Montian	60

[1] LONGWELL, C. R.: GeoTimes **2**, No. 9, 13 (1958).

Table 1 (Continued)

Era	System and Period	Series and Epoch	Stage and Age		Absolute Age
			North America	Europe	
Mesozoic	Cretaceous	Upper (Late)	No accepted classification for North America generally	Maastrichtian Campanian Santonian Coniacian Turonian Cenomanian	70 90
		Lower (Early)	These European names commonly used in North America also	Albian Aptian Barremian Hauterivian Valanginian Berriasian	
	Jurassic	Upper (Late)		Purbeckian Portlandian Kimeridgian Oxfordian	140
		Middle (Middle)		Callovian Bathonian Bajocian	
		Lower (Early)		Toarcian Pliensbachian Sinemurian Hettangian	
	Triassic	Upper (Late)		Rhaetian Norian Carnian	
		Middle (Middle)		Ladinian Anisian	
		Lower (Early)		Scythian	
	Permian	(West Texas) Ochoa Guadalupe Leonard Wolfcamp	Not established in North America	(Russia) Kazanian Kungurian Artinskian Sakmarian	220
	Carboniferous Systems: Pennsylvanian	Central North America Virgil Missouri Des Moines Atoka Morrow	Not established	(Europe) Stephanian Westphalian Upper Namurian	
	Carboniferous Systems: Mississippian	Chester Meramec Osage Kinderhook	Not established	Lower Namurian Viséan Tournaisian	

Table 1 (Continued)

Era	System and Period	Series and Epoch	Stage and Age: North America	Stage and Age: Europe	Absolute Age
	Devonian	(Eastern United States)		(Europe)	
		Bradfordian	Conewango	Famennian	
		Chautauquan	Cassadaga		
			Chemung		
		Senecan	Finger Lakes	Frasnian	270
			Taghanic		
		Erian	Tioughnioga	Givetian	
			Cazenovia	Eifelian	
			Onesquethaw		
		Ulsterian	Deerpark	Coblenzian	
			Helderberg	Gedinnian	
	Silurian	Series and Epoch		Stage and Age	
		(North America)	(Britain)		
		Cayugan	Downtonian	Not established	
Paleozoic		Niagaran	Ludlovian		
		Albion	Wenlockian		
			Valentian		
	Ordovician	Series and Epoch	Stage and Age	Series and Epoch	
		(North America)		(Britain)	
		Cincinnatian	Gamachian	Ashgillian	375
			Richmondian		
			Maysvillian		
			Edenian	Caradocian	
		Champlainian	Mohawkian	Llandeilian	
			Chazyan	Skiddavian	
		Canadian	Not established	Tremadocian	
	Cambrian	Upper (Late)	Not established	Not established	440
		Middle (Middle)			
		Lower (Early)			470

well known that during the Pleistocene, large parts of Europe and North America were covered by sheets of ice, a condition which is commonly referred to as "Pleistocene ice age". The stratigraphic evidence for the recognition of this ice age has come from many sources[1]. Most striking are the results from recent investigations of deep sea cores: such cores indicate that an abrupt change in climate may have occurred close to 11000 years ago[2, 3], but this matter is not yet entively certain. Most people assume that the ice started to melt about 20000 years ago. It is

[1] EMILIANI, C.: Ann. N. Y. Acad. Sci. **95**, 521 (1961).

[2] BROECKER, W. S., K. K. TUREKIAN and B. C. HEEZEN: Amer. J. Sci. **256**, 503 (1958).

[3] BROECKER, W. S., M. EWING and B. C. HEEZEN: Amer. J. Sci. **258**, 429 (1960).

possible to interpret the fact that frozen mammoths are still embedded in Siberian permafrost areas, as an indication that the beginning of the ice age was very abrupt, although this matter is also not yet entirely certain[1].

Ice ages have also occurred at other epochs. Best known is the Paleozoic ice age, in which South Africa, Brasil and other areas appear to have been covered by ice caps[2–5]. The reasons for the occurrence of ice ages have not yet been definitely established. Theories have been advanced, for instance, by MILANKOVITCH[6], by EWING and DONN[7, 8], EMILIANI and GEISS[9], LUNGERSGAUSEN[10] and others. A review of the various possibilities has also been given by the writer[11].

If one combines the observations on paleoglaciations, one is led to assuming as a plausible explanation, that the geographic position of the North Pole underwent changes during geologic history.

The first to investigate the climatological evidence comprehensively in this fashion was KREICHGAUER[12]. Later KÖPPEN and WEGENER[13] and KÖPPEN[14] made thorough investigations of paleoclimatic data. This yielded three attempts at a reconstruction of the polar paths which are shown in Fig. 1. The trace of the pole runs in all three attempts from somewhere near Hawaii in the Carboniferous to its present position.

Accordingly, in the Carboniferous, Western Europe and North America would have lain in an equatorial belt of rain forest. In the Permian Epoch they belonged to the adjacent dry zones so that the large deposits of salt found in these regions could be formed. At the same time, the glaciations of Brasil were replaced by forests giving rise to the formation of coal seams there. During the Mesozoic, Europe was in a dry area whereas the pole proceeded through the North East Pacific Ocean. In the Oligocene and Eocene the position of the pole caused the formation of fossil ice in Alaska and Siberia. The large loop

---

[1] FARRAND, W. R.: Science **133**, 729 (1961).

[2] BAIN, G. W.: Yale Sci. Mag. **27**, No. 5 (1953).

[3] BAIN, G. A.: Rept. Int. Geolog. Congr. Norden **21**, Pt. 12, 84 (1960).

[4] MARTIN, R.: Canad. Oil and Gas Ind. **15**, No 10 (1961).

[5] STEHLI, F. G.: Amer. J. Sci. **255**, 607 (1957).

[6] MILANKOVITCH, M.: Kanon der Erdbestrahlung und seine Anwendung auf das Eiszeitenproblem. Belgrade: Kgl. Serb. Akad. 1941.

[7] EWING, M., and W. L. DONN: Science **127**, No. 3307, 1159 (1958).

[8] EWING, M., and W. L. DONN: Science **129**, 463 (1959).

[9] EMILIANI, C., and J. GEISS: Geol. Rdsch. **46**, 576 (1959).

[10] LUNGERSGAUSEN, G. F.: Dokl. Akad. Nauk SSSR. **108**, 707 (1956).

[11] SCHEIDEGGER, A. E.: Theoretical Geomorphology. Berlin-Göttingen-Heidelberg: Springer 1961.

[12] KREICHGAUER, D.: Die Äquatorfrage in der Geologie, 1. Aufl. Steyl 1902.

[13] KÖPPEN, W., u. A. WEGENER: Die Klimate der geologischen Vorzeit. Berlin: Gebr. Bornträger 1924.

[14] KÖPPEN, W.: Meteor. Z. **57**, 106 (1940).

in the polar path during the Pleistocene is suggested by the astonishing finds of plants on the Seymour-Islands which are now covered with ice,

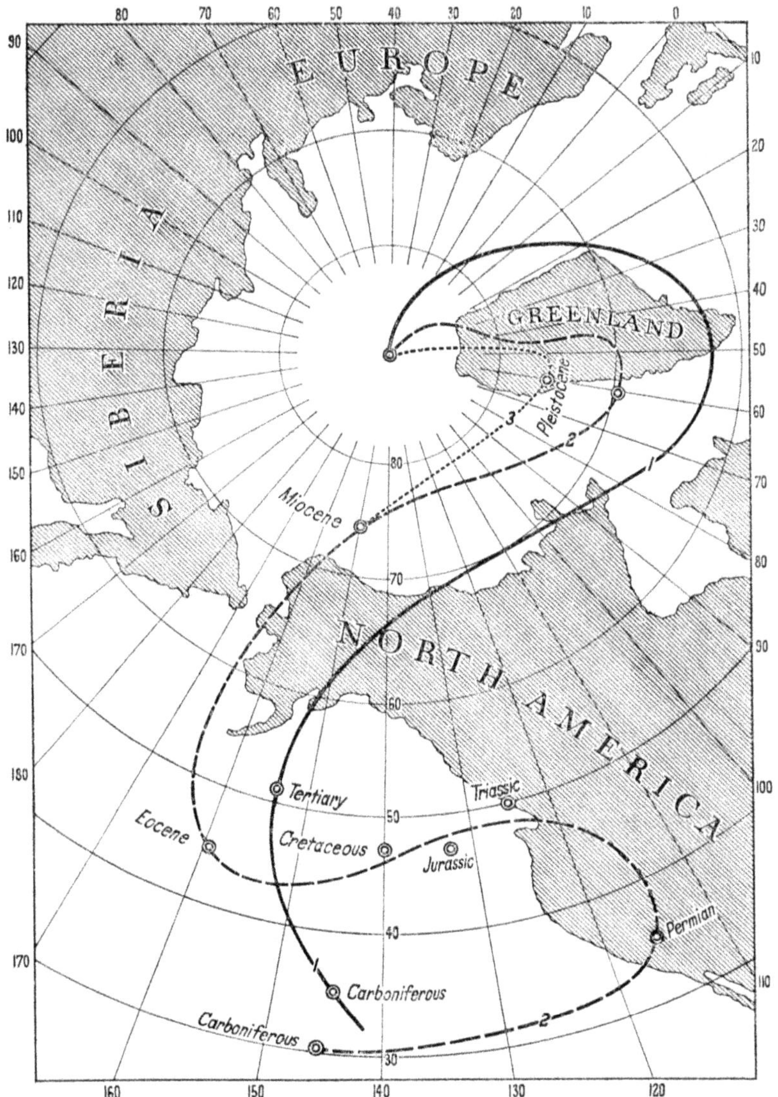

Fig. 1. Path of the North Pole as indicated by paleoclimatology. *1* after KREICHGAUER[1]; *2* after KÖPPEN and WEGENER[2]; *3* after KÖPPEN[3]. (After KÖPPEN[3])

[1] KREICHGAUER, D.: Die Äquatorfrage in der Geologie, 1. Aufl. Steyl 1902.

[2] KÖPPEN, W., A. WEGENER: Die Klimate der geologischen Vorzeit. Berlin: Gebr. Bornträger 1924.

[3] KÖPPEN, W.: Meteor. Z. **57**, 106 (1940).

and of Quarternary sediments containing fossil subtropical mollusci as far south as the Rio Negro in South America. Similarly, the much more extensive glaciation in North America as compared with that of Europe during the Pleistocene ice ages is a pointer in the same direction. The above early investigations have later essentially been corroborated by the results obtained by MA[1], KOMAROV[2], and BLACKETT[3].

It is therefore seen that it is possible to postulate a reasonably coherent path of the pole to explain various geological and climatological observations. However, as the dating of strata relative to each other in different parts of the world is not easy to achieve, it is very difficult to establish as a certainty that glaciation in one part occurred concurrently with tropical growth elsewhere. Therefore, the alteration of glaciations and tropical growth in any one area can also be explained by surmising that the climate of the *whole* Earth underwent such changes wherein the evidence in other areas might be assumed to have been obliterated for one reason or another. Nevertheless, the fact that the path of the pole as postulated by paleoclimatic investigations turns out to be more or less coherent, certainly lends considerable support to the hypothesis of polar wandering.

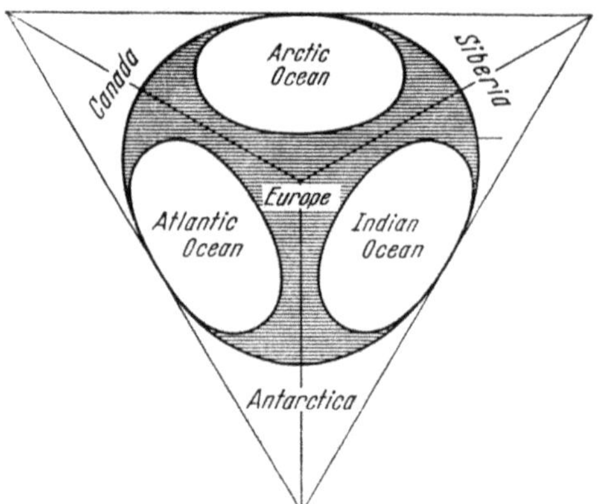

Fig. 2. The tetrahedral distribution of oceans and continents. Pacific ocean on concealed face (after HOLMES[4])

1.3. Geography of Continents and Oceans

1.31. Geometrical Arrangement. An inspection of a globe shows that the latter is mostly covered by the blue color of water. It is a fact that the land areas cover less than one-third of the surface of the Earth, the rest is covered by the sea. Many land areas form chunks of considerable size which are called *continents*. It appears that most of the continents are antipodic to oceans and, with some good will, one may say that four old continental areas have their position, roughly speaking, at the corners of a tetrahedron (see Fig. 2). In addition, the continents are all roughly triangular in shape, touching each other in the North and pointing southwards.

[1] MA, T. Y. H.: Res. Past Clim. and Cont. Drift **15**, 24pp. (1960).
[2] KOMAROV, A. G.: Priroda **1960**, No. 2, 8 (1960).
[3] BLACKETT, P. M. S.: Proc. Roy. Soc. Lond. A **263**, 1 (1961).
[4] HOLMES, A.: Principles of Physical Geology. New York: The Ronald Press 1945.

The question naturally arises whether the above features are the expression of some sort of intrinsic regularity or not. Particularly, one might think that there is a fundamental significance in the fact that continents are antipodic to oceans. However, it has been shown by EVISON and WHITTLE[1] that, if one were to take a series of areas making up one-third of the surface of a sphere, and if one were then to place these areas on the sphere in a random fashion, the most probable position of these areas would, in fact, be non-antipodic to each other (i.e. antipodic to the non-covered areas).

Fig. 3. BAKER's[2] composition of the continents. After DU TOIT[3]

A remarkable observation is that the continental structures on the Earth's surface can be made to fit together rather well like a jigsaw puzzle. The fit of the Western shore of Africa with the Eastern shore of South America is quite obvious (although even this has been questioned by JEFFREYS[4]; JEFFREYS' remarks, however, were contradicted by CAREY[5]), but the rest of the continents can also be made to fit with more or less ease. This had already been observed as early as in 1911

[1] EVISON, F. F., and R. WHITTLE: Geol. Mag. **98**, 377 (1961).
[2] BAKER, H. B.: See DU TOIT[3].
[3] DU TOIT, A. L.: Our Wandering Continents. Edinburgh: Oliver & Boyd 1937.
[4] JEFFREYS, H.: The Earth, 2nd ed. London: Cambridge Univ. Press 1952.
[5] CAREY, S. W.: Geol. Mag. **92**, 196 (1955).

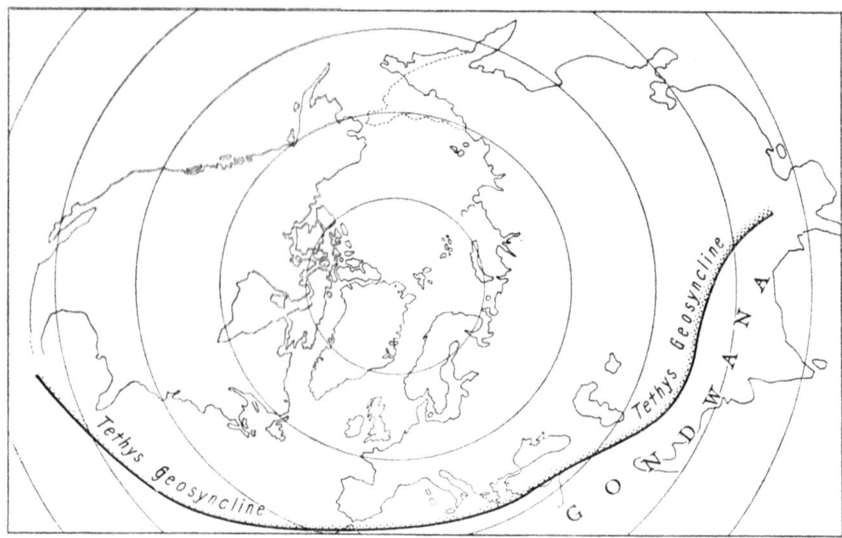

Fig. 4. Laurasia as envisaged by Du Toit[1]

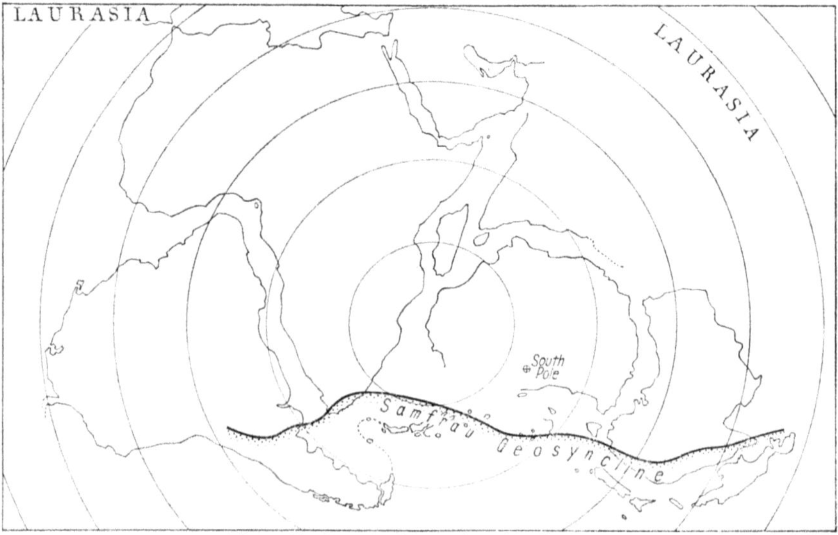

Fig. 5. Gondwanaland as envisaged by Du Toit[1]

by Baker[2] who showed the composition of the continents reproduced in Fig. 3. However, more recently (cf. Du Toit[1]) it has been the practice to fit the Earth's continents together into *two* groups, called *Laurasia* and *Gondwanaland*, rather than into one as done by Baker. Laurasia is the complex of Europe, Asia and North America, which is even at

[1] Du Toit, A. L.: Our Wandering Continents. Edinburgh: Oliver & Boyd 1937.
[2] Baker, H. B.: See Du Toit[1].

the present time not very widely dispersed; Gondwanaland is the combination of all the southern continents fitted together. In this sense, one arrives at the picture shown in Figs. 4 and 5. The present-day distribution of the pieces of Laurasia and Gondwanaland is shown in Fig. 6.

From the artificial arrangement of the continents into two big blocks, it is an easy step to postulating that the continents actually *were* formed originally as such blocks and that they subsequently "broke up" and "drifted" into their present position[1]. We shall

Fig. 6. The continental shields making up the two supercontinents of Laurasia and Gondwanaland (after Du Toit[2])

discuss the dynamical possibilities for this having occurred later, and at the present time only mention the physiographic evidence bearing thereupon as exhibited by the fit of the continents. In addition to this physiographic indication, many geological data have been collected[1-3] with the intention to find features common to the various continents which might indicate whether or not and if so, when the continents moved apart from the two original blocks. This evidence is of course somewhat problematic, but it is possible to state the following points in favor of the hypothesis of continental drift:

(a) The orogenetic activity in the southern continents is localized in a belt that can be followed continuously through Gondwanaland as the "*Samfrau*" geosyncline (see Fig. 5).

[1] Wegener, A.: The Origin of Continents and Oceans. Translated from 3rd German ed. by J. G. A. Skerl. London: Methuen 1924.

[2] Du Toit, A. L.: Our Wandering Continents. Edinburgh: Oliver & Boyd 1937.

[3] Wegmann, C. E.: Medd. om Grønland **144**, No. 7.

(b) Glaciation in the Carboniferous and Permian era seems to radiate from a point corresponding to the position of the South Pole as postulated by KÖPPEN (cf. Sec. 1.23) for that epoch, but appears to cover parts of the southern continents in such a fashion as to suggest that the latter were close together at that time.

(c) Paleobiological evidence seems to indicate that the southern continents had, even in comparatively recent times, some land connection between each other. Otherwise the simultaneous occurrence of e.g. marsupalia in South America and Australia would appear as difficult to explain. Another example of this kind is the distribution of the scorpionidae (cf. DU TOIT[1]).

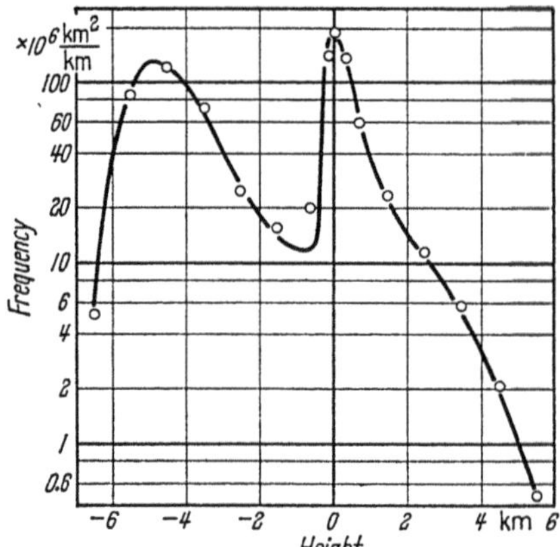

Fig. 7. The hypsometric curve

From the arguments given above it appears that, say, the pieces of Gondwanaland may have drifted some 14000 km since the Carboniferous.

Attempts have been made to determine by geodetic means whether continents are "drifting" in modern times. Unfortunately, the evidence is inconclusive.

1.32. The Hypsometric Curve. An informative way to represent the distribution of continents and oceans is obtained by calculating the percentage of the Earth's surface above or below a certain height-level. By differentiation, this leads to a statistical distribution curve of heights which indicates what percentage of the Earth's surface lies at a certain level. This distribution curve has been termed *hypsometric curve* of the Earth. It is shown in Fig. 7. The data defining the hypsometric curve (after KOSSINNA[2]) are shown in Table 2.

With the hypsomentric curve one can do some statistical analyzing. From the data of Table 2 it is at once obvious that the distribution of heights has two maxima, corresponding to the mean levels of the continents and that of the ocean floors. However, a more exact investigation of the data by JOKSCH[3] showed that the hypsometric curve is not the sum of two, but of *three* elementary distributions. The constituent distributions are logarithmico-normal defined by the equation

$$w(x) = \frac{1}{s(x-a)} \frac{1}{\sqrt{2\pi}} \exp\left\{-\frac{[\log\text{nat}(x-a)-m]^2}{2s^2}\right\}. \quad (1.32\text{–}1)$$

[1] DU TOIT, A. L.: Our Wandering Continents. Edinburgh: Oliver & Boyd 1937.

[2] KOSSINNA, E.: Die Erdoberfläche. In Handbuch der Geophysik, Bd. 2, S. 875. 1933.

[3] JOKSCH, H. C.: Z. Geophys. **21**, 109 (1955).

Table 2. *Hypsography of the Earth.* (After KOSSINNA[1])

Level (km)	Area (10^6 km^2)	%	% above level	% below level	Level (km)	Area (10^6 km^2)	%	% above level	% below level
∞			0.0	100.0	−1			37.6	62.4
	0.5	0.1				15.2	3.0		
5			0.1	99.9	−2			40.6	59.4
	2.2	0.4				24.4	4.8		
4			0.5	99.5	−3			45.4	54.6
	5.8	1.1				70.8	13.9		
3			1.6	98.4	−4			59.3	40.7
	11.2	2.2				119.1	23.3		
2			3.8	96.2	−5			82.6	17.4
	22.6	4.5				83.7	16.4		
1			8.3	91.7	−6			99.9	1.0
	105.8	20.8				5.0	1.0		
0			29.1	70.9	−∞			100.0	0.0
	43.7	8.5							

Here, a is called the vanishing point of the distribution. Instead of using the parameters m and s above, it is often convenient to use the "median" M defined by

$$\int_a^M w(x)\,dx = 0.5 \tag{1.32-2}$$

and a "normal deviation" S indicating that interval of x around the median which contains 90% of all the area underneath the distribution curve.

In the above terms, JOKSCH showed that the hypsometric curve can be represented as follows (all lengths in km):

(a) by a logarithmico-normal distribution of weight 61/100 with $a = -7.5$, $M = -4.5$ and $S = 3.7$

(b) by a logarithmico-normal distribution of weight 23/100 with $a = -0.5$, $M = 0.2$, $S = 1.2$

(c) by a logarithmico-normal distribution of weight 16/100 with $a = -7.5$, $M = 0.5$ and $S = 3.5$.

The tripartite composition of the hypsometric curve suggests that the three levels might have been created each by an individual process of the same nature.

Hypsometry refers all measured heights to the datum given by the "mean sea level". There is much geological evidence that various parts of the continents were inundated by the sea at one time or another. It is not entirely clear whether this was due to the fluctuation of the sea level itself (i.e. the amount of water in the oceans) or to a change in elevation of the continents. Both factors have probably been involved.

[1] KOSSINNA, E.: Die Erdoberfläche. In Handbuch der Geophysik, Bd. 2, S. 875. 1933.

There is evidence that even at the present time certain parts of the world are rising and others are sinking with regard to each other (cf. Sec. 1.75). This might be an indication of the occurrence of small crustal "undations".

If the total quantity of water in the hydrosphere is assumed as constant, then one can try to calculate the change in the Earth's radius that would have been necessary to cause the inundations mentioned above. EGYED[1] has done this. Using data compiled by STRAKHOV[2], he arrived at the result that the total coverage of the Earth's surface by water is *decreasing*. This would correspond to an increase of the Earth's radius during geological time. Using the present-day hypsometric curve of the Earth, EGYED obtained that an increase of the Earth's radius by 0.5 mm/year would be a satisfactory figure. However, it is by no means certain that the hypsometric curve stayed constant during geologic time. This completely obviates EGYED's argument.

1.33. A Comparison of Continents and Oceans. From the few brief remarks made thus far, it is evident that there is a fundamental geological difference between continents and oceans[3]. The uniform great depth of the ocean basins alone serves as an indication that continental and oceanic areas are not alike. Naturally, continental "structure" may continue for some distance off-shore as a "shelf", but the transition from continental to oceanic structure is rather abruptly marked by the sudden sloping of the shelf so as to reach the depths of the oceanic abysses over a short distance.

A true comparison between continents and oceans can be made only after the various geological and geophysical aspects of the two areas have been reviewed. This will be done below.

1.4. Physiography of Continental Areas

1.41. General Features. Turning first to continental areas, we note that the most conspicuous irregularities thereon are undoubtedly mountain ranges. It will of course be necessary to consider as continental mountain ranges also such occurrences as offshore island chains in the sea; the latter are nothing but the peaks of submerged mountains.

If one examines the mountains somewhat more closely, a few remarkable facts become apparent. Thus, we observe that mountains are not scattered at random over the Earth's surface, but, firstly, that they occur in ranges, secondly, that the ranges themselves form chains, and thirdly that the chains of ranges seem to form world-wide systems. The series of mountains which belong to such one world-wide system are referred to as belonging to one *orogenetic system*.

[1] EGYED, L.: Geofis. Pura Appl. **33**, 42 (1956).

[2] STRAKHOV, N. M.: Основы исторической геологии. Moscow 1948.

[3] BULLARD, E. C.: Proc. Roy. Soc. Lond., Ser. A **222**, 403 (1954).

If the above features of orogenetic systems are real, it will be necessary to obtain a physical explanation therefor. However, it is obviously not easy to prove the reality of these features: mountain chains are never perfectly regular, and therefore the existence of the features mentioned above is always based to a certain extent on intuition. Nevertheless, mathematical investigation does bear out certain regularities which will be discussed below.

Geological investigation of the rocks on the Earth's surface shows that mountain building *(orogenesis)* seems to occur in *cycles:* A mountain chain appears to be built up, and gradually it seems to be worn down by the erosive action of wind and water[1]. The final remnant of a once mountainous area after eons of detrition assumes the appearance of a *shield:* rocky knolls alternate with puddles of water.

It appears that, since the Paleozoic epoch, there have been at least two orogenetic cycles, the one referred to as the "Appalachian-Caledonian-Herzynian" cycle which took place at the end of the Paleozoic, the second called "Alpine-Himalayan-Circumpacific" cycle which started at the end of the Mesozoic and has still not yet come to an end. Sometimes, the Paleozoic cycle is counted as two but it seems preferable to count it as one for the present purpose.

In talking about an orogenetic cycle it should be understood that it is not implied that orogenesis occurred in a manner resembling a catastrophe. It seems that orogenetic belts grew at their edges, being active at different times in different parts of the world, until a new system started to develop somewhere else. UMBGROVE[2,3] was probably one of the chief proponents of such periodicity and synchronism of orogenetic activity. However, a more extreme view has been taken by STILLE[4,5], according to whom short, worldwide and synchronous orogenetic phases would alternate with long periods of acquiescence. This, however, seems to be open to criticism[6] as it is difficult to establish an exact correlation in time for orogenetic movements in widely separated regions of the Earth. Thus, as suggested above, orogenetic activity does not seem to be confined to single, short diastrophisms, but rather seems to occur during long periods of general tectonic unrest.

The development of a mountain system in an orogenetic cycle has been connected with the sequence of the occurrence of various "phases

---

[1] WEGMANN, E.: Rev. Géogr. phys. et géol. dyn., Sér. II **1**, 3 (1957).

[2] UMBGROVE, J. H. F.: The Pulse of the Earth. The Hague: M. Nijhoff Publ. Co. 1947.

[3] UMBGROVE, J. H. F.: Amer. J. Sci. **248**, 521 (1950).

[4] STILLE, H.: Grundfragen der vergleichenden Tektonik. Berlin: Gebr. Bornträger 1924.

[5] STILLE, H.: Einführung in den Bau Amerikas. Berlin: Gebr. Bornträger 1940.

[6] RUTTEN, L. M. R.: Bull. Gol. Soc. Amer. **60**, 1755 (1949).

of the orogeny"[1-8]. At the beginning of a cycle, one generally assumes that a *geosyncline*[9,10] is formed, usually thought to occur at a continental margin[11]. By this is meant that a trough would develop in the Earth's crust where the orogenetic belt would finally appear. The through would then be filled-in with sediments which would eventually be lifted up by some mechanism to form the mountains. The occurrence of a geosyncline is an entirely hypothetical assumption, but it is made plausible by the observation that, in mountain belts, the thickness of the sediments must have been tremendous.

It is not known how many orogenetic cycles occurred before the beginning of the traditional geological time scale. With the absence of fossils it becomes increasingly difficult to trace the various orogenetic movements. It is certain, however, that numerous cycles *did* occur during the last 2 billion years, the order of magnitude of their number is commonly given as eight. Before 2 billion years ago, indications are that orogenesis occurred in a different mode than that described above (cf. Sec. 2.52). A discussion of the means for arriving at the above number of cycles will be given in Sec. 2.5. In all, it is reasonable to assume that about 10 orogenetic cycles occurred altogether since the beginning of present-day type orogenesis 2 billion years ago.

1.42. Mountain Ranges. It is a striking feature[12] of many ranges of mountains and islands of our planet that they have the shape of curved arcs which at first sight appear more or less circular[13-15]. The most outstanding examples of this kind are the arcs of the Philippines, Riu Kiu, Japan, Kuriles and Aleutians. Similar features are equally found in other places of the world. The curved structure of Persia and the Himalayas is obvious, and so is the curved structure of the mountains of the Pacific Coast of British Columbia, the United States and México.

---

[1] Bemmelen, R. W. van: Mountain Building. The Hague: M. Nijhoff 1954.
[2] Bubnoff, S. N.: Byull. Mosk. Obshch. Ispyt. Prirody **63**, Otdel. Geol. **33**, No. 1, 3 (1958).
[3] Gidon, P.: Bull. Soc. Géol. France, Sér. VI, **7**, 125 (1957).
[4] Khain, V. E.: Proc. 21st Int. Geol. Congr. Norden Pt. **18**, 215 (1960).
[5] Kraus, E. C.: Geol. Rdsch. **50**, 292 (1960).
[6] Kraus, E. C.: Proc. 21st Int. Geol. Congr. Norden Pt. **18**, 236 (1960).
[7] Pavoni, N.: Vjschr. naturforsch. Ges. Zürich **105**, 181 (1960).
[8] Schmidt, E. R.: Ann. Inst. Geol. Publ. Hung. **49**, 931 (1959).
[9] Kraus, E.: Z. dtsch. geol. Ges. **106**, 431 (1954).
[10] Bleissner, M. F., and C. Teichert: Amer. J. Sci. **245**, 465 and 482 (1947).
[11] Brockamp, B.: Geologie **4**, 363 (1955).
[12] Cf. Scheidegger, A. E., and J. T. Wilson: Proc. Geol. Ass. Can. **3**, 167 (1950).
[13] Cf. Lake, P.: Geogr. J. **78**, 149 (1931).
[14] Umbgrove, J. H. F.: The Pulse of the Earth, 2nd ed. The Hague 1947.
[15] Eardley, A. J.: Proc. 8th Pac. Sci. Congr. **2**A, 677 (1956).

Let us study somewhat more closely the actual shape of the island arcs and mountain belts that LAKE[1] and others have assumed to be circular. By means of spherical trigonometry it is easy to calculate the center and radius of a circle determined by 3 points on a sphere. Thus, we shall take on each island arc 5 or more points and put the circles through each three subsequent ones. If the arcs are circular, the radii of those circles should be approximately equal and the centres more or less in the same geographical region. For getting an average center one may take the circle through the first, third and fifth points of the five given ones.

In this manner one can investigate the geographical data for several island arcs and mountain chains. The results for five such arcs are tabulated in the attached Tables 3a–e. The geographical data have been chosen for geological reasons (volcanic regions). In the island arcs there was little ambiguity as to which points should be taken; islands with great volcanoes were usually preferred if there were several possibilities. In Alaska the

[1] Cf. LAKE, P.: Geogr. J. 78, 149 (1931).

Table 3a. *The Southern Andes*

No.	Points	Latitude	Longitude
1	Maipu	34° 10′ S	69° 50′ W
2	N. Llayamas	38° 50′ S	71° 30′ W
3	S. Llayamas	46° 40′ S	73° 30′ W
4	Mt. Burney	52° 20′ S	73° 20′ W
5	Small Island NW of Cape Horn	55° 40′ S	67° 40′ W

Centers and radii of the circles through three of the above points

Circle	Center latitude	Center longitude	Radius
123	35° 30′ S	15° 10′ W	44° 20′
234	43° 0′ S	43° 20′ W	20° 30′
345	48° 50′ S	63° 10′ W	7° 20′

Table 3b. *The West Indian Arc*

No.	Points	Latitude	Longitude
1	St. George	12° 0′ N	61° 40′ W
2	St. Vincent	13° 20′ N	61° 10′ W
3	Martinique	14° 50′ N	61° 10′ W
4	Gouadeloupe	16° 10′ N	61° 40′ W
5	St. Christopher	17° 30′ N	62° 50′ W

Centers and radii of the circles through three of the above points

Circle	Center latitude	Center longitude	Radius
123	13° 35′ N	62° 55′ W	2° 0′
234	13° 50′ N	65° 50′ W	4° 30′
345	13° 50′ N	66° 0′ W	4° 50′
135	14° 0′ N	65° 40′ W	5° 30′

Table 3c. *The Aleutian Arc*

No.	Points	Latitude	Longitude
1	Komandorskii	55° 20′ N	167° 10′ E
2	Chugul	52° 0′ N	178° 10′ E
3	Mt. Unimak	54° 40′ N	164° 0′ W
4	Mt. Kukak	58° 30′ N	154° 20′ W
5	Talkeetna Mountains	62° 05′ N	148° 40′ W

Centers and radii of circles through three of the above points

Circle	Center latitude	Center longitude	Radius
123	62° 10′ N	177° 40′ W	10° 30′
234	66° 50′ N	178° 20′ E	14° 50′
345	64° 40′ N	178° 50′ E	14° 30′
135	65° 30′ N	177° 10′ W	12° 40′

Table 3d. *The Riu Kiu Arc*

No.	Points	Latitude	Longitude
1	Yaku Shima	30° 20′ N	130° 30′ E
2	Amame-o-shima	20° 20′ N	129° 20′ E
3	Sesoke-shima	26° 40′ N	128° 0′ E
4	Omoto Dake	24° 30′ N	124° 10′ E
5	Formosa	24° 20′ N	121° 10′ E

Centers and radii of circles through three of the above points

Circle	Center latitude	Center longitude	Radius
123	36° 30′ N	110° 20′ E	18° 0′
234	31° 10′ N	122° 0′ E	6° 50′
345	31° 20′ N	122° 0′ E	7° 0′
135	33° 0′ N	121° 10′ E	8° 40′

Table 3e. *The Kurile Arc*

No.	Points	Latitude	Longitude
1	Khrebet Khardimskii	56° 40′ N	161° 50′ E
2	Sopka Koryatskaya	53° 20′ N	158° 40′ E
3	Onnekotan-to crater	49° 20′ N	154° 40′ E
4	Shimushira Daka	46° 50′ N	151° 50′ E
5	Berutarube Yama	44° 30′ N	147° 0′ E

Centers and radii of circles through three of the above points

Circle	Center latitude	Center longitude	Radius
123	56° 40′ N	99° 50′ E	33° 40′
234	57° 0′ N	101° 0′ E	32° 10′
345	53° 20′ N	140° 10′ E	9° 50′
135	59° 0′ N	102° 50′ E	31° 0′

calculations were made first with another point (Mt. Russell) which, however, proved to be too much out of the range where the continuation of the arc 1234 might have been expected. Therefore, Mt. Russell may be assumed to belong to another arc.

An inspection of the above Tables 3a–e shows that the radii of the circles through three subsequent points (which approach the radii of curvature of the arc in the corresponding region) are increasing quite definitely towards one end of each arc, in a way that they become twice or three times as great as at the other end. Thus the island arcs and mountain chains are rather of the shape of a spiral than that of a circle; if one assumes them to be circles, one can, of course, reach no well defined locus for their "centers".

A further interesting feature of mountain ranges is that some appear to be associated with a significant crustal shortening taking place in their neighborhood. The amount of such crustal shortening can be estimated simply by a direct measurement of the strata in the great mountain systems. If one assumes that in a normal cross-section of a mountain range the length of the section of a stratum (which is a curved line) is equal to the length of that section before it was folded, i.e. when it was flat on the ground, one can determine how much shortening must have taken place. This, of course, assumes that the strata have undergone no deformation of area but only one of shape. Estimates of shortening obtained in this manner are quoted to be 50 to 80 km in the Appalachians, 40 to 100 km in the Rocky mountains of Canada[1]

[1] North, F. K., and G. G. L. Henderson: Alberta Soc. Pet. Geol. Guidebook **4**, 15 (1954).

and 17 km in the Coast Range of California[1]. For the Alps, HEIM[2] quotes 240 to 320 km. Compared with the assumed (unfolded) cross section of a mountain range, these values represent a shortening of up to 4:1.

1.43. Systems of Mountain Ranges. An inspection of any physiographic chart of the world shows that mountains not only occur in arcuate single ranges, but quite generally that orogenetic activity is concentrated in narrow belts. Each such belt consists of a string of arcs and reaches almost completely around the Earth. One of the most comprehensive studies of these orogenetic belts has been made by WILSON[3,4].

Accordingly, the geographical "arcs" discussed in the last Section, belong to several physiographically distinct classes. WILSON[3] discerns the following:

(i) *Single Island Arcs* which are uniformly curved chains of volcanoes with an oceanic foredeep. The Aleutians and South Sandwich Islands are examples.

(ii) *Double Island Arcs* which develop where single island arcs approach continents. In them an outer chain of islands composed of folded sedimentary rocks replaces the foredeep. Examples of such islands are Kodiak, Timor and Trinidad.

(iii) *Single Mountain Ranges* which have the same essential features as single island arcs, but due to interaction of forces and the abundance of sediments it is the volcanic arc which has grown while the foredeep has remained. The Andes of northern Chile and of Peru are examples.

(iv) *Double Mountain Ranges* which correspond to double island arcs in having a volcanic and predominantly igneous range on the concave side of a parallel range which is predominantly sedimentary. The Sierra Nevada and the Coast Range of California provide examples.

(v) *Fractured Arcs* which are complex features where only some disorganized aspects of arcuate structure exist, such as a deep ocean trench and scattered active volcanoes. If a simple arcuate pattern ever existed, it has been torn apart by faulting. The Melanesian "arcs" from the Philippines to New Zealand provide examples.

In the orogenetic belts the various types of arcs are joined together and it is a significant fact that there are only five types of junctions which are called Linkages, Common Deflections, Capped Deflections, Fractured

[1] PIRSON, L. V., and C. SCHUCHERT: Textbook of Geology. New York: Wiley 1920.

[2] HEIM, A.: Geologie der Schweiz. Leipzig 1921.

[3] WILSON, J. TUZO: Proc. Geol. Ass. Can. 3, 141 (1950).

[4] WILSON, J. TUZO: In: The Earth as a Planet, ed. Kuiper, p. 138. Chicago, Ill.: Univ. Chicago Press 1954.

Deflections and Reversed Arcs (see Fig. 8). WILSON[1] describes them as follows.

(i) *Linkages* are characterized by the extension of one of the arcs past the junction in a nearly straight line. Examples are found in the northeast Asian island arcs.

(ii) *Common deflections* are formed where two mountain arcs meet at an obtuse angle. Secondary arcs then adjoin them closely. An example is near Santiago de Chile in the Andes.

(iii) *Capped deflections* are a special case of common deflections which occur where two arcs meet at an acute angle. The secondary arc then

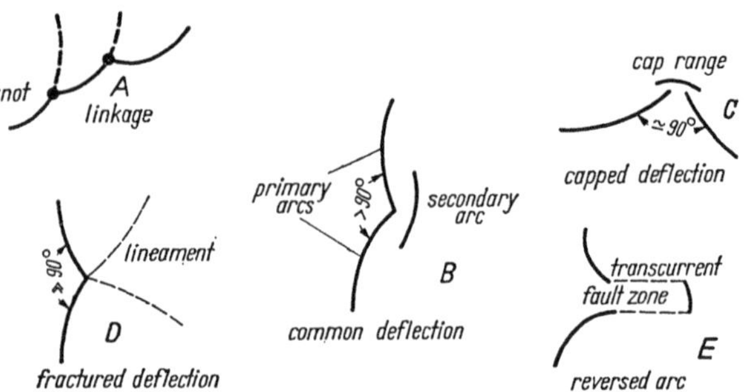

Fig. 8. The five types of junctions of arcs after WILSON[1]

is much more highly metamorphosed than in common deflections. It also is generally sharply curved. The Alps are an example.

(iv) *Fractured deflections* are formed where two mountain arcs meet at an obtuse angle if a secondary arc develops at a distance from the two arcs in conjunction so that it is separated by a wide medianland. Two conspicuous lineaments then spring from the junction, cross the medianland as shear zones and mark the end of the secondary arc. An example occurs near Seattle in the United States of America.

(v) *Reversed arcs* are found in certain instances where it appears that in a chain one arc is suddenly reversed so that it is facing the others. The ends of the arcs are connected by transcurrent fault zones. The West Indian Arc is an example.

Finally, if we are looking at the systems of mountain arcs at large, we note that the belts referred to above reach around the whole Earth. The Mesozoic-Cenozoic orogenetic systems form two large belts, one around the Pacific and the other through the Alpine and Himalayan systems to Oceania (see Fig. 9). These belts can be traced very easily.

[1] WILSON, J. TUZO: In: The Earth as a Planet, ed. Kuiper, p. 138. Chicago, Ill.: Univ. Chicago Press 1954.

They nearly follow two great circles intersecting each other at right angles. It may be suspected that the Paleozoic mountains formed a similar system, but it is not so easy to trace it owing to subsequent erosion and

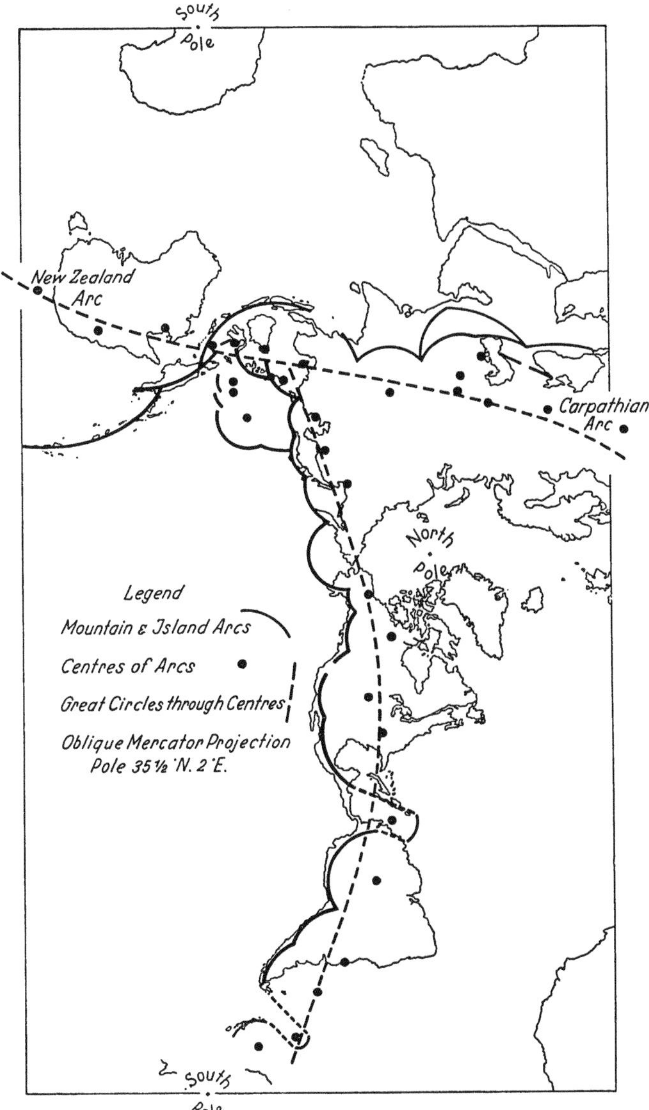

Fig. 9. The mountain and island arcs of the recent orogenetic system. After WILSON[1]

[1] WILSON, J. TUZO: In: The Earth as a Planet, ed. KUIPER, p. 138. Chicago, Ill.: Univ. Chicago Press 1954.

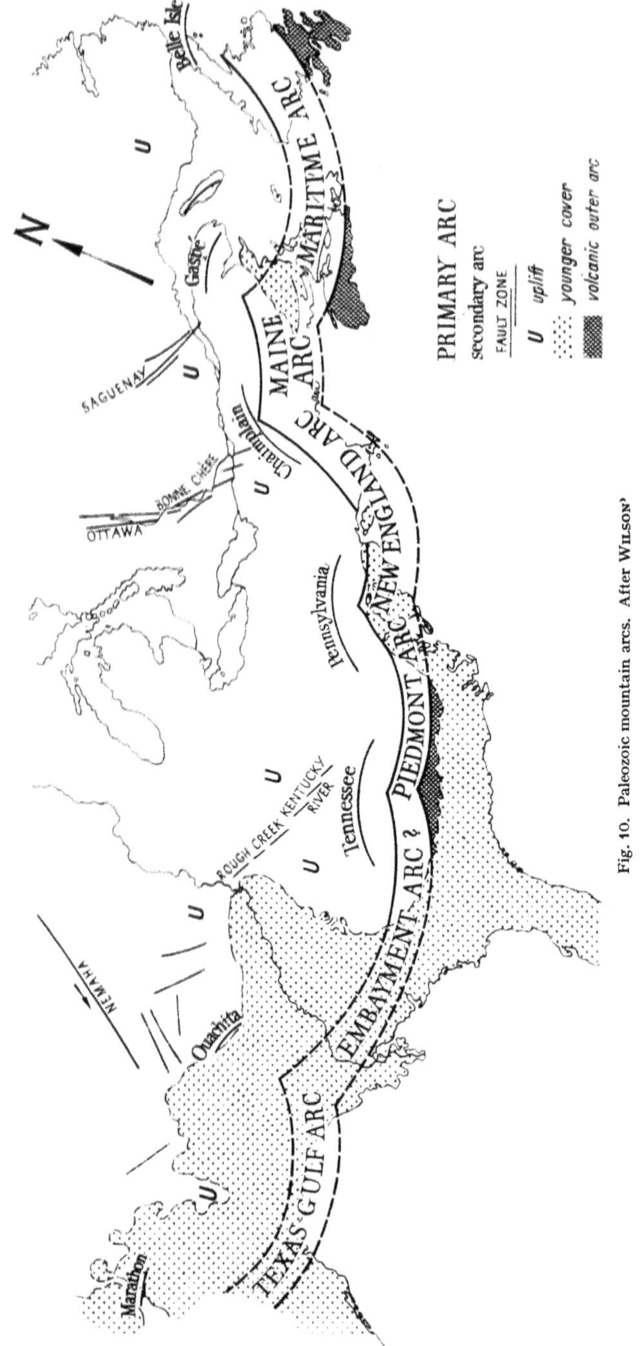

Fig. 10. Paleozoic mountain arcs. After WILSON[1]

[1] WILSON, J. TUZO: In: The Earth as a Planet, ed. KUIPER, p. 138. Chicago, Ill.: Univ. Chicago Press 1954.

recent orogenetic activity. Nevertheless, WILSON[1] has traced part of it in North America with the result shown in Fig. 10.

It thus appears that the main orogenetic activity of the Earth has been concentrated at any one time in narrow belts. The orogenetic belts form world-wide systems. The recent ones very nearly follow two great circles and the contention is that *every* orogenetic system follows this pattern; it should be noted, however, that it is very difficult to substantiate this for the older systems.

In conclusion of the present Section, it should be remarked that the view of the physiography of orogenetic systems presented above, is not the only one that has been advanced in the literature. Some of the different views will be presented in Sec. 1.45.

1.44. The Continental Shelf. The margin of continents, where oceanic and continental areas meet, are of utmost importance for geodynamic considerations[2-4]. As noted earlier, the continental structures may continue for some distance out to sea, but, eventually, a sharp drop in height occurs (continental slope), and the structure then is no longer continental. Beyond the continental slope, one may find a very deep trench[5-7] where some of the greatest ocean depths known may be reached. Just before the descent to the trench, much volcanic activity may be observed. A bathymetric chart of a typical continental margin is shown in Fig. 11.

The continental margins may be areas of considerable orogenetic activity. It has been thought that the formation of a trench might be the initial phase of the development of a geosyncline. As the trench becomes filled-in, the tremendous thicknesses of sediments found in orogenetic belts may become established. Finally, as the orogenesis occurs, it would appear that an addition is thereby created to the previously existing continent. Thus, one arrives at the idea of "*continental growth*" (see e.g. WEEKS[8]); the way in which a continental margin may grow is illustrated in Fig. 12.

1.45. Different Views of Orogenesis. A view of the geometrical characteristics of orogenetic system *different* from that presented above has been taken by various people. Thus, some authors sought to find some

[1] WILSON, J. TUZO: In: The Earth as a Planet, ed. Kuiper, p. 138. Chicago. Ill.: Chicago Univ. Press 1954.

[2] EMERY, K. O.: Proc. 8th Pac. Sci. Congr. **2**A, 810 (1956).

[3] DIETZ, R. S.: Proc. 8th Pac. Sci. Congr. **2**A, 815 (1956).

[4] DAY, A. A.: Deep Sea Res. **5**, 245 (1959).

[5] BRODIE, J. W., and T. HATHERTON: Deep Sea Res. **5**, 18 (1958).

[6] ZEIGLER, J. M., W. D. ATHEARN and H. SMALL: Deep Sea Res. **4**, 238 (1957).

[7] UDINTSEV, G. B.: Priroda **1955**, No. 12, 79 (1955).

[8] WEEKS, L. G.: Bull. Amer. Ass. Petrol. Geol. **43**, 350 (1959).

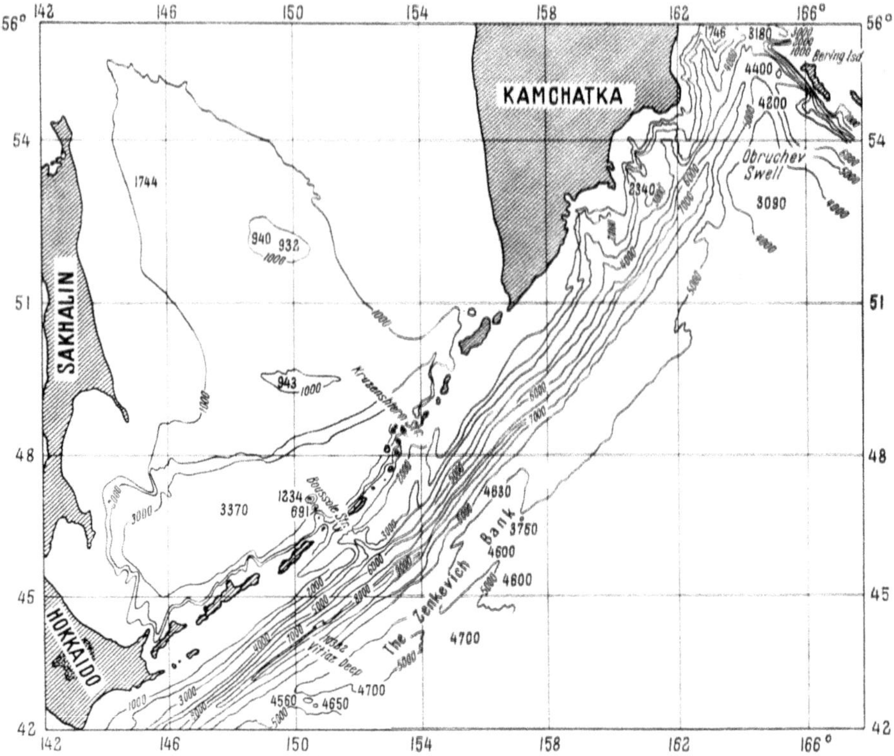

Fig. 11. Bathymetric Chart of Kurile-Kamchatka Trough. After UDINTSEV[1]. Depths in meters

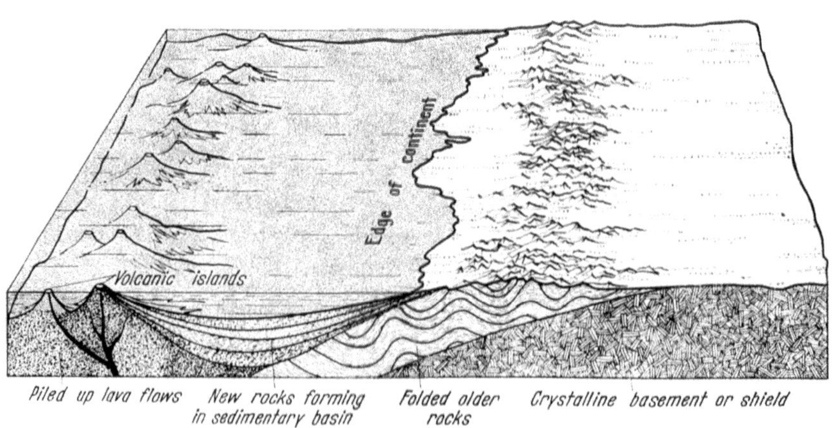

Fig. 12. The various ways in which a continent may grow, shown schematically (from SCHEIDEGGER[2])

---

[1] UDINTSEV, G. B.: Priroda **1955**, No. 12, 79 (1955).

[2] SCHEIDEGGER, A. E.: New Scientist **12**, 227 (1961).

regular linear patterns in the arrangement of orogenetic elements. Generally, these patterns have been connected with "*shear lines*" which would be part of a shear system encompassing the whole world. Accordingly, all orogenetic movements would be tangential along straight lines (i.e. great circles) connected with the overall shear pattern with only minor displacements occurring vertically. The first to seek such patterns was probably HOBBS[1]. The idea has later been taken up by many writers, particularly by VENING MEINESZ[2] and by MOODY and HILL[3]. Accordingly, a geosyncline would not at all consist of a downwarping of the crust, but of a break (transcurrent fault, see Sec. 1.61) with a substantial horizontal and minor vertical movement. This yields, superficially, the appearance of a trench. Island arcs would fit into such a pattern by regarding them as appropriate parts of X's rather than as curves.

The assumptions, then, of this type of interpretation are that there exists a shear pattern common to the entire crust of the Earth. The major elements of this shear pattern would be large transcurrent faults which cause a wholesale polygonal segmentation of the outer crust of the Earth. The basic pattern would have been formed very early in the history of the Earth.

In this connection, one also might mention some investigations of BROCK's[4, 5] according to whom the whole Earth is divided into a mosaic-pattern of stable elements *(tesserae)*. The boundary lines of these elements would correspond to the shear patterns mentioned above, but the lines would separate blocks which are subject to *vertical* motions in opposite directions rather than blocks which are subject to horizontal motions in opposite directions. The lines, therefore, might be called *hinge lines* rather than shear lines. Again, these lines are assumed to be very old so that, again, the basic pattern of the Earth's physiography would have been formed very early in its history.

It is not quite clear, however, how the persistence of a primeval line pattern can be upheld in the light of the known large changes in the location and geographic arrangement of orogenetic activity that occurred during the history of the Earth.

Orogenetic activity has also been connected with the formation of basins rather than with the formation of mountains. The statistics of the geometrical properties of basins shows some remarkable features. Thus, FAIRBRIDGE[6] noted that 290 basins occupy 55×10^6 km^2 or 36%

[1] HOBBS, W. H.: Bull. Geol. Soc. Amer. **22**, 123 (1911).
[2] VENING MEINESZ, F. A.: Trans. Amer. Geophys. Un. **28**, 1 (1947).
[3] MOODY, J. D., and M. J. HILL: Bull. Geol. Soc. Amer. **67**, 1207 (1956).
[4] BROCK, B. B.: Trans. Geol. Soc. South Afr. **59**, 150 (1956).
[5] BROCK, B. B.: Trans. Geol. Soc. South Afr. **62**, 326 (1959).
[6] FAIRBRIDGE, R. W.: Comm. No. 107, Assoc. Int. Seism., U.G.G.I., Assemb. Gén. Toronto, 1957.

of the continental areas of the world. The average size of a basin is 180000 km², the average thickness of the sediments therein is 1800 m. The average depth of basins is 5100 m and the volume of sediments in the unfolded areas of the world is 2.8×10^8 km³. DALLMUS[1] suggested a mechanism of basin evolution, essentially based on the "crustal shortening" that has to be connected with any form of subsidence.

Another view of orogenesis has been taken by GLANGEAUD[2] who paid particular attention to the symmetry (in cross-section) properties of mountain chains. Upon classifying mountain chains into mono-liminary and bi-liminary ones, he proceeded to describe the orogenetic physiography of each.

Finally, we note a view of orogenetic features that has been introduced by CAREY[3]. The latter author proposed that all orogenesis occurs originally along straight lines, and that the corners and arcs evident on the Earth's surface are formed by a subsequent rotating motion owing to continental drift. For an orogenetic system which has thus been flexed into an arc, CAREY introduced the term *orocline*. By straightening out the "oroclines" on the Northern Hemisphere, CAREY arrived at a reconstruction of Laurasia similar to that postulated earlier by DU TOIT on totally different grounds (cf. Sec. 1.31).

1.5. Physiography of Oceanic Areas

1.51. General Remarks. Knowledge of the morphology of the ocean bottom has been accumulating only recently. Whereas the physiography of continental areas, including that of the shelves and of adjacent island arcs, has been the object of the study of geologists for a long time, the investigation of the true ocean bottom had to await the advent of refined measuring techniques, such as echo sounding, deep sea coring etc.

As noted ealier, the total area that is covered by the sea, is approximately two thirds of the Earth's surface. However, all those regions that are either continental shelves or that are situated between continents and marginal island arcs (including their fore-deeps) are usually considered in conjunction with orogenetic processes that occur on the continents. Separating these areas from the total area covered by the sea, one still ends up with over half of the Earth's surface as "true" oceanic area. It therefore becomes evident that the study of oceanic features is of extreme importance in the discussion of geodynamics.

In the course of modern oceanographic investigations, it turned out that the deep sea is not just a bottomless abyss, but that there is a

[1] DALLMUS, K. F.: In: Habitat of Oil, p. 883ff., publ. by Amer. Assoc. Petrol. Geol., 1958.

[2] GLANGEAUD, L.: Rev. géogr. phys. géol. dyn. (2), **1**, 200 (1957).

[3] CAREY, S. W.: Pap. & Proc. Roy. Soc. Tasmania **89**, 255 (1955).

rather varied topography. The major structural elements that have been discovered are (i) abyssal plains, (ii) mid-ocean ridges, (iii) smaller features such a guyots, midocean islands, and (iv) archipelagic aprons. Finally, (v) extensive fracture zones have been discovered.

There are a series of books[1-4] and many papers[5-15] in existence dealing with the general aspects of the oceanic areas of the world. Detailed investigations of specific areas (i.e. the North Atlantic[16], the Indian Ocean[17], the Pacific Ocean[18], the Mediterranean[19], the Caribbean[20] and, particularly, the Arctic[21-25]) are even more numerous.

We shall discuss below each of the oceanic features mentioned above with emphasis on those characteristics that might have a bearing upon the understanding of geodynamic processes.

1.52. Abyssal Plains. The abyssal plains are usually thought to be the least disturbed areas of the oceans. In their original state, they show a hilly relief of roughly 300 meters (particularly in the Pacific), but many parts are smooth. Their average depth of submersion is approximately 5000 to 6000 meters[26]. At one time, it had been thought[27] that

---

[1] Shepard, F. P.: Submarine Geology. New York 1948.

[2] Bourcart, J.: La géographie du fond des mers. Paris: Payot 1949.

[3] Kuenen, P. H.: Marine Geology. New York: Wiley 1950.

[4] Heezen, B. C., M. Tharp and M. Ewing: The Floors of the Oceans. Geol. Soc. Amer. Spec. Pap. 65 (1959).

[5] Lees, G. M.: Proc. Roy. Soc. Lond., Ser. A **222**, 400 (1954).

[6] Bullard, E. C.: Mem. Proc. Manchester Lit. & Phil. Soc. **97**, 47 (1956).

[7] Adams, R. D.: Nature, Lond. **180**, 778 (1957).

[8] Elmendorf, C. H., and B. C. Heezen: Bell Syst. Techn. J. **36**, 1047 (1957).

[9] Segre, A. G.: Coll. Int. Centre Nat. Rech. Sci. **83**, 53 (1959).

[10] Laughton, A. S.: Endeavor **18**, 178 (1959).

[11] Heezen, B. C.: Submarine Topography. In: McGraw-Hill Encyc. Science & Tech. 216 (1960).

[12] Menard, H. W.: Trans. Amer. Geophys. Un. **41**, 274 (1960).

[13] Arrhenius, G.: Oceanography (Amer. Ass. Advan. Sci.) 129 (1961).

[14] Ewing, M., and M. Landisman: Oceanography (Amer. Ass. Adv. Sci.) 3 (1961).

[15] Smith, D. T.: Min. J. February-March 1961.

[16] Tolstoy, I.: Bull. Geol. Soc. Amer. **62**, 441 (1951).

[17] Lisitsyn, A. P., i A. B. Zhivago: Izv. Akad. Nauk SSSR, Ser. Geograf. **1958**, No. 2, 9 (1958).

[18] Menard, H. W.: Experientia **15**, 205 (1959).

[19] Segre, A. G.: Geol. Rdsch. **47**, 196 (1958).

[20] Heezen, B. C.: Trans. Sec. Caribb. Geol. Conf. 12 (1960).

[21] Fisher, R. L., A. J. Carsola and G. Shumway: Deep Sea Res. **5**, 1 (1958).

[22] Hope, E. R.: J. Geophys. Res. **64**, 407 (1959).

[23] Gordienko, P. A., i. A. F. Lakhtionov: Izv. Akad. Nauk SSSR, Ser. Geograf. **1960**, No. 5, 22 (1960).

[24] Treshnikov, A. F.: Priroda **1960**, No. 2, 25 (1960).

[25] Volkov, P.: Morskoi Flot **1961**, No. 3, 35 (1961).

[26] Kuenen, P. H.: Marine Geology. New York: Wiley 1950.

[27] Umbgrove, J. H. F.: The Pulse of the Earth. The Hague: Nijhoff 1947.

the Atlantic abyssal plains were different from the Pacific abyssal plains. This, however, does not seem to be the case. Since the original relief of abyssal plains appears to have been hilly, there is some doubt as to whether they really can be considered as "undisturbed".

The abyssal plains are covered with sediments. Exhaustive studies have been made of the nature and composition of these sediments[1-8]. It turns out that pelagic sedimentation rates are very slow, of the order of 0.2 cm per millenium[9]. Even at this rate, the average tickness of sediment solids on the ocean floors should be 1 to 3 km (accumulation over a billion years or so); however, it appears that the actual sediment thickness is only 0.5 km and it is therefore seen that there is here a discrepancy to be explained. A careful investigation of the physical properties of pelagic sediments by HAMILTON[10] yielded the result that consolidation and lithification of the lower sediment layers can be held responsible for their seeming disappearance.

1.53. Mid-Ocean Ridges. The most important features of the topography of the ocean bottom are mid-ocean ridges. These are huge "mountain" ranges rising above the abyssal plains, sometimes (in the form of islands) reaching to the surface of the sea. It has been conjectured as early as 1956 by EWING and HEEZEN[11] that the system of mid-ocean ridges, with some exceptions, is a continuous one, reaching all around the globe. This conjecture has subsequently been proven to be correct[12-15]; the presently established picture of the world-wide ridge system is shown in Fig. 13. It is interesting to note that the ridges follow almost everywhere closely the median lines of the ocean basins.

Tracing the ridge system, we note that in the Atlantic, there is the Mid-Atlantic Ridge[16] which has been well-known for some time. It is

[1] NAFE, J. E., and C. L. DRAKE: Geophysics **22**, 523 (1957).

[2] SUTTON, G. H., H. BERCKHEIMER and J. E. NAFE: Geophysics **22**, 779 (1957).

[3] BROECKER, W. S., K. K. TUREKIAN and B. C. HEEZEN: Amer. J. Sci. **256**, 503 (1958).

[4] SHEPARD, F. P.: Science **130**, 141 (1959).

[5] HAMILTON, E. L.: J. Sed. Petrol. **30**, 370 (1960).

[6] HEEZEN, B. C. *et al.*: C. R. Acad. Sci., Paris **251**, 410 (1960).

[7] ARRHENIUS, G.: Pelagic Sediments. In: The Sea, Ideas and Observations, ed. Goldberg. New York: Interscience 1961.

[8] ERICSON, D. B. *et al.*: Bull. Geol. Soc. Amer. **72**, 193 (1961).

[9] DIETRICH, G., u. K. KALLE: Allgemeine Meereskunde. Berlin: Bornträger 1957.

[10] HAMILTON, E. L.: Bull. Geol. Soc. Amer. **70**, 1399 (1959).

[11] EWING, M., and B. C. HEEZEN: Amer. Geophys. Monogr. **1**, 75 (1956).

[12] MENARD, H. W.: Bull. Geol. Soc. Amer. **69**, 1179 (1958).

[13] MENARD, H. W.: Experientia **15**, 205 (1959).

[14] HESS, H. H.: Preprints. Int. Oceanogr. Congr., Washington 33 (1959).

[15] EWING, M., and B. C. HEEZEN: Science **131**, 1677 (1960).

[16] TOLSTOY, I.: Bull. Geol. Soc. Amer. **62**, 441 (1951).

continuous from the Arctic to Tristan da Cunja; thence it swings around to the south of Africa and connects with the Madagascar and St. Paul Ridges in the Indian Ocean. Finally, it continues south of Australia and connects with the principal ridges in the Pacific Ocean[1,2]. The ridges of the Arctic Ocean, i.e. the Lomonosov Ridge and the Marvin Ridge seem to be detached from the world-wide system mentioned above[3–5].

Best investigated is the Mid-Atlantic Ridge complex[6,7]. It is up to 2000 km wide (1500 is an average) and can be visualized as a broad

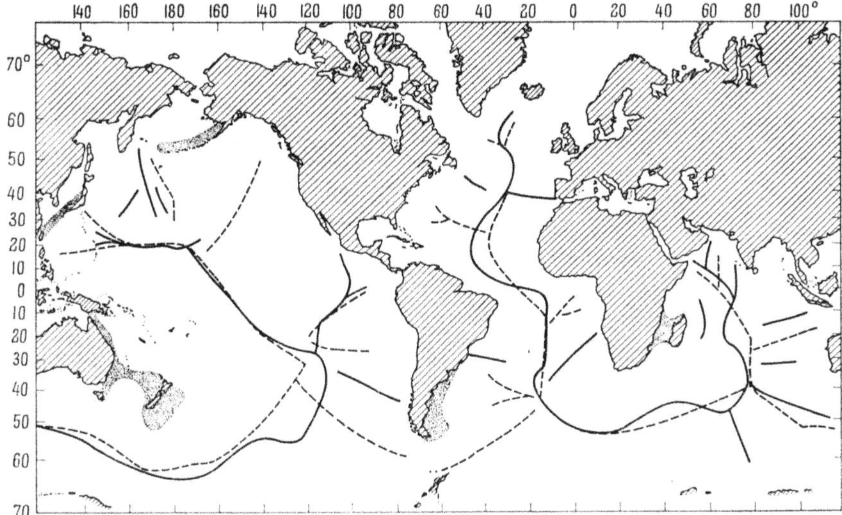

Fig. 13. World-wide distribution of mid-ocean ridges (solid lines); the dotted lines indicate the position of the medians of the ocean basins. After MENARD[8]

swell of varied topography. It is characterized by a high central zone (Main Range) consisting of parallel ridges following the general trend of the Main Range. The flanks of the Ridge are characterized by a succession of flats which have been called terraces. These seem to be areas of great sedimentary deposition. Finally, between the terraced zone and the adjacent abyssal plain, one may usually recognize a mountainous area standing out as a distinct physiographic unit. A typical profile across the Mid-Atlantic Ridges is shown in Fig. 14.

[1] SHUMWAY, G.: J. Geol. **62**, 573 (1954).
[2] SHUMWAY, G.: Deep Sea Res. **4**, 250 (1957).
[3] CRARY, A. P., and N. GOLDSTEIN: Deep Sea Res. **4**, 185 (1957).
[4] BURKHANOV, V. F.: Priroda **1957**, No. 5, 21 (1957).
[5] GAKKEL', YA. YA.: Priroda **1958**, No. 4, 87 (1958).
[6] TOLSTOY, I.: Bull. Geol. Soc. Amer. **62**, 441 (1951).
[7] ELMENDORF, C. H., and B. C. HEEZEN: Bell Syst. Techn. J. **36**, 1047 (1957).
[8] MENARD, H. W.: Experientia **15**, 205 (1959).

An interesting feature of mid-ocean ridges is that there appears to be a definite rift[1], approximately 100 km wide, at their crest. Depths in this rift exceed the maximum depths at the adjacent sides of the ridge out to 160 km or more.

Regarding the nature of the sea bottom on mid-ocean ridges, ELMENDORF and HEEZEN[2] state that, on the Mid-Atlantic Ridge, the steeper slopes are probably bare rock, and that the sediment removed from these slopes probably is deposited in the inter-range basins. If cores are

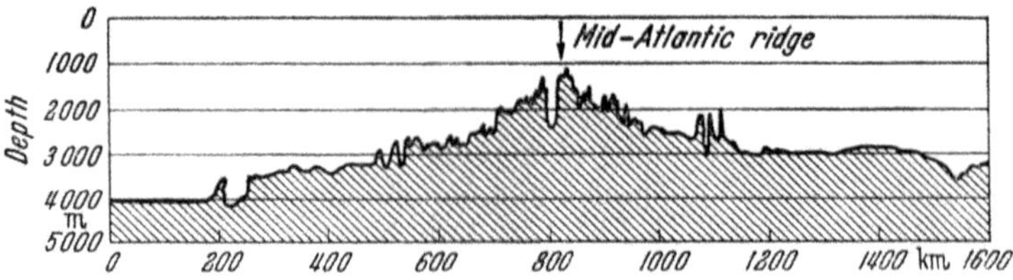

Fig. 14. Typical profile of the Mid-Atlantic Ridge (after ELMENDORF and HEEZEN[2])

taken, one usually finds globigerina ooze, but the sides of the ridges are often covered by red clay. On the Lomonosov Ridge, cores have been taken which revealed glacial and interglacial deposits reminiscent of deposits on the present continent of Asia. It thus appears that, if not over its whole extent, then at least in a considerable part of its greatest heights, now submerged at a mean depth of 1000 m below sea level, the Lomonosov Ridge was above water (BURKHANOV[3]). Evidence that other ridges were higher in recent times, although not above water, has also been mentioned by HESS[4].

The age of the formation of the ridges is uncertain. On the Mid-Atlantic Ridge, most of the rocks (mostly as evident on islands) are Quaternary, but occasionally, Tertiary material is found. None are older than Tertiary (HESS[4]). BURKHANOV[3] claims that the Lomonosov Ridge is of Mesozoic age, but it is not quite clear on what evidence this claim is based.

From the above remarks, it appears that mid-ocean ridges are somewhat like oceanic counterparts of continental orogenetic belts. It is not certain, however, whether they are folded like mountains (evidencing crustal shortening) or whether they have slowly been building up due to fissuring of the oceanic crust by tensional forces and the consequent oozing out of material from the deeper parts of the Earth. The existence of the axial rift would seem to favor the latter hypothesis (see the discussion by KOCZY and BURRI[5]).

[1] HILL, M. N.: Deep Sea Res. **6**, 193 (1960).
[2] ELMENDORF, C. H., and R. C. HEEZEN: Bell Syst. Techn. J. **36**, 1047 (1957).
[3] BURKHANOV, V. F.: Priroda **1957**, No. 5, 21 (1957).
[4] HESS, H. H.: Proc. Roy. Soc. Lond., Ser. A **222**, 341 (1954).
[5] KOCZY, F. F., and M. BURRI: Deep Sea Res. **5**, 7 (1958).

1.54. Smaller Features in Basins. Finally, we turn our attention to the remaining oceanographic features mentioned in Sec. 1.51. These features are mostly types of smaller irregularities that may occur within the confines of the oceans.

One of these is the *sea mount*. Sea mounts of various shapes and sizes are found in many basins. If they reach to the surface of the sea,

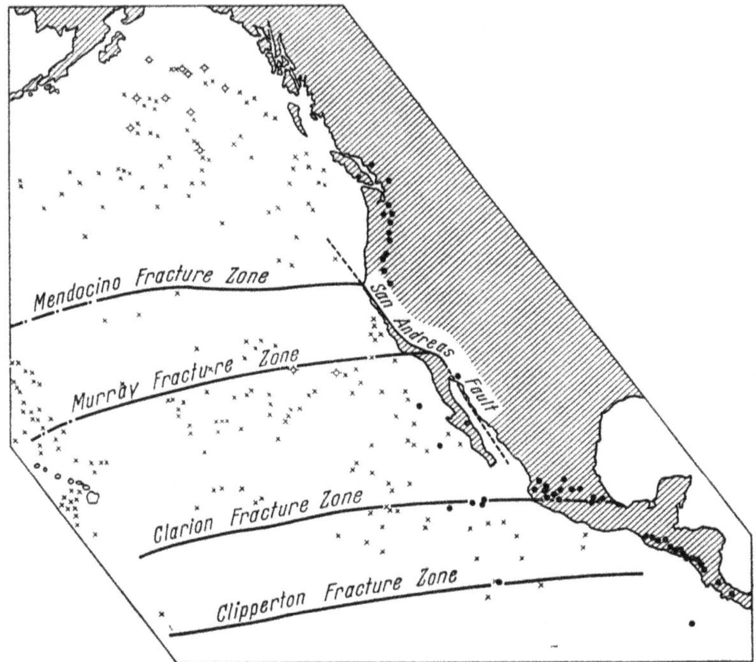

Fig. 15. Fracture zones off the Pacific Coast of North America (after MENARD[1]).

they give rise to the development of coral atolls[2]. Some of the sea-mounts are flat-topped; they are then called *guyots*. The flat top seems to indicate that these sea mounts once reached to the surface of the water and that wave action levelled their tops. They represent, thus, drowned ancient islands. They are particularly numerous in some parts of the Pacific[3, 4].

The above observations point to a relative subsidence of many features in the ocean. Coral atolls seem to point in the same direction; for, on a subsiding sea mount, under favorable climatic conditions, coral

[1] MENARD, H. W.: Bull. Geol. Soc. Amer. **66**, 1149 (1955).

[2] WIENS, J. H.: Ann. Ass. Amer. Geogr. **49**, No. 1, 31 (1959).

[3] DIETZ, R. S.: New Scientist **5**, 14 (1959).

[4] HAMILTON, E. L.: Sunken Islands of the Mid-Pacific Mountains. Geol. Soc. Amer. Memoir No. 64 (1956).

growth could just keep up with subsidence. From a hole drilled on Bikini, KUENEN[1] estimated a rate of subsidence (before glacial times) of a little less than 0.5 cm/century. It would thus appear that sea mounts and atolls are of volcanic origin. After their formation, they start sinking owing to the extra weight caused by their presence.

As noted, sea mounts are particularly numerous in the Pacific. A series of them is concentrated within a number of "fracture zones" off the shores of North America. The Pacific fracture zones were first discovered by MENARD[2-4]; later, such fracture zones were also found elsewhere[5,6]. A drawing of MENARD's fracture zones is given in Fig. 15. Within these fracture zones, the topography, apart from featuring an accumulation of sea mounts, is characterized by deep, narrow troughs, asymmetrical ridges and escarpments.

Most of the Pacific Ocean bottom is hilly with an average relief of about 300 m. However, part of it is smooth, and it is found that almost all the smooth parts are concentrated around islands or sea mounts, and have the form of "archipelagic aprons"[7,8]. These aprons seem to consist of volcanic rocks several kilometers thick. This material buries the original hilly relief of the sea floor so as to form a smooth plain. The total volume of material contained in the archipelagic aprons in the Pacific has been estimated by MENARD[7] as 4×10^6 km^3.

1.6. Physiography of Faults and Folds

1.61. Faults. The strata deposited by sedimentation cannot generally be expected to have persisted in their original position throughout the ages. The effects of orogenesis have had a profound influence on their appearance. The most common deformations of strata are faults and folds.

Starting with *faults*, one can say that these are fracture surfaces along which the rocks have undergone a relative displacement. They occur in parallel or subparallel systems which have usually a wide lateral distribution.

Physically, a fault is determined by first stating its position on the Earth (by giving latitude and longitude) and secondly by giving its "character" by fixing (a) the *fault plane* and (b) the direction of relative

[1] KUENEN, P. H.: Marine Geology. New York: Wiley 1950, see p. 467.
[2] MENARD, H. W.: Bull. Geol. Soc. Amer. **66**, 1149 (1955).
[3] MENARD, H. W., and R. L. FISHER: J. Geology **66**, 239 (1958).
[4] MENARD, H. W.: Bull. Geol. Soc. Amer. **70**, 1491 (1959).
[5] PANOV, D. G.: Izv. Akad. Nauk SSSR, Ser. Geol. **1958**, No. 9, 84 (1958).
[6] HERSEY, J. B., and M. S. RUTSTEIN: Bull. Geol. Soc. Amer. **69**, 1297 (1958).
[7] MENARD, H. W.: Bull. Amer. Ass. Petrol. Geol. **40**, 2195 (1956).
[8] EMERY, K. O.: J. Geol. **68**, 464 (1960).

motion of the two sides of the fault thereon[1]. It is thus seen that, in order to determine the character of a fault, one needs *three* parameters, two determining the position of the fault plane, and one determining the direction of the motion vector. Sometimes more parameters are given, but then constraints must apply between them.

Most representations of faults aim at fixing (a) the fault surface, and (b) the direction of motion. The fault surface can be determined e.g. by giving the direction of strike (i.e. the line of intersection of the fault surface with a horizontal plane) together with the dip (i. e. the angle between the fault plane and a horizontal plane) as shown in Fig. 16 Alternatively, the dip direction and dip can be given. The direction of motion can be determined e.g. by giving the angle between the motion vector and the strike direction of the fault. This angle is called "slip angle". Alternatively, introducing more than three parameters for the determination of a fault, the direction of the motion vector can be given by stating its azimuth (referring to the northward direction) and its inclination to the horizontal. Another way to give the direction of the motion-vector is by stating strike and dip, or dip direction and dip, of a plane which is orthogonal to the motion-vector. The latter plane is commonly called "auxiliary plane" of the fault.

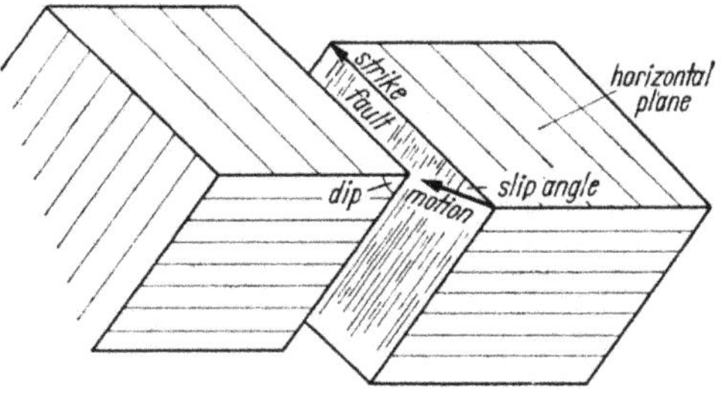

Fig. 16. A fault

The terminology of faults has been coined by geologists[2, 3]. It is customary to call a fault transcurrent if the slip angle (that is the angle between strike and motion vector) is smaller than 45°. Transcurrent faults may be further classified as "dextral" or "sinistral". Imagine oneself standing on one side of the fault, and looking at the other side. If the other side appears to have moved to the right, the fault is dextral, if it appears to have moved to the left, the fault is sinistral. If the slip angle is greater than 45°, one also differentiates between two cases: normal faults, if the (over-) hanging wall moves downwards, and reversed faults or thrusts if the hanging wall moves up.

Fig. 17 will serve to illustrate the geologists' conception of faulting.

Phenomena related to faults are *dykes*. Dykes are, in the main, nearly vertical fissures between 3 and 30 metres wide that have been

[1] SCHEIDEGGER, A. E.: Bull. Seism. Soc. Amer. **47**, 89 (1957).
[2] HOLMES, A.: Principles of Physical Geology. London: T. Nelson 1944.
[3] LENSEN, G. J.: New Zeal. J. Geol. Geophys. **1**, 307 (1958).

infilled with some intrusive material. Flow structures in dykes exhibiting the infilling process are not uncommon[1]. If the dykes have not been filled-in, they are called *joints*. The two sides of a dyke appear to have moved apart in a direction normal to the fissure as if under tension, such that there is neither a lateral nor a vertical dislocation. Dykes are much less common than faults and they are also somewhat restricted in distribution.

A similar cause of the origin as of dykes (i.e. tensional stresses) may also be postulated for the great *rift valleys* in the world (the Rhine

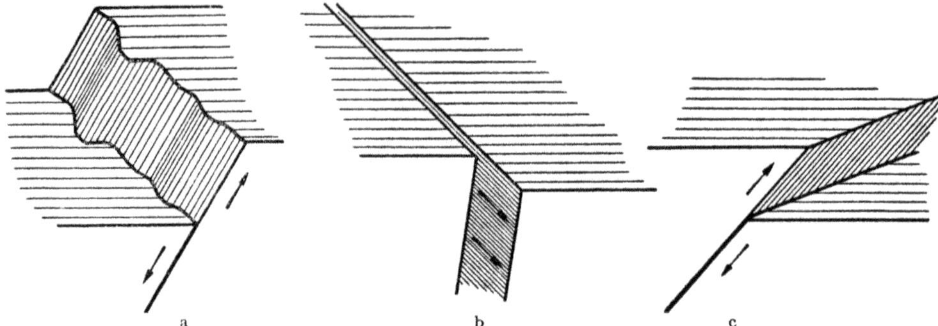

Fig. 17a—c. The major types of faults: (a) normal fault; (b) transcurrent fault; (c) reversed fault

Graben, the African Rift Valleys etc.). These "grabens" may be taken as the expression of a giant rift in the surface of the Earth[2].

1.62. Folds. In the formation of mountain ranges, it is evident that not faulting, but a process properly called *folding* is of prime importance. It appears that the strata in such mountain ranges have been contorted to a fabulous extent[3]. It almost looks as if some supernatural giant took originally level strata, pulled them up and folded them over at his will, until they may become recumbent[4]. Naturally, the action of water and wind will erode some of the folded materials and the physiographic appearance of a mountain range is therefore one of high peaks and deep valleys. Nevertheless, the continuity of the folded strata can be traced from peak to peak and the original (i.e. undisturbed by erosion) position of the layers can be reconstructed. The horizontal distances over which the strata are folded over may be up to several score kilometers. The Western Alps, in a profile from Lausanne over the Dent Blanche to the plain of Piemont, give a classical example.

[1] Spry, A.: Geol. Mag. **90**, 248 (1953).

[2] Wheeler, R. R.: Bull. Geol. Soc. Amer. **70**, 667 (1959).

[3] Cf. e.g. Holmes, A.: Principles of Physical Geology. New York: The Ronald Press Co. 1945.

[4] Sitter, L. U. de: Bull. Soc. Belge Géol. **66**, No. 3, 352 (1957).

The nomenclature of folds has again been created by geologists[1]. If the strata are bent upwards into an archlike form, the arrangement is termed *anticlinal*; if the strata are bent downwards into a through-like form, the arrangement is termed *synclinal*. The two sides of a fold are called its *limbs*. The bisecting plane of the two limbs is termed the *axial plane* of the fold. Finally, the *axis* of the fold is the intersection of the axial plane with the uppermost stratum. The angle of the axis with the horizontal plane is called its *pitch* (see Fig. 18).

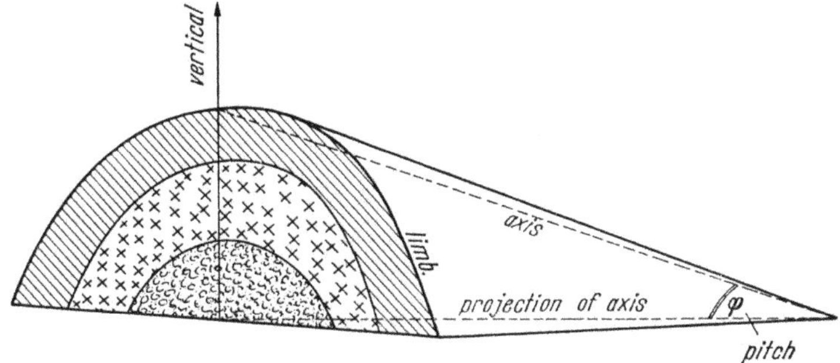

Fig. 18. Schematic drawing of an anticlinal fold

It would thus appear that, in order to determine a fault, one has to give strike and pitch of its axis, and the dips of both of its limbs. However, this is not sufficient. There must be an additional parameter which is vaguely referred to by geologists as "intensity". It refers to the *radius of curvature* at the crest of the fold.

It is therefore seen that one needs the following parameters in order to fix a fold geometrically: (i) strike and (ii) dip of the axial plane, (iii) the angle of aperture of the fold, i.e. the angle which the two limbs form, (iv) the pitch of the axis and (v) the radius of curvature at the crest. These are altogether five parameters which are needed in addition to the geographical co-ordinates of the point of the fold under consideration. Naturally, another equivalent set could equally well be used.

1.63. Patterns of Faults and Folds. As noted above, faults (as well as folds) often occur in the form of systems.

Patterns of faults have been discovered in many parts of the world. All of these are connected with geodynamic processes. Some of the patterns have been interpreted in terms of shear lines, others in terms of arrangements of joints[2-4]. It will surprise no-one to find such

[1] Cf. e. g. HOLMES, A.: Principles of Physical Geology. New York: The Ronald Press Co. 1945.

[2] HALLER, J.: Schweiz. mineral. petrogr. Mitt. **37**, 11 (1957).

[3] HODGSON, R. A.: Bull. Amer. Ass. Petrol. Geol. **45**, 1 (1961).

[4] HODGSON, R. A.: Bull. Amer. Ass. Petrol. Geol. **45**, 95 (1961).

patterns in tectonically active areas, but striking patterns of joints have been found by MOLLARD[1, 2] even in geotectonically quiet regions of the world, such as on the Canadian Prairies, in Venezuela and in West Pakistan. These patterns are particularly evident if one inspects air photographs, as the drainage of an area tends to accentuate any existing features[1–4]. For the process causing fault patterns, SONDER[5] coined the term *rhegmagenesis*.

The various *patterns of folds* are nothing but what has been called earlier "orogenetic systems". Thus, for a discussion of fold patterns, the reader is referred to the earlier sections of this book.

1.7. Physiography of Some Special Features

1.71. Meteor Craters. Finally, we turn our attention to special physiographic features upon the Earth's surface. One type of such features are crater-like holes in the ground, caused by the impact of meteors.

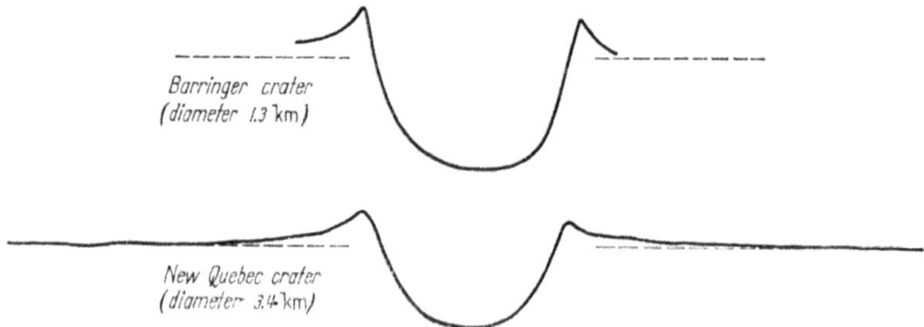

Fig. 19. Crater profiles (after MILLMAN[6]). Vertical scale five times horizontal scale

Striking examples of such craters have been found in Arizona[7] (Barringer Crater), in Northern Quebec[6] (New Quebec Crater) and elsewhere[8, 9]. Typical profiles of these craters are shown in Fig. 19.

It had not always been established that such craters actually were caused by meteors. One of the hypotheses advanced in earlier times was that these craters were of volcanic origin. However, it is now established

---

[1] MOLLARD, J. R.: Oil in Canada **9**, No. 40, 26 (1957).

[2] MOLLARD, J. R.: Proc. Second Williston Basin Conf. 109 (1959).

[3] EGYED, L.: Gerlands Beitr. Geophys. **66**, 271 (1957).

[4] HAMAN, P. J.: Lineament Analysis on Aerial Photographs. Publ. by Author. Calgary, Alta. 1961.

[5] SONDER, R. A.: Mechanik der Erde. Stuttgart: E. Schweizerbart 1956.

[6] MILLMAN, P. M.: Publ. Dom. Obs. Ottawa **18**, No. 4, 59 (1956).

[7] BARRINGER, D. M.: Proc. Acad. Nat. Sci. Philadelphia **66**, 556 (1914); **76**, 275 (1924). — NININGER, H. H.: Arizona's Meteorite Crater; Publ. Amer. Meteorite Museum, Sedona, Ariz. 1956.

[8] DIETZ, R. S.: J. Geol. **67**, 496 (1959).

[9] NININGER, H. H.: Science **130**, 1251 (1959).

beyond doubt (by drilling and by other direct means) that these craters were created by the impact of celestial objects.

This result has prompted one to search for other features on the Earth's surface that might have been caused by meteoritic impact. One of the characteristics of the craters is that they are almost perfectly circular. The hypothesis has therefore been advanced that such features as the Gulf of St. Lawrence[1] or the arc of the Lesser Antilles[2] were created by the impact of huge meteorites. This is, of course, somewhat doubtful[3], but it is certain that many other smaller circular features have been formed in this fashion. In this instance, one is peaking of "fossil" craters[4-8]. The Earth has been stricken by meteors in the distant past; hence a hole resulted in the ground which was subsequently filled-in with detritus and other foreign material. Further changes may have occurred during glaciation, and at the present time, all that is seen is perhaps a change in the appearance of the vegetation due to the difference in soil above the originally shattered and unshattered area.

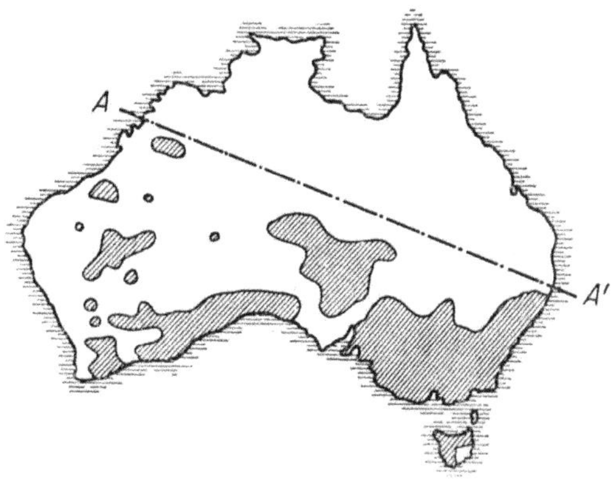

Fig. 20. Areas in Australia where tektites were found. No tektites were found above the line AA'. After VARSAVSKI[9]

A phenomenon related to meteoritic impact may be the occurrence of *tektites*. The latter are glassy pebbles found in strangely specific areas of the world (see Fig. 20); one of the contentions as to their origin is that they came from the Moon. The physiography of the occurrence of tektites has been studied in a big monograph by BAKER[10]; shorter descriptions have also been given by other authors[9, 11].

---

[1] WILLMORE, P. L., and A. E. SCHEIDEGGER: Trans. Roy. Soc. Can. **50**, Ser. III, Can. Comm. Oceanogr. 21 (1956).

[2] SCHULZ-WEIDNER, W.: Umschau **55**, No. 23, 713 (1955).

[3] In fact, the Gulf of St. Lawrence, e.g., simply seems to be a basin (cf. loc. cit.[1]).

[4] BEALS, C. S., G. M. FERGUSON and A. LANDAU: J. Roy. Astron. Soc. Can. **50**, 203 (1956).

[5] BEALS, C. S.: Sky and Telescope **16**, No. 11 (1957).

[6] INNES, M. J. S.: J. Roy. Astron. Soc. Can. **51**, 235 (1957).

[7] BEALS, C. S.: Nature, Lond. **181**, 559 (1958).

[8] MILLMAN, P. M.: Publ. Dom. Obs. Ottawa **24**, No. 1 (1960).

[9] VARSAVSKI, C. M.: Geochim. et Cosmochim. Acta **14**, 292 (1958).

[10] BAKER, G.: Tektites. Mem. Nat. Mus. Victoria, Melbourne, No. 23 (1959).

[11] O'KEEFE, J. A., B. E. SHUTE: Aerospace Eng. **20**, No. 7, 26 (1961).

1.72. Boudinage. A particularly intriguing feature is the occurrence of boudins. The term "boudinage structure" was introduced by LOHEST[1]. It is French meaning "sausage structure" and refers to a fractured sheet of rock wedged in between non-fractured rocks, or, in geological terms, to a fractured competent layer situated between two incompetent layers. Each fragment of the fractured rock often looks like a sausage, and the whole array resembles in many cases a chain of sausages. Hence the name of these structures.

Thus, boudins are oblong bodies; they have been described e.g. by RAMBERG[2] and DE SITTER[3]. Their thickness varies[2] from about 1 cm to approximately 20 meters. The shortest boudins are little longer than thick; the largest are many times longer than thick.

The shape of boudins may be quite varied. Some of them are rectangular bodies with sharp corners, others are barrel-shaped or lenticular, and finally some of them have an appearance as if they would have been twisted around their central axis.

Boudinage structures are quite common in gneissic and regional metamorphic areas where the rocks are well exposed. They are also abundant in some low-grade schist regions. Usually, of course, only a two-dimensional cross-section of the structures is exposed in rock outcrops so that the three-dimensional arrangement of the boudins is not always easy to infer.

A physiographic appearance similar to that of boudinage strcutures is exhibited by pinch-and-swell structures. In the latter the wedged-in "compentent" layer is not completely fractured but seems to pinch out and swell up regularly in a pattern which is also sausage-like. Such structures are very common in conformable pegmatites and in quartz veins. From the close physiographic similarity of such pinch-and-swell structures with boudinage one might expect that the physical explanation of the two phenomena should also be similar.

1.73. Diapirs. An interesting phenomenon is the existence of diapirs (domes). Generally, a dome is a fold which is "anticlinial" in every direction, viewed from its crest. This can also be stated by saying that the "axis" of the fault degenerates into a point.

One usually calls domes only such features which are caused by intrusions from below[4-6]. The intrusive material may be salt, gypsum or some other material capable of plastic flow. The strata above the intrusive material are pushed upward and represent the "dome".

---

[1] LOHEST, M.: Ann. Soc. Géol. Belg. **36** B, 275 (1909).
[2] RAMBERG, H.: J. Geol. **63**, 512 (1955).
[3] SITTER, L. U. DE: Geol. en Mijnb. **20**, 277 (1958).
[4] TRUSHEIM, F.: Z. dtsch. geol. Ges. **109**, 111 (1957).
[5] FEELY, H. W., and J. L. KULP: Bull. Amer. Ass. Petrol. Geol. **41**, 1802 (1957).
[6] HEIM, A.: Ecl. Geol. Helv. **51**, 1 (1958).

Of particular interest is the case when the intrusive material reaches the surface of the Earth. One speaks in such cases of "piercement

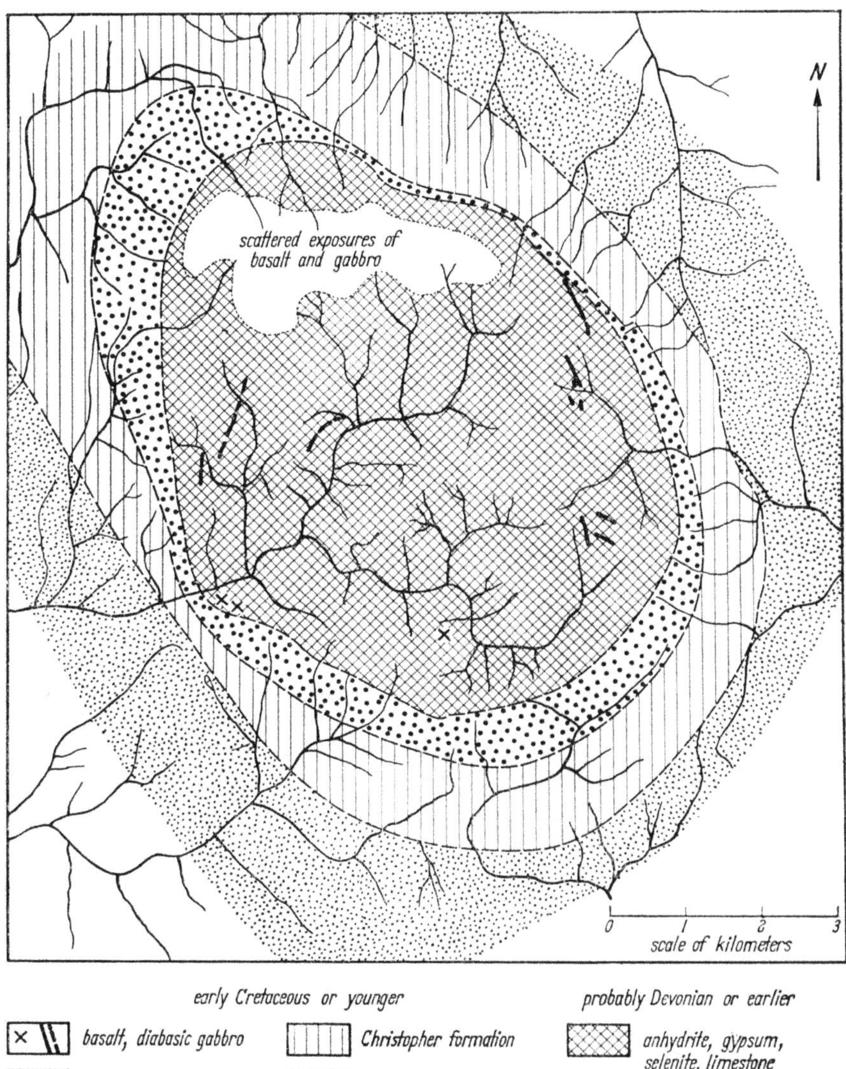

Fig. 21. Geological sketch of the Isachsen Dome in the Canadian Arctic. After HEYWOOD[1]

domes". The physiographic appearance is then that of a circular structure (see Fig. 21). A number of such structures have been discovered in the Canadian Arctic and were described by HEYWOOD[1]. A particularly

[1] HEYWOOD, W. W.: Trans. Canad. Inst. Min. Met. **58**, 27 (1955).

beautiful example is the Isachsen Dome on Ellef Ringnes Island (see Fig. 21). In this particular dome, the core-rocks are gypsum, anhydrite, limestone and basalt. The intrusion of this material has upwarped the overlying sediments and caused faults radial and tangential to the outer contact of the dome. The sedimentary rocks are even sharply upturned and overturned in many places. The effects of the diapirism are not apparent farther away than 3 to 5 km from the structure.

1.74. Volcanoes. A different type of circular structure that occurs on the surface of the Earth is found in connection with volcanism. It is well known that volcanism produces cone-like mountains of lava and ash which have one or more craters at the summit. Gas and lava is found to pour out from these craters in various proportions, often in cataclysmic spasms. Physiographic descriptions of volcanic eruptions have been given, for instance, by RUE[1], by RITTMANN[2], and by COTTON[3]. The magma erupting when a volcano is active seems to come from chambers not far below[4].

The best-known volcano is undoubtedly Mount Vesuvius as it has been observed and investigated since the time of antiquity. Its basis is circular with a diameter of approximately 16 km. It rises gently to an elevation of about 595 meters above sea level and hence abruptly to the two summits of the mountain. One is Monte Somma, the other Mount Vesuvius proper. The cone of Vesuvius rises at an inclination of about 30° to about 1300 m above sea level. It consists mostly of a loose accumulation of cinders. Throughout history, periods of activity have alternated with periods of acquiescence.

The above picture of Mount Vesuvius is fairly typical for an average volcano. Specific morphological studies of volcanoes have been made in many parts of the world[5-10]. A drawing of a typical volcano is shown in Fig. 22.

[1] RUE, E. A. DE LA: L'homme et les volcans. Paris: Gallimard 1958.

[2] RITTMANN, A.: Vulkane und ihre Tätigkeit. 2. Aufl. Stuttgart: Ferdinand Enke 1960.

[3] COTTON, C. A.: Volcanoes as Landscape Forms, 2nd ed. Christchurch: Whitcombe & Tombs Ltd. 1952.

[4] GORSHKOV, G. S.: Bull. Volc. (2) **19**, 103 (1948).

[5] JAGGAR, T. A.: Origin and Development of Craters. Geol. Soc. Amer. Memoir No. 21 (1947).

[6] GROVER, J. C.: Geograph. J. **123**, 298 (1957).

[7] MACHADO, F.: Atlantida **2**, 225, 305 (1958).

[8] CASTELLO BRANCO, A. DE, *et al.*: Le volcanisme de l'île de Faial et l'éruption du volcan de Capelinhos. Memoria No. 4, Serviços Geológicos de Portugal, Lisboa 1959.

[9] VLODAVEC, V. I.: Die Vulkane der Sowjetunion. Gotha: VEB Geogr.-Kart. Anst. 1954.

[10] SATO, H.: Distribution of Volcanoes in Japan. Proc. Intern. Geog. Un. Reg. Conf. of 1957, p. 184 (1959).

A particularly detailed study of the morphology of volcanoes has been made by JAGGAR[1] in Hawaii. The actual craters are very similar to explosion pits, except that they are elevated above their surroundings.

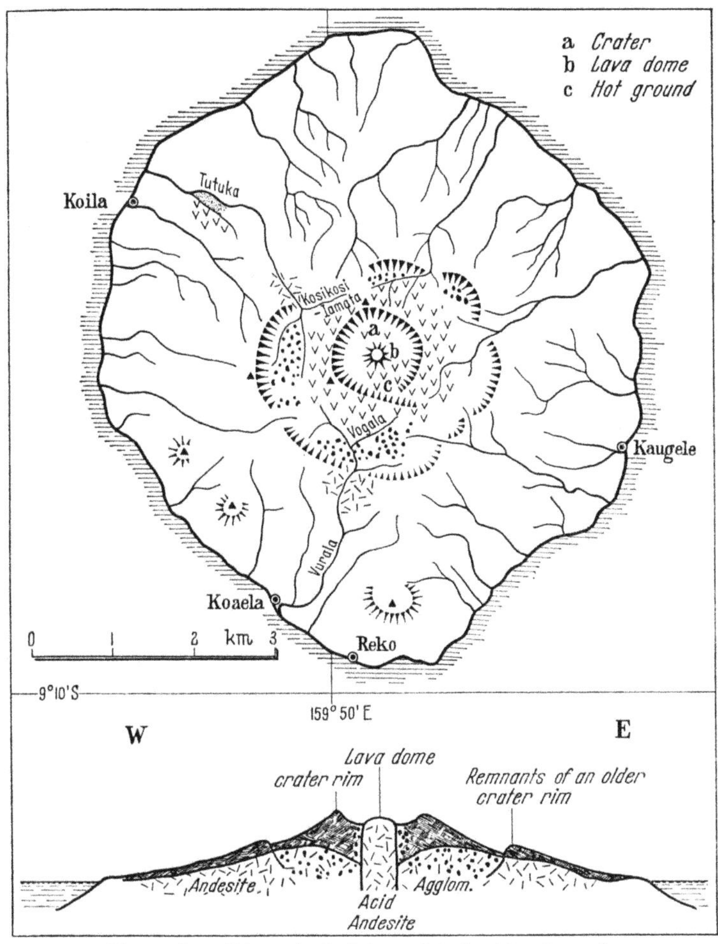

Fig. 22. Savo Volcano, in the Solomon Islands. After GROVER[2]

In a real grand-style volcanic eruption, the resulting crater may be up to 8 km in diameter. A profile of the crater resulting from the eruption of Krakatoa in 1883 is shown in Fig. 23.

Volcanoes exist in many parts of the world. There is a general accumulation of them in orogenetic belts. There are, however, orogenetic

[1] JAGGAR, T. A.: Origin and Development of Craters. Geol. Soc. Amer. Memoir No. 21 (1947).

[2] GROVER, J. C.: Geograph. J. **123**, 298 (1957).

belts without any volcanoes (such as the Alps) and there are non-orogenetic areas with prominent volcanoes (such as the Belgian Congo). Volcanoes do thus not lack a certain randomness regarding their geographical distribution.

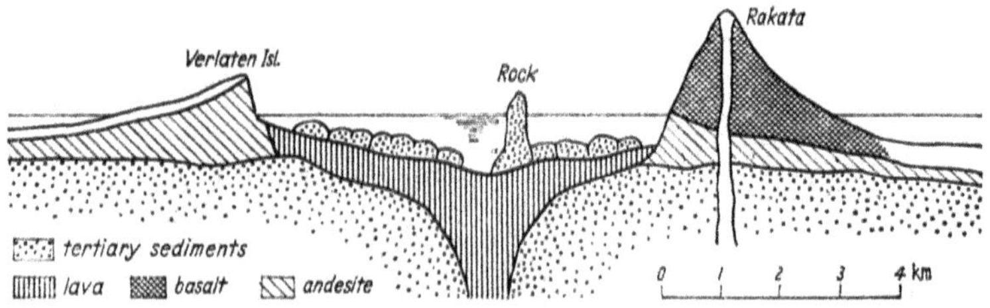

Fig. 23. Cross-section through Krakatoa after its eruption. After NEUMAYR and SÜESS[1]

1.75. Vertical Displacements. There is some conclusive evidence that, in addition to tectonic movements leading to horizontal shifts in the Earth's surface (orogenetic movements), vertical displacements (often called "*epeirogenetic movements*") are also taking place. Studies proving this have been particularly undertaken in the vicinity of Scandinavia[2-13]. In that area, the evidence for vertical displacements (in this case: an uplift) is threefold: geologic, mareographic, and geodetic.

The *geologic* evidence is based upon an analysis of the strata in question. From the existence of fine-grained sediments such as silt, clay and mud, which commonly occur at the bottom of the sea, it is inferred that the central part of the area was depressed below sea level at the end of the Pleistocene ice age. Later on, it must have gradually risen, because remnants of forest and bog vegetation become apparent. It thus becomes possible to date the time of transition of any one spot from the sea-bottom to dry land. DE GEER[2] was a pioneer in such studies which were continued by SAURAMO[3]. From his results, SAURAMO concluded that the largest uplift took place near the Gulf of Bothnia; and that since 6800 B.C. the land has risen there by about 250 meters.

[1] NEUMAYR, M., u. F. E. SÜESS: Erdgeschichte, 3rd ed. Stuttgart: Bibliogr. Inst. 1920.

[2] GEER, C. J. DE: C. R. Int. Geol. Congr. Stockholm **2**, 849 (1910).

[3] SAURAMO, M.: Fennia **66**, No. 2, 3 (1939).

[4] GUTENBERG, B.: Bull. Geol. Soc. Amer. **52**, 750 (1941).

[5] KUKKAMÄKI, T. J.: Veröff. Finn. Geodät. Inst. No. 26, 120 (1939).

[6] KÄÄRIÄINEN, E.: Veröff. Finn. Geodät. Inst. No. 42 (1953).

[7] WEGMANN, E.: Geol. Rdsch. **43**, 4 (1955).

[8] SAXOV, S.: Dansk. Geol. Fören. Medd. **13**, No. 6, 518 (1958).

[9] BIRKENMAJER, K.: Proc. 21st Int. Geolog. Congr. Norden **21**, 281 (1960).

[10] GUDELIS, V.: Collect. Papers, 19th Int. Geogr. Congr. Vilnius 201 (1960).

[11] HONSKALO, T.: Geophysica, Hels. **7**, 117 (1960).

[12] KÖSTER, R.: Proc. 21st Int. Geolog. Congr. Norden **18**, 89 (1960).

[13] USHAKOV, S. A.: Dokl. Akad. Nauk SSSR **133**, 205 (1960).

The *mareographic* evidence is based upon tide-level gauge readings. From these the change of sea-level and the change in elevation of the

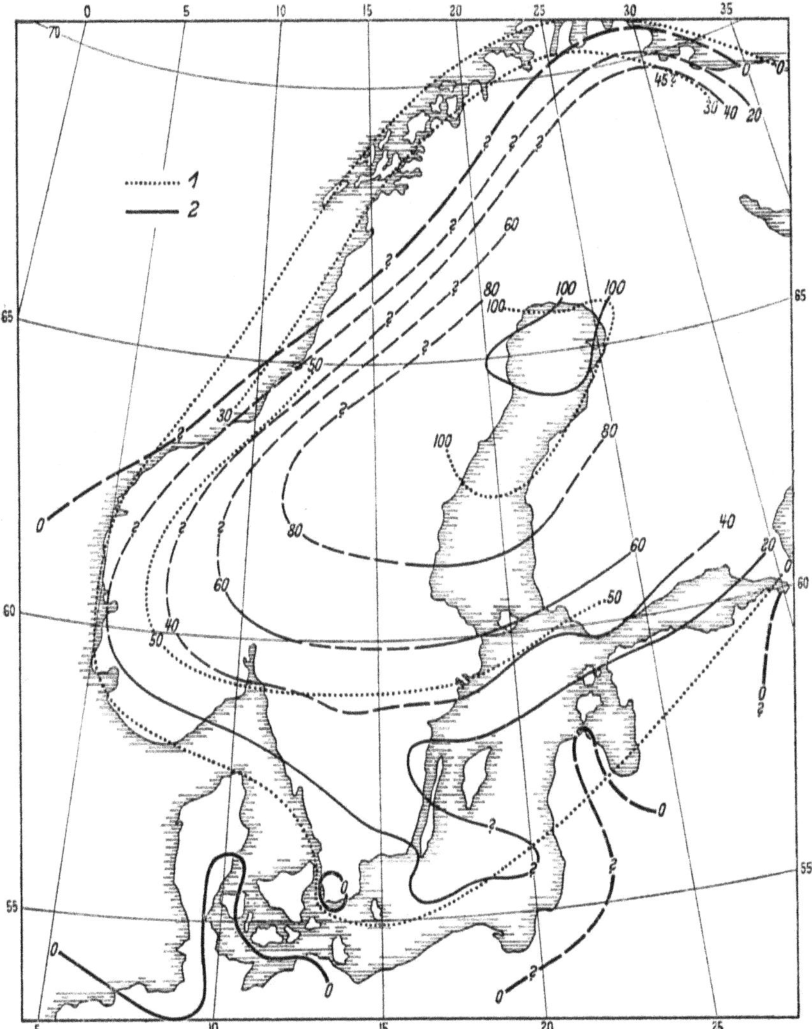

Fig. 24. The uplift of Fennoscandia, after GUTENBERG[1]. *1* lines of uplift (in meters); *2* lines of uplift velocity (in cm/century)

coast can be calculated. The results of such investigations have been summarized by GUTENBERG[1]. From them, GUTENBERG constructed a map of Fennoscandia which is shown in Fig. 24.

[1] GUTENBERG, B.: Bull. Geol. Soc. Amer. **52**, 750 (1941).

Finally, the *geodetic* evidence[1,2], based on precise levelling measurements, corroborates the above findings.

In Scandinavia, the occurrence of the uplift has commonly been interpreted as a rebounding after the disappearance of the ice put there by the most recent glaciation. However, this interpretation has recently been questioned by Russian investigators[3-6] who claim that the uplift occurred also *before* the onset of the last ice age. The fact that vertical movements of the Earth's surface have also been discovered in other parts of the world[7-10] may or may not be a corroboration of this claim.

---

[1] Kukkamäki, T.: Veröff. Finn. Geodät. Inst. No. 26, 120 (1939).

[2] Kääriäinen, E.: Veröff. Finn. Geodät. Inst. No. 42 (1953).

[3] Zhelin, G. A.: Izv. Akad. Nauk Eston, SSR., Ser. Mekh. Fiz. Mat. Nauk **1957**, **6**, No. 3, 205 (1957).

[4] Meshcheryakov, Yu. A.: Priroda **1958**, No. 9, 15 (1958).

[5] Meshcherikov (Meshcheryakov), J. A. (Yu. A.): Pub. Bur. Centr. Séism. Int. A., **20**, 261 (1959).

[6] Gzovskii, M. V., V. N. Krestnikov i G. I. Reisner: Izv. Akad. Nauk SSSR, Ser. Geofiz. **1959**, No. 8, 1147 (1959).

[7] Valentin, H.: Verh. Dtsch. Geographentag. 1955, **29**, 148 (1955).

[8] Hori, S.: Japan J. Geol. Geogr. **28**, 1 (1957).

[9] Svoboda, K.: Vermessungstechn. **7**, No. 8, 221 (1959).

[10] Greene, G. W.: U. S. Geol. Surv. Prof. Pap. **400** B, 275 (1960).

II. Geophysical Data Regarding the Earth

2.1. The Layering of the Earth

2.11. Earthquakes and Seismic Waves. Seismology, the study of earthquakes, has yielded some very pertinent information about the structure of the Earth. Earthquakes are shocks that occur within the Earth. It appears that these shocks originate each in a region which is small compared with the Earth as a whole; this region can be regarded as a point for most purposes and is referred to as the *focus* of the earthquake. The point directly above the focus on the Earth's surface is termed the *epicenter* of the earthquake.

After an earthquake has occurred, one observes effects at seismic stations throughout the world. Such stations are equipped with *seismographs*, instruments designed to amplify and register any tremors of the Earth's surface in their vicinity. The seismograph writes a *seismogram*, a line related to the motion of the Earth in any one chosen direction. Any change of amplitude or frequency in the seismogram of an earthquake is called a *phase*. The principal phases in the seismogram of an earthquake have been called P, S and L and it has been established that they represent the first onsets of compressional, transverse bodily and surface waves (cf. Sec. 3.21), respectively. A typical seismogram is shown in Fig. 25.

One of the principal outcomes of observational seismology has been the recognition that it is possible to regard "phases" as travelling along "rays". It is thus possible to construct travel-time tables for the various phases and to trace their paths through the interior of the Earth. The travel-time of a phase does not depend appreciably on the location of the epicenter and the station, but only on the epicentral distance and the depth of the focus.

From the above description, it is obvious that seismology yields data about the Earth in two ways: first, one can analyse the effect that the interior of the Earth has upon elastic wave transmission, and second, one can analyze the occurrence and the mechanism of the earthquakes themselves. Treatises on the various aspects of seismology have been written e.g. by MACELWANE[1],

---

[1] MACELWANE, J. B.: When the Earth Quakes. Milwaukee: Bruce 1947.

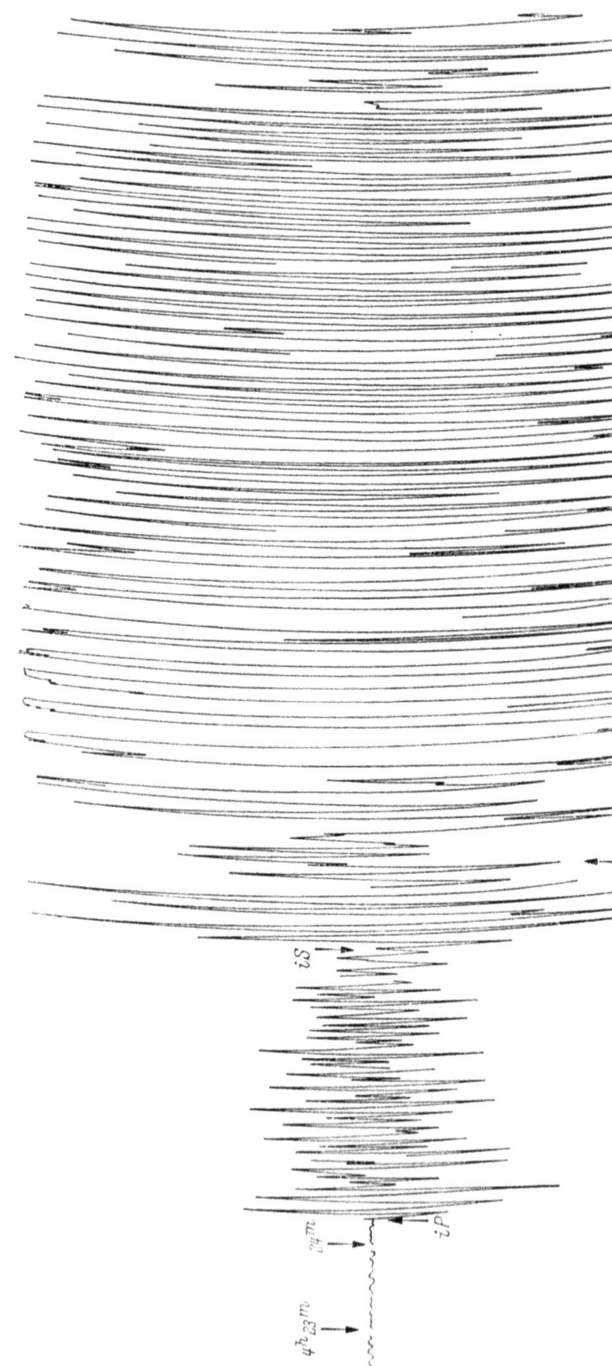

Fig. 25. Typical seismogram, showing P, S and L phases (earthquake of 28. Dec. 1908, epicenter near Messina; record written on a Wiechert seismograph at Hamburg, Germany). After NEUMAYR and SÜESS[1]

[1] NEUMAYR, M., u. F. E. SÜESS: Erdgeschichte 3rd ed. Stuttgart: Bibliogr. Inst. 1920.

LEET[1], SAVERENSKII and KIRNOS[2], RICHTER[3], BULLEN[4] and others[5].

We shall turn our attention first to the facts that may be gleaned from the transmission of seismic waves through the Earth; this yields much information about the Earth's interior. It is clear that it is immaterial whether the source of the seismic wave energy is natural or artificial (conventional or nuclear explosions). The discussion of the nature of seismic foci will be relegated to Sec. 2.2.

2.12. The Basic Division of the Earth into Layers. The study of the records written by natural or artificial earthquakes at seismic stations forms a subject of vast complexity. The distortion of the various seismic phases at discontinuities follows complex laws and many theoretical investigations of these laws have been undertaken. These theoretical investigations have been reviewed in the books on seismology mentioned in Sec. 2.11, and the reader is referred thither for the details.

The main result of the studies of elastic wave transmission within the Earth is that there are two fundamental discontinuities which divide the interior of the Earth into three principal layers called the *crust*, the *mantle*, and the *core*.

The most shallow of these discontinuities has been termed *Mohorovičić discontinuity* (after its discoverer[6]); it lies at depths of from 5 to 60 km beneath the surface of the Earth. On it, the velocity of longitudinal elastic waves jumps from some low value to a uniform value of 8.1 km/sec. This discontinuity divides the crust from the mantle of the Earth.

The second main discontinuity is the core-boundary at a depth of 2900 km beneath the surface of the Earth (discovered by OLDHAM[7] and GUTENBERG[8]). On it, transverse elastic waves disappear, from which it has been inferred that the Earth's core is liquid. However, it has recently come to light that there may be an *inner core* within the core which could again be solid (cf. Sec. 2.14).

The basic division of the Earth into 3 layers, viz. core, mantle, and crust, is fundamental in any study of geodynamics. As noted above,

---

[1] LEET, L. D.: Earth Waves. Cambridge (Mass.): Harvard Univ. Press 1950.

[2] SAVARENSKII, E. F., i. D. P. KIRNOS: Элементы сейсмологии и сейсмометрии. Moscow: Gos. Iz-vo Tekh.-Teoret. Lit. 1955.

[3] RICHTER, C. F.: Elementary Seismology. San Francisco: Freeman & Co. 1958.

[4] BULLEN, K. E.: An Introduction to the Theory of Seismology, 2nd ed. London: Cambridge Univ. Press 1959.

[5] BYERLY, P.: Seismology. New York: Prentice-Hall 1942.

[6] MOHOROVIČIĆ, A.: Jb. met. Obs. Zagreb für 1909 **9**, Pt. 4, 1 (1910).

[7] OLDHAM, R. D.: Quart. J. Geol. Soc. **62**, 456 (1906).

[8] GUTENBERG, B.: Nachr. Ges. Wiss. Göttingen, math.-phys. Kl. **1914**, 1, 125 (1914).

the evidence for this division is chiefly based upon elastic wave propagation studies through the Earth's interior.

2.13. Crustal Studies. We have noted above that it is the Mohorovičić Discontinuity which defines the lower boundary of the crust. It is a most significant result of seismic investigations to have established that the crust thus defined is about 5 km thick beneath oceans and about 35 km thick beneath continents (the bounday of the continents is thereby taken as the 1000 meter depth line).

A confirmation of the broad structure of the Earth's crust as outlined above has been obtained by an analysis of the dispersion of seismic surface waves originating from earthquakes. This dispersion depends on the thickness (and other parameters) of the surface layer, and hence the latter can be inferred from empirically determined dispersion curves. The values obtained in this fashion are entirely in line with those quoted above.

Many detailed studies have been made of the position of the Mohorovičić discontinuity in various areas of the world. Since oceans and continents show a different depth of this discontinuity, let us consider the various studies of continental and of oceanic areas separately.

Turning first to continents, we note that in plains areas, the depth of the Mohorovičić discontinuity is usually about 35 km below the surface. In mountainous areas, it has been claimed that the Mohorovičić discontinuity is depressed so that the corresponding mountains would have *roots*. MINTROP[1,2] has made a study of all the seismic evidence available regarding the possibility of a depression of the Mohorovičić discontinuity (i.e. of the existence of roots) beneath the Alps and beneath the Sierra Nevada. He maintains that such roots do, in fact, *not* exist and that earlier results[3] to the contrary are based on a faulty interpretation of seismic data. However, MINTROP introduces some high velocity layers below the Mohorovičić discontinuity and his visualization of the structure of the Earth's crust is not at all in accordance with the commonly accepted view.

Some of the above difficulties may be resolved by noting that, in addition to the Mohorovičić discontinuity, there are other seismic discontinuities in the Earth's crust. The most important of these is the *Conrad*[4] *discontinuity* where the seismic velocity jumps from approximately 6.1 to 6.4–6.7 km/sec. It exists at varying depths in continental areas and is supposed to separate a "granitic" from a "basaltic" (or

[1] MINTROP, L.: Ann. Geofis., Roma **5**, 163 (1952).

[2] MINTROP, L.: Geol. Rdsch. **41**, 67 (1953).

[3] GUTENBERG, B.: Bull. Geol. Soc. Amer. **54**, 473 (1943).

[4] CONRAD, V.: Laufzeitkurven des Tauernbebens vom 28. Nov. 1923. Wien: Mitt. Erdb. Komm. No. 59 (1925).

"intermediate") layer; not too much chemical significance, however, should be attached to these designations. In the light of the existence of such a Conrad discontinuity, the problem of mountain roots takes on a different slant. It is conceivable that mountains may have roots in the Conrad disontinuity rather than in the Mohorovičić discontinuity. Instances where this has been claimed to be the case have been reported

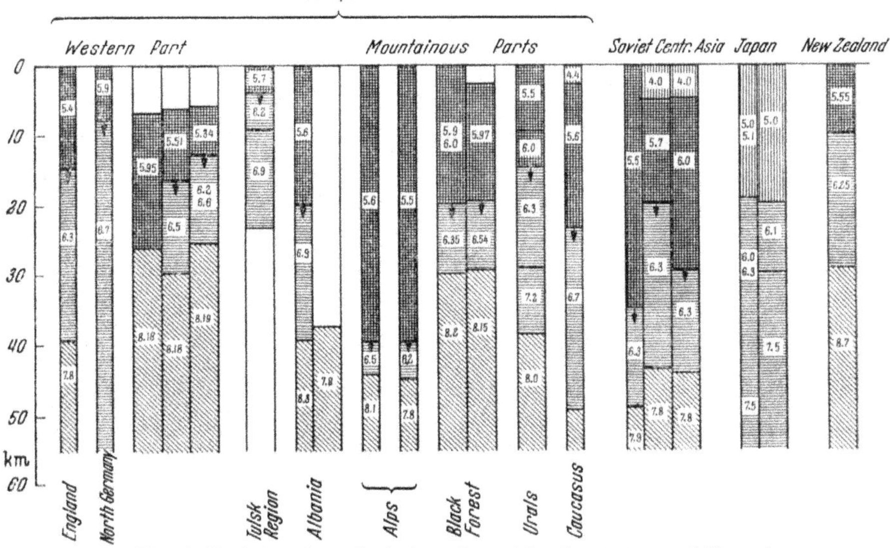

Fig. 26. Typical continental seismic sections. After SAVARENSKII and KIRNOS[1]

from Central Asia[2] and from California[3]. The "mountain roots" may therefore possibly be roots in the "granitic" layer, roots in the "basaltic" layer, or may in certain instances not exist at all.

A series of typical continental seismic velocity profiles is shown in Fig. 26.

Turning now to *oceanic areas*, we recall again the remarkable result that the Mohorovičić discontinuity is, beneath oceans, at a depth of only about 5 km below the solid surface. Two typical oceanic sections are shown in Fig. 27. The first of these sections (Type A) applies to normal deep sea areas and the second (Type B) to areas where an archipelagic apron is present.

No really satisfactory seismic sections across mid-ocean ridges seem to have been published, so that the structure of the crust in the vicinity

[1] SAVARENSKII, E. F., i D. P. KIRNOS: Элементы сейсмологии и сейсмометрии. Moscow 1955.

[2] KOSMINSKAYA, I. P., G. G. MIKHOTA i YU. V. TULINA: Izv. Akad. Nauk SSSR, Ser. Geofiz. **1958**, No. 10, 1162 (1958).

[3] GUTENBERG, B.: Geol. Rdsch. **46**, 30 (1957).

of the ridges is still somewhat of a mystery, although EWING and EWING[1] seem to have evidence that there may be a root beneath the Mid-Atlantic Ridge. A cross-section of the Mid-Atlantic Ridge as postulated by these authors is shown in Fig. 28.

The results of detailed investigations made in the neighborhood of some oceanic islands are more reliable. WOOLLARD[2] has compiled and collated the available data. A typical section as postulated by him is shown in Fig. 29 (incorporating also gravity data; cf. Sec. 2.3).

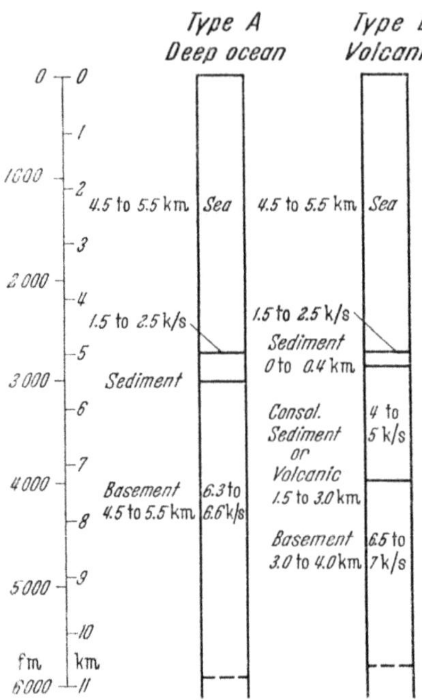

Fig. 27. Two typical sections of the Earth's crust in oceanic areas; (A) in an abyssal plain, (B) in area where an archipelagic apron is present (after GASKELL[5])

Most interesting are the regions where the oceanic and the continental types of crust meet, i.e. the continental margin. Much seismic work has been done in such regions, but it turned out to be rather difficult to trace the Mohorovičić discontinuity exactly. The available information has been collected by OFFICER et al.[3]; an abstract of this compilation is shown in Fig. 30, presenting a cross section across a typical continental margin with adjacent island arc and foredeep. Below volcanoes in island arcs, GORSHKOV[4] has discovered magma chambers by seismic investigations. These are at a depth of 50–60 km.

2.14. The Interior of the Earth. We are now turning our attention to the Earth's interior, i.e. to the regions below the Mohorovičić discontinuity. Again, the study of seismograms yields much information with regard to the elastic wave transmission properties of the interior of the Earth[6]. The investigations bearing upon this problem have been made

[1] EWING, J., and M. EWING: Bull. Geol. Soc. Amer. **70**, 291 (1959).
[2] WOOLLARD, G. P.: Proc. Roy. Soc. Lond., Ser. A **222**, 361 (1954).
[3] OFFICER, C. B.: Bull. Geol. Soc. Amer. **68**, 359 (1957).
[4] GORSHKOV, G. S.: Bull. Volcan. **19**, 103 (1958).
[5] GASKELL, T. F.: Proc. Roy. Soc. Lond., Ser. A **222**, 341 (1954).
[6] BULLEN, K. E.: An. Introduction to the Theory of Seismology. London: Cambridge Univ. Press 1959.

by various people. JEFFREYS obtained a graph of the two fundamental wave-velocities with depth; it is presented (solid line) in Fig. 31.

As noted earlier, a principal discontinuity exists in the Earth's interior at a depth of 2900 km, characterized by the fact that no shear waves

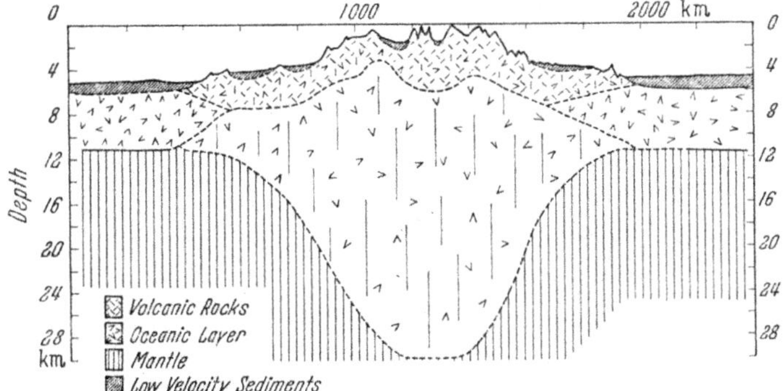

Fig. 28. Structure section across the Mid-Atlantic Ridge south of Azores (after EWING and EWING[1])

can be transmitted across it. This has given rise to the assumption that this discontinuity defines the boundary of a *liquid core* of the Earth. A further discontinuity in the neighborhood of 5000 km depth suggests the existence of an *inner core* within the core. Speculations have been that this inner core is again solid. This is, however, not yet very certain.

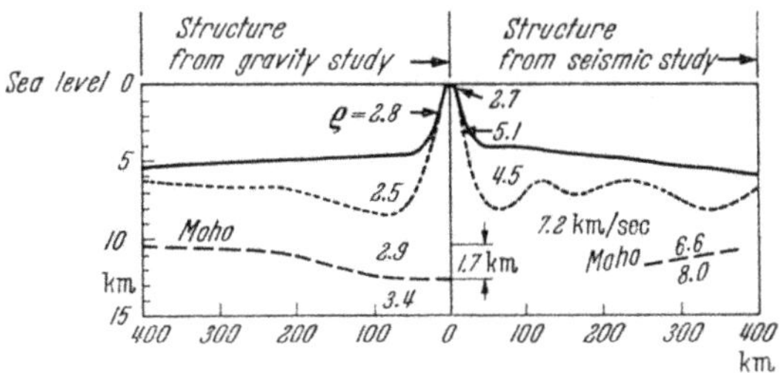

Fig. 29. Interpretation of gravity and seismic studies across a typical oceanic island (Bermuda) in terms of crustal structure (after WOOLLARD[2])

An alternative curve of P and S velocities has been deduced by GUTENBERG[3]. It is also shown in Fig. 31 (dotted line). It will be observed that in most essential points, JEFFREYS' and GUTENBERG'S curves are in substantial agreement.

The existence of a further discontinuity at a depth of about 900 km has been postulated by BIRCH upon chemical grounds. It will be discussed more fully in Sec. 2.81.

---

[1] EWING, J., and M. EWING: Bull. Geol. Soc. Amer. **70**, 291 (1959).

[2] WOOLLARD, G. P.: Proc. Roy. Soc. Lond., Ser. A **222**, 361 (1954).

[3] GUTENBERG, B.: Trans. Amer. Geophys. Un. **32**, 373 (1951).

JEFFREYS' values for the P and S velocities shown in Fig. 31 can be used to estimate the density distribution within the Earth[1]. The equation of equilibrium at distance r from the center of the Earth postulates

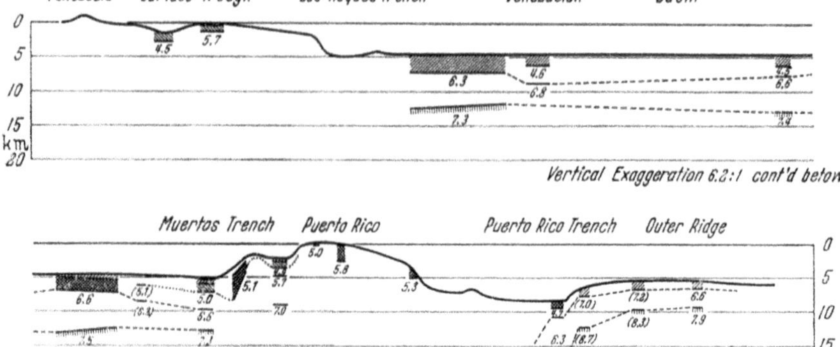

Fig. 30. Typical cross section across a continental margin, with adjacent island arc and foredeep (north-south from Caribbean to Atlantic). After OFFICER *et al.*[2]

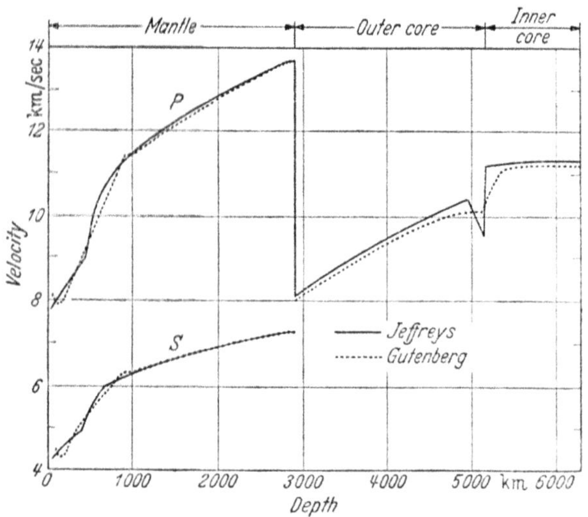

Fig. 31. P and S velocity distributions in the Earth's interior. Solid line after JEFFREYS, dotted line after GUTENBERG[3] (after BULLEN[1])

---

[1] Cf. e.g. BULLEN, K. E.: An Introduction to the Theory of Seismology. London: Cambridge Univ. Press 1959.

[2] OFFICER, C. B. *et al.*: Bull. Geol. Soc. Amer. **68**, 359 (1957).

[3] GUTENBERG, B.: Trans. Amer. Geophys. Un. **32**, 373 (1951).

(after BULLEN[1])

$$dp/dr = -g\varrho = -\varkappa m\varrho/r^2 \tag{2.14-1}$$

where p is the pressure, g the local gravitational pull, ϱ the density, m the mass of the matter inside a sphere of radius r and $\varkappa$ the gravitational constant. Assuming adiabatic conditions in a homogeneous substance, we have

$$\frac{k}{\varrho} = \frac{dp}{d\varrho} = c_p^2 - \frac{4}{3}c_s^2 \tag{2.14-2}$$

where k is the (adiabatic) incompressibility and c_p, c_s denote the local P and S velocities, respectively. This result is a consequence of infinitesimal elasticity theory [cf. Eq. (3.21-17/18) and (3.21-6c)]. Hence

$$\frac{d\varrho}{dr} = -\frac{\varkappa m\varrho}{r^2\left(c_p^2 - \frac{4}{3}c_s^2\right)}. \tag{2.14-3}$$

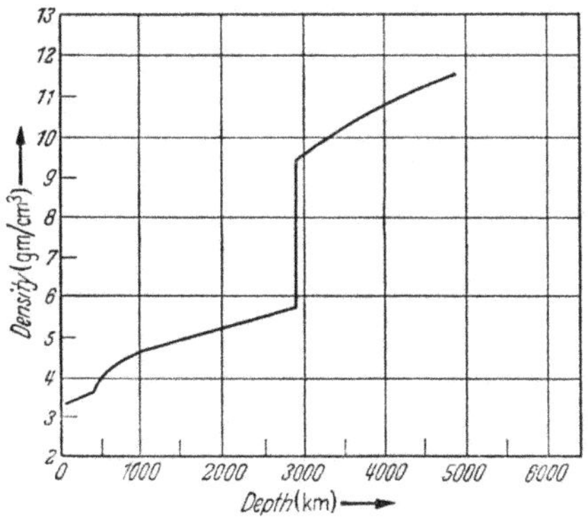

Fig. 32. Density distribution in the Earth's interior (after BULLEN[1])

This equation can be evaluated numerically, but there is some difficulty in determining the analytical continuation across the surfaces of discontinuity. Fortunately, it turns out that various reasonable assumptions that can be made do not influence the result very much; thus, the density variation is probably pretty close to that shown in Fig. 32, as calculated by BULLEN[2].

If one collects all the information presented above, one arrives at a *model* of the Earth. In essence, the above discussion refers to BULLEN's model; other models have been suggested by MOLODENSKII. In constructing Earth models, the Earth is split into various layers with homogeneous composition and with more or less homogeneous physical properties. BULLEN's (principal) model is shown in Fig. 33.

Turning now to some of the regions in detail, we note that it has been found that a low-velocity layer exists in the Earth's upper mantle (region B); this low-velocity layer is often called *asthenosphere channel*. GUTENBERG[3] was probably the first to postulate this channel as early

[1] Cf. e. g. BULLEN, K. E.: An Introduction to the Theory of Seismology. London: Cambridge Univ. Press 1959.

[2] Inhomogeneities may affect BULLEN's calculations. See BULLARD, E. C.: Verh. K. Ned. Geol. Mijnb. Genoot. **18**, 23 (1957).

[3] GUTENBERG, B.: Z. Geophys. **2**, 24 (1926).

as in 1926. He substantiated his claim in a series of later papers[1-5]; the low velocity channel was subsequently also found by other people[6-9]. Thus, it is fairly well established that the velocity begins to decrease near the Mohorovičić discontinuity and stays below the value immediately below the Mohorovičić discontinuity to a depth of almost 200 km.

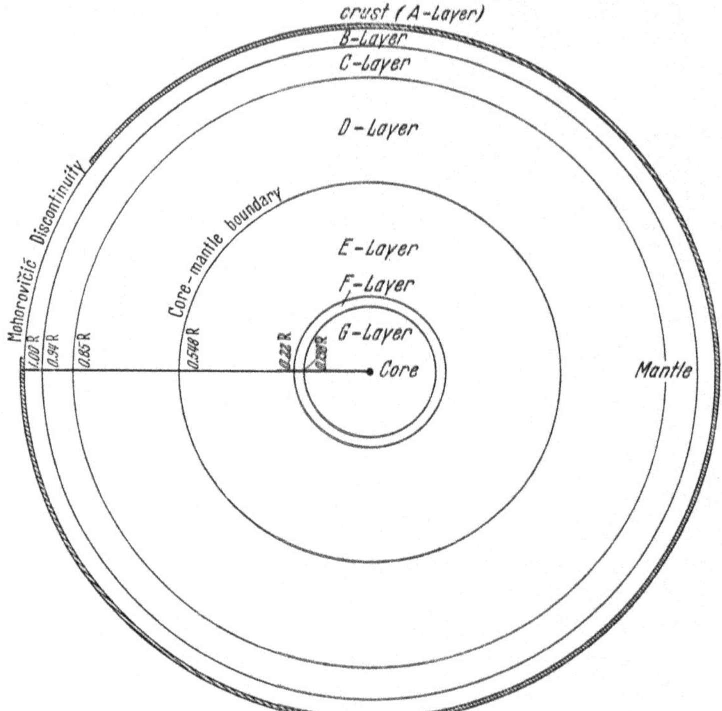

Fig. 33. BULLEN's principal Earth model

There may even be some regional variation in this inhomogeneity of the Earth's mantle[10]. The detailed S and P velocities across the low velocity layer, as postulated by GUTENBERG, are shown in Fig. 34.

[1] GUTENBERG, B.: Bull. Geol. Soc. Amer. **65**, 337 (1954).
[2] GUTENBERG, B.: Bull. Geol. Soc. Amer. **66**, 1203 (1955).
[3] GUTENBERG, B.: Trans. Amer. Geophys. Un. **39**, 486 (1958).
[4] GUTENBERG, B.: Geophys. J. Roy. Astr. Soc. **2**, 348 (1959).
[5] GUTENBERG, B.: Ann. Geofis., Roma **12**, 439 (1959).
[6] SHIROKOVA, E. I.: Izv. Akad. Nauk SSSR., Ser. Geofiz. **1959**, 1127 (1959).
[7] VESANEN, E. *et al.*: Geophysica, Helsinki **7**, 1 (1959).
[8] LEHMANN, I.: Geophys. J. Roy. Astr. Soc. **4**, 124 (1961).
[9] MACDONALD, G. J. F., and N. F. NESS: J. Geophys. Res. **66**, 1865 (1961).
[10] BULLEN, K. E.: Trans. Amer. Geophys. Un. **35**, 838 (1954).

The next of BULLEN's layers (the C layer) has always been a region of some controversy. We have already mentioned the contention of BIRCH who noted that a chemical transition must take place at this depth. In addition, studies of the heat flow also indicate that some rather interesting phenomena must occur at this depth. However, since the reasoning is mostly based upon thermodynamic arguments, we shall refer to these investigations in their proper context.

Finally, we turn our attention to the *core*. As noted earlier, there are indications that there is a solid inner core within the outer liquid core. This was first postulated by LEHMANN[1] and seems to have been substantiated later on. The question, however, is not entirely settled as of yet.

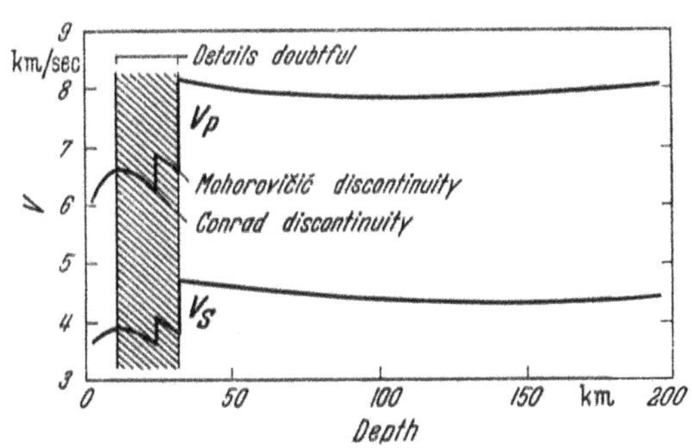

Fig. 34. GUTENBERG's low velocity layer in the Earth's upper mantle (V_p for P-, V_S for S-waves). After GUTENBERG[2]

2.2. Focal Studies

2.21. General Remarks. We shall now turn our attention to the analysis of the occurrence and the mechanism of the earthquakes themselves. The investigation of the occurrence of earthquakes leads to *seismicity* studies, the investigation of the focal mechanism leads to *magnitude* studies and *fault* plane studies.

We shall discuss these topics in their turn.

2.22. Seismicity Studies. Seismicity studies have as their objective a definition of the geographical distribution of earthquakes. In order to achieve this aim, it is necessary to make epicenter determinations of as many shocks as is possible and to plot the latter on a map. It is obvious that this can be done only for such shocks which have occurred since the introduction of reasonably accurate instrumentation, i.e. since about 1904. One of the most accurate collections of data bearing upon this subject has been compiled by GUTENBERG and RICHTER[3].

One of the outcomes of seismicity studies is that the foci of earthquakes may occur at various levels, to a depth of about 700 km. It is therefore convenient to separate earthquakes according to their depth

[1] LEHMANN, I.: Publ. Bur. Centr. Séism. Int. A **14**, 3 (1936).

[2] GUTENBERG, B.: Bull. Geol. Soc. Amer. **66**, 1203 (1955).

[3] GUTENBERG, B., and C. F. RICHTER: Seismicity of the Earth and Associated Phenomena. Princeton: Princeton Univ. Press 1949. [Second edition 1954.]

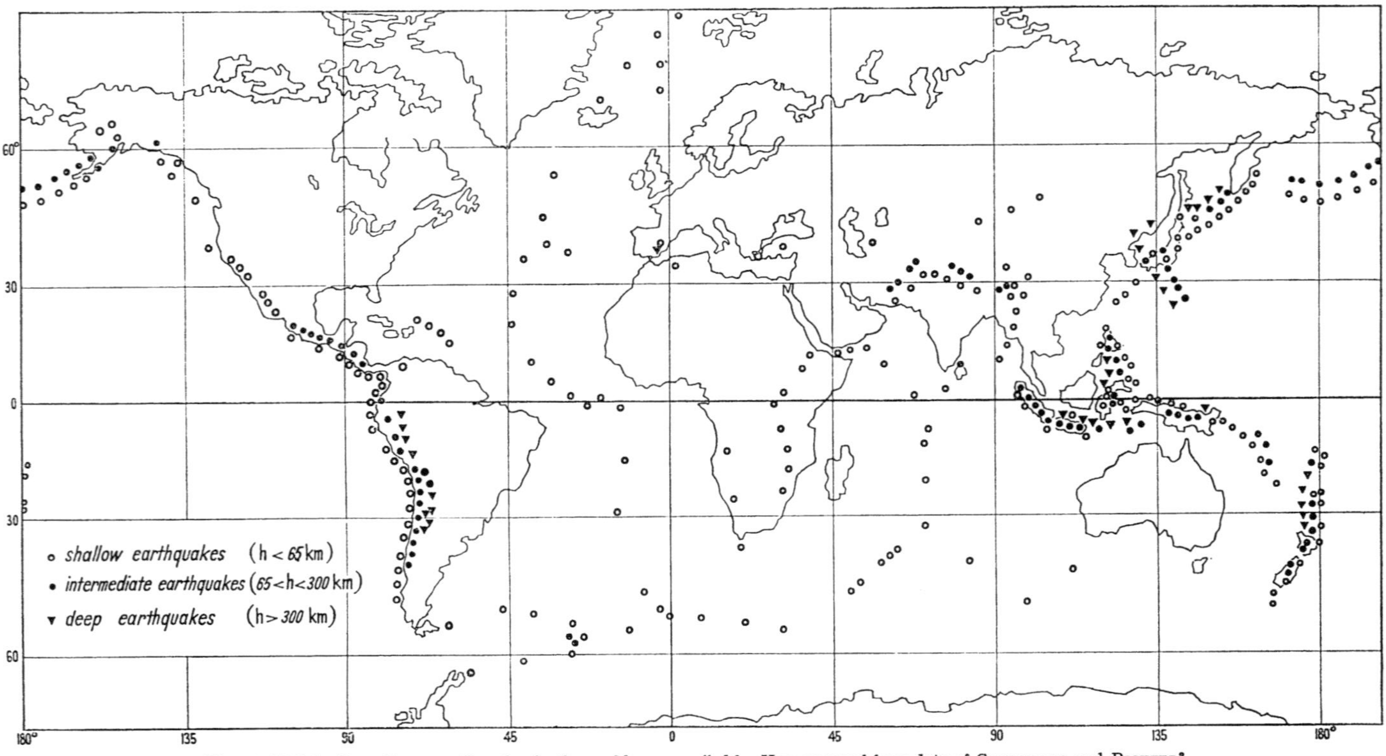

Fig. 35. Distribution of large earthquakes in the world, as compiled by HIERSEMANN[1] from data of GUTENBERG and RICHTER[2]

[1] HIERSEMANN, L.: Freiberger Forschungsh. C **24** (1956).

[2] GUTENBERG, B., and C. F. RICHTER: Seismicity of the Earth and Associated Phenomena. Princeton: Princeton Univ. Press 1949. [Second edition 1954.]

of focus; a common classification of earthquakes is into *shallow* ones (depth of focus less than 65 km), *intermediate* ones (depth of focus between 65 and 300 km) and *deep* ones (depth of focus more than 300 km). A distribution map of the large shallow and deep shocks as compiled by HIERSEMANN[1] from data by GUTENBERG and RICHTER is shown in Fig. 35. An inspection of this map makes it at once obvious that the earthquakes are distributed mainly in the zones of recent orogenetic activity.

Seismicity studies have been extended, particularly by Russian authors[2-4], to the seismic regionalization of earthquake areas. This per-

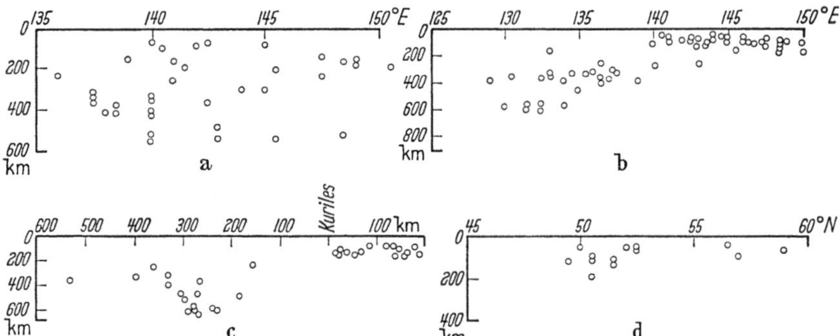

Fig. 36a—d. Vertical cross section through epicentral zones of the northern Pacific Coast: (a) Mariana Islands area; (b) Japan area; (c) Kurile Islands area; (d) Aleutian Islands area. After KOGAN[5]

mits one to assign quantitatively an earthquake-risk number to any one region which is particularly important from an engineering standpoint.

GUTENBERG and RICHTER postulated that the distribution of foci beneath orogenetic (mountain and island) arcs is characteristic of the latter. Accordingly, earthquakes would occur in localized zones which are almost planar, dipping at some intermediate angle (30 to 60°) into the Earth (see Fig. 40). However, this view has recently been challenged by some Russian scientists who claim that the pattern of distribution of earthquake foci beneath an orogenetically active island arc is not as simple as had been envisaged by GUTENBERG and RICHTER. It thus appears that the "plane" postulated by GUTENBERG and RICHTER is the lower boundary of earthquake foci beneath an island arc, but that the whole sector above it is more or less filled-in with foci. Fig. 36 shows the results of KOGAN's[5] investigations in this regard.

---

[1] HIERSEMANN, L.: Freiberger Forschungsh. C **24** (1956).
[2] BELOUSOV, V. V.: Izv. Akad. Nauk SSSR., Ser Geofiz. **1954**, No. 3, 209 (1954).
[3] RIZNICHENKO, YU. V.: Ann. Geofis., Roma **12**, 227 (1959).
[4] GUBIN, I. E.: Закономерности сейсмических проявлений на территории Таджикистана. Moscow: Iz-vo Akad. Nauk SSSR. 1960.
[5] KOGAN, S. D.: Dokl. Akad. Nauk SSSR. **101**, 63 (1955).

A further remark concerns the observation that there seems to be a lack of earthquake foci at a depth of 475 km. GUTENBERG[1] has investigated this question from the energy-standpoint and found that there is a minimum of energy-release at that depth.

2.23. Magnitude Studies. After an earthquake occurs, it is a natural desire to try to ascribe to it a certain value representative of its intensity. The original intensity scales were set up by investigating the destructiveness of earthquakes upon human structures. It is obvious, however, that a classification achieved in this manner depends not only on the intrinsic severity of the earthquake, but also on the type and number of man-made structures that happen to be in the epicentral area. Modifications of the scale have been attempted by specifying more exactly the type of structures affected, the type of ground they stand on etc. In spite of these efforts, intensity scales arrived at in this manner remain at best mostly descriptive.

Any serious attempts to assign a value to the severity of an earthquake, therefore, must be based upon an analysis of instrumental records. Such a value, called *magnitude*, has been defined empirically by RICHTER[2]. Accordingly, the relation between magnitude M of two earthquakes at a standard epicentral distance of 100 km, and for the maximum recorder trace amplitude B is

$$M_1 - M_2 = \log B_1 - \log B_2 \qquad (2.23\text{–}1)$$

where the seismogram is supposed to have been written on a standard torsion seismometer of free period 0.8 sec, static instrumental magnification of 2800 and damping ratio of 50:1. The zero of the scale is defined by setting $M=3$ for $B=1$ mm.

The above definition of earthquake magnitude was devised for the analysis of local shocks in southern California. In order to assign magnitudes to more distant shocks and to shocks observed elsewhere, various different scales have been devised, all based upon measurements of certain "standard" amplitudes in seismograms, in conjunction with an equation of the type shown in (2.23–1). The possibilities have been summarized by GUTENBERG and RICHTER[3]. Accordingly, a magnitude M can be determined, in addition to the Richter-magnitude outlined above, from the amplitudes of surface waves for shallow teleseisms, and from the amplitude-to-period ratio of body waves for teleseisms, shallow and deep-focus. The scales, which are all based on Eq. (2.23–1), were originally adjusted to coincide at a magnitude of 7, but it was subsequently

[1] GUTENBERG, B.: Vening Meinesz Anniversary Volume (publ. by Kon. Ned. Geolog.-Mijnbowkundig Genot.), p. 165. 1957.

[2] RICHTER, C. F.: Bull. Seism. Soc. Amer. **25**, 1 (1935).

[3] GUTENBERG, B., and C. F. RICHTER: Ann. Geof., Roma **9**, 1 (1956).

found impossible to have them coincide everywhere. It has therefore been suggested[1] to use as a final "unified" magnitude m a certain average of these various magnitudes. This ambiguity, of course, is a reflection of the fact that analogous parts of the seismograms of two earthquakes are not proportional, not even on identical instruments. In view of this it must be conceded that the term "magnitude" does not yield an absolute characterization of an earthquake; rather it indicates the amplitude of a particular part of a seismogram which that earthquake has produced on a particular instrument.

An unambiguous indication of the intensity of an earthquake would be obtained if it were possible to give the amount of energy released by it. Unfortunately, it is very difficult to do this. In fact, the only indication of the energy released in an earthquake is obtained by an inference from the above-mentioned, incompletely defined magnitude scales.

Attempts to correlate magnitude with earthquake energy E started in 1942 when GUTENBERG and RICHTER[2] correlated the amplitude expectable in a particular part of a seismogram with the total energy released as estimated from the total energy flux going through one station, calculated from the seismic trace by using elasticity theory. The originally proposed connection between magnitude and energy has since been modified several times[3], and the relationship at present[3] suggested is

$$\log E = 9.4 + 2.14M - 0.054M^2 \qquad (2.23\text{–}2)$$

where E is the energy in ergs and M the magnitude.

Eq. (2.23–2) is a quadratic relationship between $\log E$ and M. It may be seen, however, that a linear relationship is a sufficient approximation in a significant range. In particular, if the "unified" magnitude m mentioned above (which is a particularly weighted mean of the other magnitudes) is used, one can express the energy relationship as follows

$$\log E = 5.8 + 2.4\,m. \qquad (2.23\text{–}3)$$

This relationship implies essentially

$$E \sim B^{2.4} \qquad (2.23\text{–}4)$$

However, it should be noted that LATTER *et al.*[4] stated, from the analysis of atomic bomb blasts, that

$$E \sim B \qquad (2.23\text{–}5)$$

would be a better formula.

[1] GUTENBERG, B., and C. F. RICHTER: Ann. Geof., Roma **9**, 1 (1956).
[2] GUTENBERG, B., and C. F. RICHTER: Bull. Seism. Soc. Amer. **32**, 163 (1942).
[3] GUTENBERG, B., and C. F. RICHTER: Bull. Seism. Soc. Amer. **46**, 105 (1956).
[4] LATTER, A. I., E. A. MARTINELLI and E. TELLER: Phys. Fluids **2**, 280 (1959).

With regard to the various magnitudes mentioned in the course of the present discussion, the corresponding relationships with energy have been compiled by GUTENBERG and RICHTER[1] and are shown in Table 4.

Table 4. *Comparison of Various Earthquake Magnitude Scales and their Relationship with Energy.* (After GUTENBERG and RICHTER[2])

Richter magnitude M	3.0	4.0	5.0	6.0	7.0	8.0	9.0
Teleseismic magnitude M (mean from surface and body waves)	2.4	3.6	4.7	5.8	6.8	7.9	8.9
"Unified" magnitude m	4.0	4.7	5.4	6.1	6.8	7.5	8.1
log energy (ergs)	15.4	17.2	18.9	20.5	22.1	23.7	25.2

In view of the ambiguity of defining a magnitude in the first place, the various postulated relationships are presumably equally ambiguous. At any rate, it is possible to obtain a rough idea of the order of magnitude (within a factor of 10 or so) of the energy released in a given earthquake, by studying the amplitudes produced in certain parts of its seismograms. It is doubtful indeed whether more can be hoped for. Obviously, it must be expected that the frequency-spectrum of energy release is highly characteristic of individual earthquakes. It is therefore very questionable whether a comparison of the energy released, by, say, two earthquakes could be obtained simply by comparing analogous parts of the corresponding seismograms, which is the procedure inherent in using magnitudes for energy-determination. A better procedure would be to calculate the energy directly from the total seismic trace which an earthquake produces at a station.

In a strained elastic body, the strain is proportional to the square root of the elastic energy stored. It is therefore possible to interpret the energy released in an earthquake in terms of a corresponding *strain release.* If the cumulative strain release of all earthquakes of the world is plotted, it may thus be possible to obtain an idea of the corresponding total strain-build-up in the world. This has been done by BENIOFF[3] taking

$$\text{Strain} \sim \sqrt{\text{Energy}} \qquad (2.23\text{–}6)$$

for shallow and deep earthquakes separately (see Figs. 37a and 37b). In viewing BENIOFF's figures it should be recalled that the constants in the magnitude-energy relationship are not very well established; in fact, BENIOFF used an older form of that relationship rather than that given

[1] GUTENBERG, B., and C. F. RICHTER: Ann. Geofis., Roma **9**, 1 (1956).
[2] GUTENBERG, B., and C. F. RICHTER: Bull. Seism. Soc. Amer. **46**, 105 (1956).
[3] BENIOFF, H.: Geol. Soc. Amer. Spec. Pap. **62**, 61 (1955).

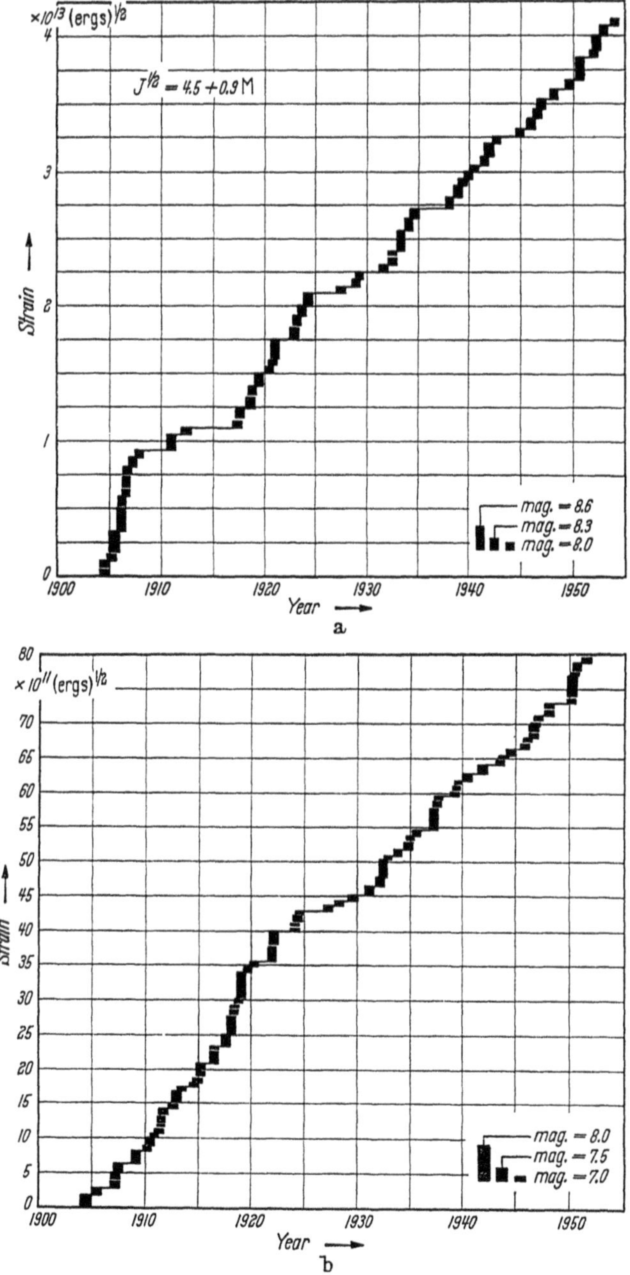

Fig. 37a and b. Strain release (a) in the world's shallow earthquakes and (b) in the world's deep earthquakes. After BENIOFF[1]

[1] BENIOFF, H.: Geol. Soc. Amer. Spec. Pap. **62**, 61 (1955).

in (2.23–3). Thus, only the general shape of the curves, and not the particular values represented by them, is of significance. It is seen that there appears to be a more or less continuous creation of strain in deep and shallow earthquakes alike. However, the release of this strain in the upper layers appears to occur in periods of high activity which alternate with periods of relative acquiescence, whereas in the deep layers the release of strain is more or less steady.

The above theory of earthquake magnitudes is essentially that developed at the California Institute of Technology. Other people have also worked on the subject; a partial list of references is given below[1–9].

2.24. Fault Plane Studies. Investigations into the mechanism at the focus of an earthquake resulted into what is called *fault plane studies*. Such investigations were initiated some thirty years ago, but it has been only recently that they have been made on a sufficient number of earthquakes to permit world-wide effects to become recognized.

We have already stated that the cause of a seismic disturbance is confined to a relatively small region within the Earth which is termed the "focus" of the earthquake. Within the focus, a mechanical event must take place such as the sudden occurrence of failure of the material. Such a phenomenon would of necessity be connected with displacements within the focal region which, in turn, would cause seismic waves. The nature of the seismic waves originating from the focal region, therefore, sould be an indication of the displacements occurring therein. In this instance, it cannot be expected that it will be possible to infer every detail of the focal mechanism from seismic waves, but rather the general behavior of the focal region as a whole. As a first approximation, therefore, it seems reasonable to enclose the focal region within a sphere (termed "focal sphere")[10] and to study the motion of the surface of this sphere as it may be inferred from seismic evidence. The size of the focal sphere must be such that it encloses the whole region of the focus in which mechanical deformation may have taken place during the earthquake, but that it is still small compared with the size of the whole Earth.

---

[1] Bisztricsany, E. de: Z. Geophys. **24**, 154 (1958).
[2] Filippo, D. di, L. Marcelli: Pub. Bur. Centr. Séism. Int. A **20**, 17 (1959).
[3] Iida, K., and H. Aoki: J. Earth. Sci., Nagoya **4**, No. 2, 63 (1956).
[4] Kogan, S. D.: Izv. Akad. Nauk SSSR., Ser. Geofiz. **1959**, No. 9, 1372 (1959).
[5] Panner, N.: Československ. Akad. Věd. Stud. Geophys. et Geod. **3**, No 3, 242 (1959).
[6] Savarenskii, E. F.: Ann. Geofis., Roma **12**, 369 (1959).
[7] Savarenskii, E. F. *et al.:* Izv. Akad. Nauk SSSR., Ser. Geofiz. **1960**, 633 (1960).
[8] Solov'ev, S. L.: Izv. Akad. Nauk SSSR., Ser. Geofiz. **1956**, No 3, 357 (1956).
[9] Solovyov, S. L.: Publ. Bur. Centr. Séism. Int. A **20**, 39 (1959).
[10] Scheidegger, A. E.: Trans. Roy. Soc. Can. **49**, Ser. III, Sec. 4, 65 (1955).

The various phases of a seismic disturbance, as has also already been pointed out, may be thought to have travelled along curved paths. The direction by which these phases leave the focal region can be calculated from travel-time tables. Thus, to each phase observed at any one seismic station corresponds a point on the focal sphere, viz. the point where the direction which the phase travelled when starting out from the focus intersects the focal sphere. If the motion near the focal sphere corresponding to each phase can be calculated from the observed motion at the station,

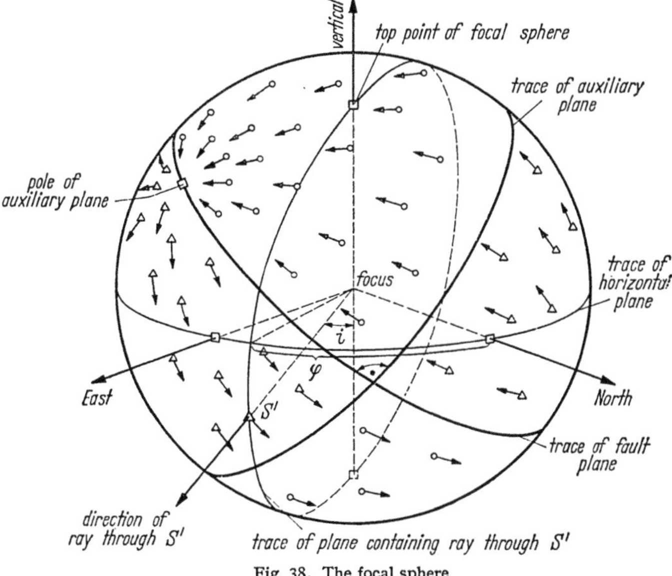

Fig. 38. The focal sphere

the geometrical change of the surface of the focal sphere during the occurrence of the earthquake considered can be inferred. Finally, it is a problem of mechanics to determine the possible failure patterns inside the focal sphere that might produce the observed displacements on the surface.

The problem is thus to determine, from seismic evidence, the displacement pattern on the focal sphere. This can be achieved by means of the hypothesis of conservation of phase signs[1]. This hypothesis can best be demonstrated by referring the reader to Fig. 38 representing the focal sphere. The vector of the displacement at each point of the surface of the focal sphere can be split into its radial and tangential (to the surface of the sphere) component. It is easy to see that these components, if the mode of dislocation of the focal sphere is assumed to correspond to that of an orange which is sliced down the middle with one part being shifted over the other, are as depicted in Fig. 38. In this figure, a radial

[1] SCHEIDEGGER, A. E.: Bull. Seism. Soc. Amer. **47**, 89 (1957).

component "out" from the center of the focal sphere is denoted by O, a radial component "in" by Δ. The tangential components are shown as arrows. The tangential components may further be split into a component lying in the plane of the ray through the point in question (called *SV*) and another one at right angles to it (called *SH*). The hypothesis of conservation of phase signs can then be stated as follows: "the directions (signs) of the phases *P*, *SH* and *SV* arriving at a seismological observatory, are identical to the directions on the focal sphere in that point through which the ray has passed".

It is thus evident that the radial displacements of the focal sphere can be determined from *P* phases, whereas for the tangential displacements an analysis of *S* waves is necessary. It so happens that it is much easier to read *P* phases from seismograms than *S* phases, and therefore the direction of the radial displacement of the focal sphere can be obtained for many more points than the direction of the tangential displacement. In general, one has thus to be content with the knowledge of the radial displacements only. From these, however, it cannot be decided which of the solidly drawn two planes (in Fig. 38) is the true "fault" plane, and which is the plane orthogonal to it (the latter is often termed "auxiliary plane" of the fault). "Fault plane solutions" obtained from *P*-phases only, are therefore ambiguous to this extent.

The technique of obtaining fault plane solutions from seismic data has been developed by Japanese workers on the subject[1]. It has later been perfected by Byerly[2] and by Keilis-Borok[3]. Solutions have been produced by many researchers, mostly Japanese, Canadian and Russian[4]. At the beginning of 1959, fault plane solutions of a total of 336 earthquakes have been available and the result of these investigations is summarized in Tables 5 to 8[5]. For this purpose, the world has been divided into 9 regions as shown in Fig. 39. Table 5 gives for the 9 regions, as well as for the world as a whole, the percentage of tensional and compressional earthquakes and the ratio of the two; the percentage of transcurrent and dipslip earthquakes and the ratio of the two; and the average slip angle (i.e. the angle of the motion vector with the horizontal).

[1] Cf. Kawasumi, H.: Publ. Bur. Centr. Séism. Int. A **15** (pt. 2), 258 (1937) for a summary.

[2] Byerly, P.: Geol. Soc. Amer. Spec. Pap. **62**, 75 (1955).

[3] Cf. Keilis-Borok, V. I.: Trav. Sci. Assoc. Int. Séismol., U.G.G.I. **19** (1956); for a summary in English.

[4] A *summary* of the fault plane solutions available in the literature up to 1957 has been given by the writer in Trans. Roy. Soc. Can., l.c.; in Publ. Dom. Obs. Ottawa **18**, No. 3 (1957), and in Geofis. Pura Appl. **38** (1957); and by Schäffner, H. J.: Freiberger Forschungsh. C **63** (1959).

[5] After Scheidegger: Bull. Seism. Soc. Amer. **49**, 337 (1959). Of the 336 earthquakes many were very small, so that only 265 were used for the calculation of the Tables.

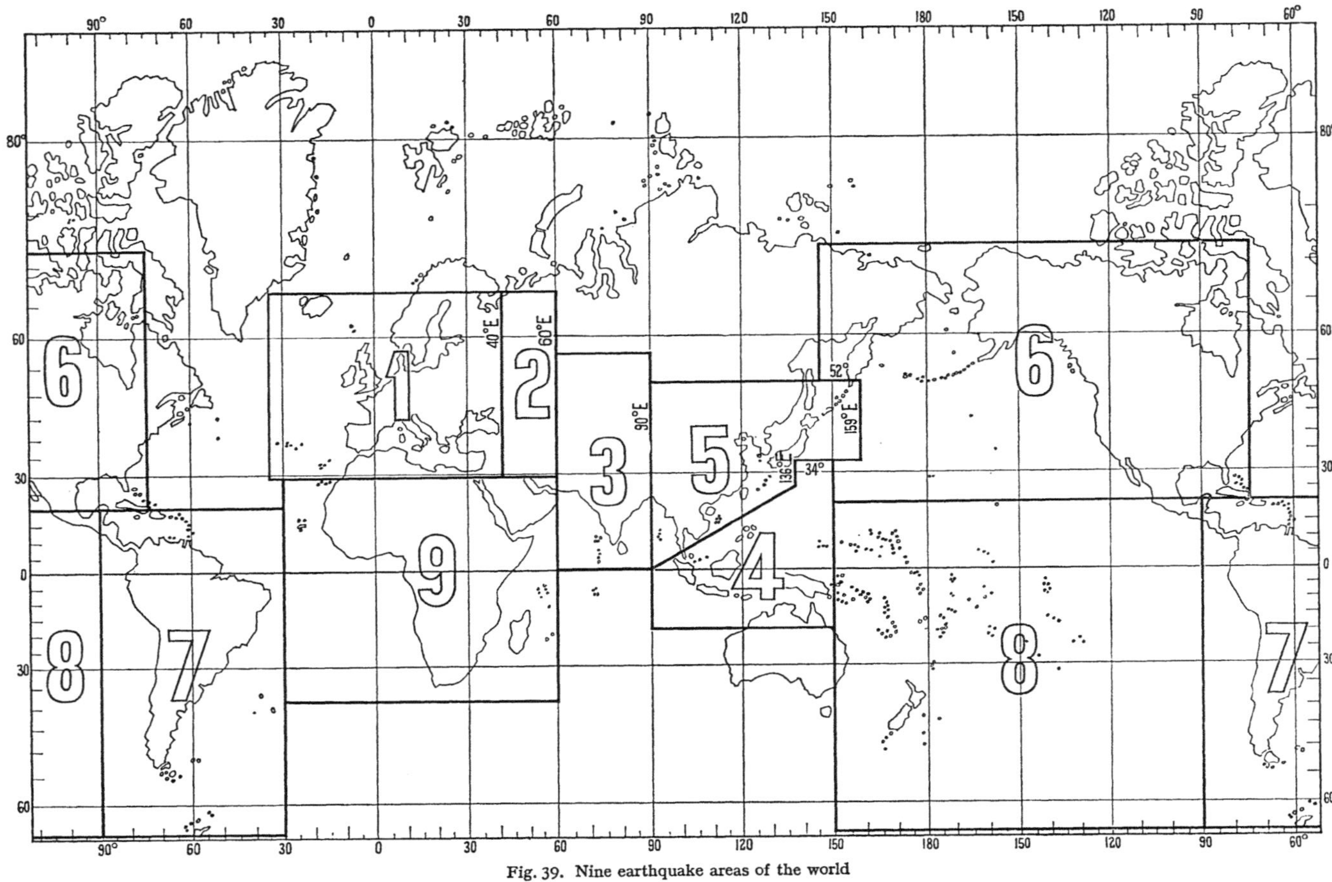

Fig. 39. Nine earthquake areas of the world

It is also possible to analyse the distribution of faulting with depth (see Table 6). If this be done, a very peculiar situation occurs at $0.03 R$: it is seen that the ratio of the number of tension to pressure earthquakes as well as the ratio of the number of strike-slip to dip-slip earthquakes becomes a minimum. The slip angle, consequently, has a maximum at that depth.

The earthquakes analyzed in Table 5 have then ben split into "shallow" and "deep" ones. Therein, an earthquake was deemed "shallow" if the depth of focus was less or equal to 140 km, all others were deemed "deep". The results for "shallow" earthquakes are shown in Table 7, the results for the "deep" earthquakes are shown in Table 8.

It may be remarked that the calculation of the various averages in the Tables is in fact, somewhat crude. The earthquakes that have been analyzed in the literature contain shocks of many different magnitudes, and it would appear that in the calculation of averages one should weight the earthquakes accordingly. Much as this would have been desirable,

Table 5. *All Earthquakes*

Region	No. of earth-quakes	Per cent			Ratio ten. to pres.	Per cent			Ratio st. sl. to dip sl.	Avg. slip angle
		Pres.	Ten.	Ind.		Strike slip	Dip slip	Ind.		
1	17	17.6	52.9	29.4	3.000	88.2	11.8	0.0	7.500	23
2	12	58.3	41.7	0.0	0.714	12.5	87.5	0.0	0.143	67
3	14	64.3	28.6	7.1	0.444	28.6	60.7	10.7	0.470	64
4	59	40.7	45.8	13.6	1.125	47.4	50.8	1.7	0.933	45
5	73	27.4	16.4	56.2	0.600	75.3	22.6	2.1	3.333	26
6	32	40.6	59.4	0.0	1.462	76.6	21.9	1.6	3.500	30
7	13	23.1	53.8	23.1	2.333	92.3	3.8	3.8	24.000	19
8	43	65.1	25.6	9.3	0.393	94.1	3.5	2.3	27.000	14
9	2	50.0	50.0	0.0	1.000	100.0	0.0	0.0	∞	26
World	265	40.8	35.8	23.4	0.880	68.8	28.9	2.3	2.386	32

Table 6. *Distribution with Depth*

Depth	No. of earth-quakes	Per cent			Ratio ten. to pres.	Per cent			Ratio st. sl. to dip sl.	Avg. slip angle
		Pres.	Ten.	Ind.		Strike slip	Dip slip	Ind.		
0.00 R	143	37.7	37.7	24.5	1.000	78.7	20.6	0.7	3.814	25
0.01 R	27	55.6	22.2	22.2	0.400	63.0	31.5	5.6	2.000	35
0.02 R	9	44.4	44.4	11.1	1.000	66.7	33.3	0.0	2.000	38
0.03 R	26	61.5	27.0	11.5	0.438	34.6	55.8	9.6	0.621	55
0.04 R	6	50.0	33.3	16.7	0.667	91.7	8.3	0.0	11.000	21
0.05 R	20	35.0	20.0	45.0	0.571	62.5	32.5	5.0	1.923	39
0.06 R	4	25.0	25.0	50.0	1.000	50.0	50.0	0.0	1.000	36
0.07 R	7	28.6	57.1	14.3	2.000	57.1	42.9	0.0	1.333	45
0.08 R	7	28.6	71.4	0.0	2.500	50.0	50.0	0.0	1.000	50
0.09 R	10	20.0	60.0	20.0	3.000	60.0	40.0	0.0	1.500	37
0.10 R	6	33.3	50.0	16.7	1.500	75.0	25.0	0.0	3.000	31

Table 7. *Shallow Earthquakes*

Region	No. of earth-quakes	Per cent			Ratio ten. to pres	Per cent			Ratio st. sl. to dip sl.	Avg. slip angle
		Pres.	Ten.	Ind.		Strike slip	Dip slip	Ind.		
1	16	12.5	56.2	31.2	4.500	90.6	9.4	0.0	9.667	22
2	12	58.3	41.7	0.0	0.714	12.5	87.5	0.0	0.143	67
3	5	20.0	60.0	20.0	3.000	40.0	50.0	10.0	0.800	55
4	25	60.0	36.0	4.0	0.600	66.0	34.0	0.0	1.941	32
5	46	23.9	6.5	69.6	0.273	76.1	21.7	2.2	3.500	22
6	31	38.7	61.3	0.0	1.583	79.0	19.4	1.6	4.083	29
7	11	27.3	63.6	9.1	2.333	90.9	4.5	4.5	20.000	21
8	31	67.7	25.8	6.4	0.381	95.2	4.8	0.0	19.667	15
9	2	50.0	50.0	0.0	1.000	100.0	0.0	0.0	∞	26
World	179	40.8	35.8	23.5	0.877	75.7	22.9	1.4	3.305	27

Table 8. *Deep Earthquakes*

Region	No. of earth-quakes	Per cent			Ratio ten. to pres.	Per cent			Ratio st. sl. to dip sl.	Avg. slip angle
		Pres.	Ten.	Ind.		Strike slip	Dip slip	Ind.		
1	1	100.0	0.0	0.0	0.000	50.0	50.0	0.0	1.000	44
2	..	....	....	....	....	....	....	....	....	..
3	9	88.9	11.1	0.0	0.125	22.2	66.7	11.1	0.333	70
4	34	26.5	52.9	20.6	2.000	33.8	63.2	2.9	0.535	55
5	27	33.3	33.3	33.3	1.000	74.1	24.1	1.8	3.077	40
6	1	100.0	0.0	0.0	0.000	0.0	100.0	0.0	0.000	78
7	2	0.0	0.0	100.0	....	100.0	0.0	0.0	∞	5
8	12	58.3	25.0	16.7	0.428	91.7	0.0	8.3	∞	12
9	..	....	....	....	....	....	....	....	....	..
World	86	40.7	36.0	23.2	0.866	54.6	41.3	4.1	1.324	43

Table 9. *Comparison of Deep and Shallow Earthquakes*

Region	Ratio of tension to pressure		Ratio of strike slip to dip slip		Slip angle	
	Deep	Shallow	Deep	Shallow	Deep	Shallow
3.	0.125	3.000	0.333	0.800	70	55
4.	2.000	0.600	0.535	1.941	55	32
5.	1.000	0.273	3.077	3.500	40	22
8.	0.428	0.381	∞	19.667	12	15
World	0.886	0.877	1.324	3.305	43	27

such a weighting procedure is quite impracticable. The whole question of magnitude is still not yet quite settled; and of many shocks implicit in our Tables, magnitude determinations have never even been attempted. The only thing that could be done, therefore, was to give equal weight to every possible fault plane solution; double weight, of course, has been attached to fault plane solutions where S analysis has been carried out and it is therefore known which is the fault- and which is the auxiliary plane, so that every earthquake has the same weight. However, some of the very small shocks have been omitted for the calculation of the Tables.

The gist of the information contained in Tables 5 to 8 may now be summarized as follows:

(i) The usual case is that in which earthquakes represent transcurrent (strikeslip) faulting[1].

(ii) There are exceptions to the usual case where, in rather small and well-defined regions, the faulting in earthquake foci is normal or reversed rather than transcurrent. One such area is the Pamir Knot, the other Micronesia and Indonesia. These regions appear rather small on a world-wide scale.

(iii) Regarding the dip-slip compoment of the faults, there is no evidence of a level of no strain. Thrust and tension features seem to balance each other.

One can also attempt a more direct comparison between deep and shallow earthquakes. The result of this attempt is shown in Table 9. It is only in Regions 3, 4, 5 and 8 that a proper comparison can be made, since only there do we have enough deep-focus earthquakes (as defined above; i.e. focal depth 140 km or more) to permit our obtaining any significant results. An inspection of Table 9 shows that there is a pretty good correlation between the results from deep and shallow earthquakes in any one area, as well as in the world as a whole. This seems to indicate that, whatever it is that causes earthquakes, acts in a similar fashion at small and at great depths. At any rate, there does not seem to be a definite break in the failure pattern at some intermediate depth. Unfortunately, the number of fault-plane solutions is still far too small to permit one to make an analysis in terms of smaller depth-steps (say: 0.01R) for all the nine earthquake areas under consideration.

2.3. Gravity Data

2.31. Gravity and Gravity Anomalies. We are turning now our attention towards the *gravity field* of the Earth.

The average value of the Earth's gravity acceleration

$$g = 980 \text{ cm/sec}^2 \qquad (2.31\text{–}1)$$

yields a means to estimate the Earth's mass and hence its mean density. The latter turns out to be

$$\varrho_{\text{mean}} = 5.5 \text{ g/cm}^3. \qquad (2.31\text{–}2)$$

[1] This seems to have been noted first by J. H. Hodgson: Geofis. Pura Appl. **32**, 31 (1955).

It is very significant that this is much higher than the density of the rocks in any of the accessible parts of the Earth. The above value, therefore, gives rise to much speculation about the composition of the Earth's interior.

It turns out that the gravity value upon the Earth's surface is not constant, but depends on the geogrraphical location of the point under consideration. There are two effects which *a priori* may be assumed to affect gravity: The Earth's rotation, and the altidude at which gravity is being measured. The effects of the *Earth's rotation* will be discussed more fully in Sec. 4.12 where the theoretical equilibrium figure of the Earth will be calculated. The gradient of the potential function W given there yields the theoretical value of gravity at sea level as a function of the latitude φ. The formula for the "normal" gravity value currently internationally adopted is the following:

$$g = 978.049\,(1 + 0.005\,2884 \sin^2\varphi - 0.000\,0059 \sin^2 2\varphi)\ \mathrm{cm/sec^2}. \qquad (2.31\text{–}3)$$

The second effect, that of *altitude*, can be taken care of as follows. For a spherical mass M, the attraction (neglecting the centrifugal force) is at a distance a from its center

$$g = \varkappa M/a^2 \qquad (2.31\text{–}4)$$

where $\varkappa$ is the gravitational constant. Hence the change of g for a change of altitude is

$$\partial g/\partial a = -\,2g/a = -\,0.3086\ \text{milligal/meter} \qquad (2.31\text{–}5)$$

where 1 gal equals 1 $\mathrm{cm/sec^2}$.

The above formula is valid for a change in altitude if this change occurs in free air. Hence gravity "anomalies" calculated by means of (2.31–5) [as referred to the "theoretical" value obtained from (2.31–3)] are called *free-air anomalies.*

However, in most cases, if gravity is measured on the surface of the Earth, it would appear as more logical to make a correction for the attraction of the rocks between the gravity station and the level of reference. If these rocks are approximated by an infinite horizontal slab, the induced gravity change is

$$\Delta g = 2\pi\varkappa\varrho h \qquad (2.31\text{–}6)$$

where ϱ is the density of the slab and h its thickness. For an average value of the density ($= 2.67\ \mathrm{g/cm^3}$) one obtains

$$\Delta g = 0.1119\ \text{milligal/meter}. \qquad (2.31\text{–}7)$$

Combination of the corrections (2.31–5) and (2.31–7) yields, if applied to any measured gravity value, the latter's *Bouguer reduction.* Further corrections can be applied to acount for the topographical shape of the surrounding terrain[1].

The reduced gravity values, compared with the "normal" values, can be represented in a map of gravity anomalies. These anomalies are able to give some indications regarding the mass-distribution in the Earth's crust.

2.32. Distribution of Gravity Anomalies. Isostasy. If the Bouguer anomalies are calculated for various parts of the world, then it becomes at once obvious that they tend towards large negative values with increasing elevations. In fact, it appears that the free air reduction leads to smaller anomalies than the Bouguer reduction. One can interpret this in terms of *isostasy*. By this term is meant that the mountain ranges are supported by "roots" of low density from below, as floating masses upon a denser substratum. There are two hypotheses by which this effect could be achieved. According to PRATT[2] (cf. GARLAND[1]), the assumed mass deficiency would consist of an anomaly in density extending to constant depth, whereas according to AIRY[3], the density of the floating matter would be assumed as constant and the latter would therefore have to extend to a depth varying with surface elevation. It is now generally the Airy hypothesis which is favored. Based upon it, HEISKANEN[4] has published tables for isostatic reduction of gravity values which is supposed to reduce to zero all such values which correspond to proper isostatic adjustment. Gravity anomalies obtained by isostatic reduction are called *isostatic anomalies*.

It is quite certain that the isostatic theory gives a proper interpretation of many gravity observations. There are, however, notable exceptions. These show that isostatic adjustment can at best be expected to be achieved on a broad scale; it certainly does not hold for small-scale features.

Of large-scale deviations from isostasy one should mention the Fennoscandian shield. An explanation suggested is that isostatic adjustment has not yet been achieved since the melting of the ice cover which was present there during the most recent (Pleistocene) ice age (cf. Sec. 1.23).

Another characteristic pattern of gravity anomalies is connected with island arcs. A typical gravity profile across an island arc is shown in Fig. 40. It thus appears that the actual "active" (volcanic) part of an

[1] GARLAND, G. D.: Handbuch der Physik, Bd. 47, S. 202. Berlin-Göttingen-Heidelberg: Springer 1956.

[2] PRATT, J. H.: Phil. Trans. Roy. Soc. Lond. **149**, 745 (1859).

[3] AIRY, G. B.: Phil. Trans. Roy. Soc. Lond. **145**, 101 (1855).

[4] HEISKANEN, W.: Publ. Isostatic Inst. **2** (1938).

island arc is connected with positive isostatic anomalies, whereas the foredeeps show large negative ones. TSUBOI[1] has developed a method for the direct calculation of the depth of the interface between "crust" and "mantle" (corresponding to a density difference of 0.4 g/cm³) which, according to the Airy hypothesis, is supposed to cause the gravity anomalies. He found that this assumed interface (which may perhaps be

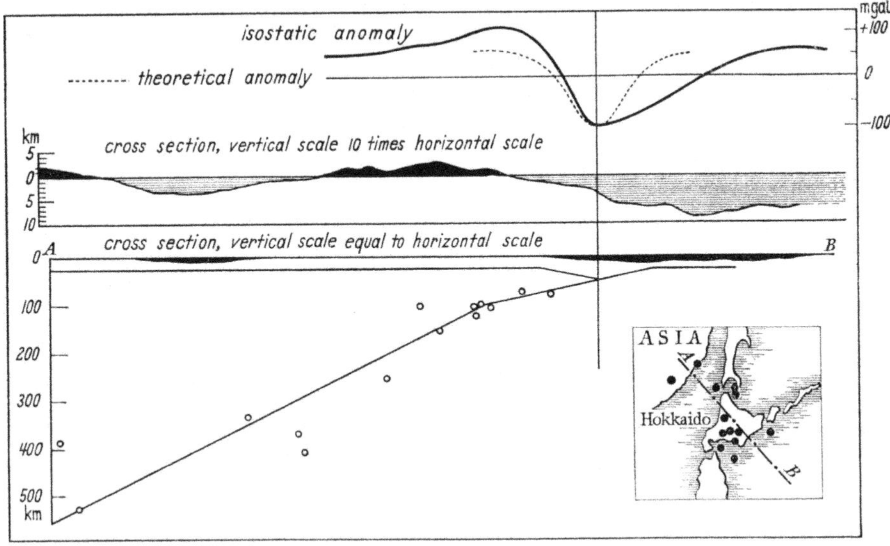

Fig. 40. Gravity anomalies and vertical distribution of the foci of deep earthquakes (after GUTENBERG and RICHTER[2])

identified with the Mohorovičić discontinuity) must dip rather sharply at the margin of continents, at an angle ϑ which is remakably constant from place to place so that

$$\tan\vartheta = 0.1\,. \qquad (2.32\text{–}1)$$

Major gravity anomalies have also been discovered over the African Rift Valleys[3]; this seems to suggest that the latter are indeed rifts in the Earth's crust.

A pattern of negative gravity anomalies similar to that of Fennoscandia is present in the interior of India. The explanation of this is not clear as it is certain that central India was not glaciated during the last ice age. The difficulty thus encountered may be collated with a

[1] TSUBOI, C.: Vening Meinesz Aniversary Volume (publ. by Kon. Ned. Geolog.-Mijnbowkundig Genot.), p. 287. 1957.

[2] GUTENBERG, B., and C. F. RICHTER: Seismicity of the Earth and Associated Phenomena. Princeton: Princeton Univ. Press 1949. (Second ed. 1954.)

[3] BULLARD, E. C.: Phil. Trans. Roy. Soc. Lond. A **237**, 237 (1938).

contention by LYUSTIKH[1] who claims that the anomalies of Fennoscandia are not caused by isostatic effects either.

The above observations refer to continental areas. Turning now to the sea, it may be noted that there exists in the oceans (especially over the abyssal plains) a consistent gravity anomaly of approximately 20 to 30 milligal[2]. This seems to indicate that there is an excess of mass beneath the oceans. It is quite possible that this statement, in fact, puts matters in the wrong light. Since the area of the oceans is much greater than the area covered by continents, one should probably regard the gravity values over oceans as "normal" and say that there is a definite mass deficiency beneath continental areas.

The above statements have recently been challenged by WORZEL[3] who claims that they are actually based on biassed data. Most of the gravity values hitherto reported from the "deep" sea were, in fact, obtained in anomalous areas. According to WORZEL[3], the true gravity anomaly in abyssal plains is −20 milligal. This, if correct, would appear to make the continents heavier than the oceans and put the whole theory of isostasy in doubt.

The gravity profiles that have been run across mid-ocean ridges show no very well defined patterns. Most of the traverses go across areas of volcanic islands which cause large and rather erratic anomalies; only a few do not. An interpretation of one of the latter for a typical area (Bermuda) in terms of crustal structure[4] (also making use of seismic data) is shown in Fig. 29. In general, there seems to be a slight decrease in gravity over the ridges as compared to the slopes on either side of it[5]. The pattern delineated above has been supported by observations of CRARY and GOLDSTEIN[6] over the Marvin Ridge in the Arctic. Attempts at calculating isostatic anomalies from the gravity profiles on mid-ocean ridges have been singularly unsuccessful[5].

Finally, a word may be said with regard to depressed areas, particularly with regard to the connection between isostasy and some theories of the origin of geosynclines. Since sediments are less dense than the substratum, it is quite clear that geosynclines cannot be *caused* by the weight of the sediments that are deposited there[7] (at least if the theory of isostasy holds). Notwithstanding this simple argument, statements

---

[1] LYUSTIKH, E. N.: Trudy Geofiz. In-ta Akad. Nauk SSSR. No. 38 (165) (1957).

[2] DIETRICH, G., u. K. KALLE: Allgemeine Meereskunde. Berlin: Bornträger 1947.

[3] WORZEL, J. L.: Publ. Inst. Geodesy, Ohio State Univ. No. **7**, 65 (1957).

[4] WOLLARD, G. P.: Proc. Roy. Soc. Lond., Ser. A **222**, 361 (1954).

[5] HESS, H. H.: Proc. Roy. Soc. Lond., Ser. A **222**, 341 (1954).

[6] CRARY, A. P., and N. GOLDSTEIN: Deep. Sea Res. **4**, 185 (1957).

[7] HOLMES, A.: Principles of Physical Geology. London: Nelson 1947.

to the contrary are found in the literature. The negative gravity anomalies[1, 2] found in deep ocean trenches also indicate that there is a phenomenon other than isostasy which causes the depressed areas.

In conclusion, it may be stated that the analysis of gravity anomalies is of great use in conjunction with seismic work for the establishment of crustal sections.

2.4. Underground Stresses[3]

2.41. General Remarks. It is clear that the present-day appearance of the Earth's surface with all its mountains and other irregularities, must be the result of the action of forces. It stands to reason that such forces are always present within the Earth's crust; if a direct means could be found to ascertain these stresses, an interesting body of information could be assembled. This has recently been possible.

The geologic stress state within the Earth's crust cannot be entirely arbitrary. It is obvious that there cannot be a pressure or a tension normal to the Earth's surface and that there cannot be a shearing force parallel to it. The latter condition implies that the normal to the surface is one of the principal directions of stress at or near the surface (cf. Sec. 7.12.). Thus, except in strongly folded areas, one principal direction of stress is nearly vertical, the other two are horizontal. The relative magnitudes of the principal stresses characterize the possible standard geological stress state. These states are described in detail in Sec. 7.12.

If one wants to determine the geological stresses present in the Earth's crust, one has to investigate the cases where these stresses might become manifest. One particular type of manifestation occurs in natural and artificial openings, such as mines, caves, tunnels, wells, etc. Such openings modify the geologic stress state in their immediate vicinity. What is observed within the openings is the effect of the modified stress state. It will be necessary, in special cases, to refer to this modified stress state.

The modification of the geologic stress state due to an opening in the earth is usually calculated by assuming that the material of the earth is elastic. On a free opening, there can be no normal pressures and tangential stresses present. Because of the fact that the equations of elasticity theory are linear, it is possible to calculate the influence of

---

[1] Vening Meinesz, F. A.: Gravity Expeditions at Sea. Publ. Ned. Geod. Comm. **111** (1941).

[2] See also Talwani, M. *et al.*: J. Geophys. Res. **66**, 1265 (1961), also J. Geophys. Res. **64**, 1545 (1959).

[3] This Section after A. E. Scheidegger: J. Alta. Soc. Petrol. Geol. **9**, 287 (1961).

the opening in question upon a stress state in which only one principal stress is different from zero, and all other components of the stress tensor are zero. Then, three perpendicular stresses may simply be superposed upon each other to yield the influence of the opening upon a general tri-axial geological stress state.

We shall refer to the calculations indicated above in the proper context. However, let us note here that general problems of elasticity where cavities are involved, have been reviewed, for instance, by SOUTHWELL and GOUGH[1], by TIMOSHENKO and GOODIER[2], by GREEN and ZERNA[3] and by MUSKHELISHVILI[4]. Special investigations into the stress concentration at the margin of holes have been made by SAWIN[5], by LEVIN[6] and by LOWENGRUB[7]. Applications of elasticity theory to rock cavities have been discussed by TERZAGHI and RICKART[8]. We shall refer to these investigations as the need arises.

Other manifestations of tectonic stresses also occur in connection with seismic effects, and with the creation of geological surface faulting. The mapping of faults can lead to a recognition of the stresses present, by applying ANDERSON'S *theory of faulting*, described in detail in Sec. 7.1.

We shall discuss the various possibilities for determining the underground stresses in their turn below.

2.42. Direct Measurements. First of all, attempts have been made to measure underground stresses directly, mostly in mines, by means of a variety of stress-meters.

Stress-meters are generally extensometers which measure changes in length. If the elastic constants of the material (rock) are known, it is then possible to determine the stress changes that induce the changes in length recorded by the extensometer. If the elastic constants of the stress-meter itself are adjusted properly, then conditions can be chosen in which the elastic constants of the rock do not have too great an influence upon the calculated stress changes from the measured length

---

[1] SOUTHWELL, R. V., and H. J. GOUGH: Phil. Mag. **1**, 71 (1926).

[2] TIMOSHENKO, S., and N. GOODIER: Theory of Elasticity, 2nd ed. New York: McGraw-Hill 1951.

[3] GREEN, A. E., and W. ZERNA: Theoretical Elasticity. Oxford: Clarendon Press 1954.

[4] MUSKHELISHVILI, N. I.: Some Basic Problems of the Mathematical Theory of Elasticity. Transl. from Russian. Groningen: P. Noordhoff 1953.

[5] SAWIN, G. N.: Spannungsverteilung am Rande von Löchern. Berlin: Verlag Technik 1956.

[6] LEVIN, E.: J. Appl. Mech. **27**, 283 (1960).

[7] LOWENGRUB, M.: Quart. Appl. Math. **19**, 119 (1961).

[8] TERZAGHI, K., and T. RICKART: Géotechnique, Lond. **3**, No. 2, 57 (1952).

changes. Various types of stress meters have been described, for instance, by HAST[1], POTTS[2] and MAY[3].

As noted above, the stress-meters only measure changes of stresses. In order to determine the actual stresses present, e.g. in a mine wall, one must measure the stress change starting from the stressed condition of the rock in situ to the completely unstressed state of the same rock. This may be done by inserting the stress meter into a small bore-hole into the wall of the mine, and then cutting the whole assembly, stress meter plus a chunk of rock into which it had been inserted, loose from the mine wall. In the cut-out piece of rock, the stresses may be assumed as zero so that the stress change measured during the cutting operation is indicative of the stresses originally present in the mine wall. These stresses can therefore be calculated.

One difficulty is immediately apparent in connection with the above method: If the stresses are measured in a mine wall, what is obtained are truly the stresses in that mine wall and not the geologic stresses which were present before the mine was dug. The latter do have to be determined in a further calculation in which the connection between the stresses near the mine and the originally present geologic stresses is introduced[4-8].

Some results of a determination of tectonic stresses from mine wall measurements have been reported, e.g. by TALOBRE[9]. The main result is that the horizontal stresses cannot be explained by simply assuming that the rock pressure is essentially that due to the weight of the overburden, with the lateral pressures adjusting themselves in conformity with POISSON's number of the material. Thus, indications are that in most cases, one is faced with a truly tri-axial stress state.

Other attempts at direct stress-measurements have been made, based on seismic investigations[10,11]: The presence of stresses changes the wave-transmittance properties of rocks, and thus it ought to be possible to infer the stresses from the seismic wave velocities. However, such studies are still in their infancy.

[1] HAST, N.: The Measurement of Rock Pressures in Mines. Stockholm, Sveriges Geologiska Undersökning, 183 pp. (1958).

[2] POTTS, E.: Quart. Colo. Sc. Mines **53**, No. 3 (1957).

[3] MAY, A. N.: Trans. Canad. Inst. Min. Met. **63**, 497 (1960).

[4] BOSHKOV, S.: Trans. Canad. Inst. Min. Met. **59**, 264 (1956).

[5] GIBSON, W. L.: Trans. Canad. Inst. Min. Met. **59**, 141 (1956).

[6] ROUX, A. J. A. *et al.:* Trans. Canad. Inst. Min. Met. **59**, 19 (1956).

[7] CORTLETT, A. V., and C. L. EMERY: Trans. Canad. Inst. Min. Met. **62**, 186 (1959).

[8] PANEK, L. A.: Mining Eng. **13**, 282—285 (1961).

[9] TALOBRE, J.: Geol. u. Bauw. **25**, 148 (1960).

[10] KUNDORF, W., u. D. ROTTER: Z. Geophys. **27**, 35 (1961).

[11] MEISSER, O.: Abh. dtsch. Akad. Wiss. Berlin, Kl. Bergbau No. 1, 17 (1961).

2.43. Manifestations in Wells. Of the holes that have been artificially made in the Earth's crust, oil wells form a significant proportion. It turns out that the tectonic stresses can be calculated from the response of such oil wells to fracturing operations.

A well-fracturing operation consists in building up the pressure inside the well until the formation adjacent to the well experiences a break-down. The aim in performing such fracturing operations is to increase the flow of the underground fluids into the well. If the pressure as a function of time be measured at the bottom of the well bore (where the fracture occurs), one obtains generally a typical curve, an example of which is shown in Fig. 41. The original fluid pressure is P_0, which

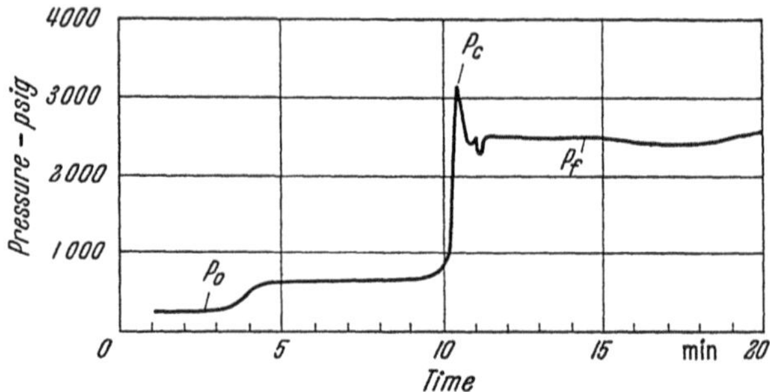

Fig. 41. Typical bottom-hole pressure curve

also is the pressure in the well until the latter is made to increase. At P_c the formation fractures and thereafter settles down to a "plateau" P_f at which fluid is flowing from the well into the fracture.

Several attempts to describe the theory of well fracturing have been reported in the literature; of these, the best known is probably that of HUBBERT and WILLIS[1]. These authors assumed that the well consists of a very long cylinder. The pressure inside the well, then, applies to the walls of the cylinder, whence it is possible to calculate the stresses in the vicinity of the well from elasticity theory by a superposition of the geologic stress state with the stress state due to the pressure inside the well. For a single principal stress S, the stresses around a cylindrical cavity of radius have been given, e.g. by MILES and TOPPING[2], as follows:

$$\left.\begin{aligned} \sigma_r &= \frac{S}{2}\left\{1 - \frac{a^2}{r^2} + \left(1 - 4\frac{a^2}{r^2} + 3\frac{a^4}{r^4}\right)\cos 2\vartheta\right\} \\ \sigma_t &= \frac{S}{2}\left\{1 - \frac{a^2}{r^2} - \left(1 + 3\frac{a^4}{r^4}\right)\cos 2\vartheta\right\} \\ \tau &= -\frac{S}{2}\left(1 + \frac{2a^2}{r^2} - 3\frac{a^4}{r^4}\right)\sin 2\vartheta \end{aligned}\right\} \qquad (2.43\text{–}1)$$

[1] HUBBERT, M. K., and D. G. WILLIS: Trans. AIME **210**, 153 (1957).
[2] MILES, A. J., and A. D. TOPPING: Trans. AIME **179**, 186 (1949).

where the symbols have the meaning indicated in Fig. 42. For a triaxial stress state one simply has to superpose the stresses for the three principal directions. In conformity with the fundamental geometry of the cylindrical well model, the stress state near the bore is a plane state; the vertical stresses are not affected by the pressure in the well. Since this is obviously unrealistic, one has to assume that the vertical stress induced is equal to the difference of the well pressure and original formation pressure. As to the fracture condition, one generally assumes that the rock has no tensile strength at all so that the emergence of a tensile stress anywhere would cause formation breakdown.

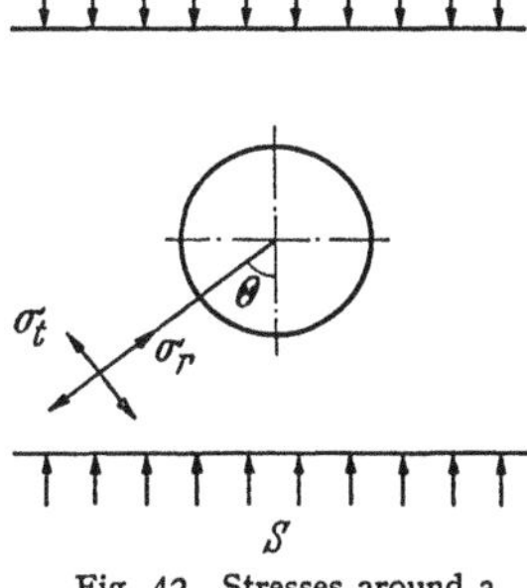

Fig. 42. Stresses around a cylindrical hole

From well fracturing data, it should thus be possible, upon the basis of Eq. (2.43–1), to calculate the geologic stress state in the vicinity of a well. The pressures P_f and P_c of the pressure curve yield two equations for the determination of the geological stresses from bottom-hole data. The pressure P_c is the critical pressure at which fracture occurs, which can be expressed by stating that one of the tangential stresses at the well bore must become zero. The pressure P_f is that pressure which is just able to keep a fracture open that may have been formed. Thus, from the two conditions, it is possible to obtain two relations for the geologic stresses. A third condition is obtained if one assumes that the average vertical pressure must be equal to the weight of the overburden above. The latter is generally taken as 1 lb./sq. inch/foot. Thus, it ought to be possible to determine the geological stress state from well fracturing data using the above formulas. This has been attempted by the writer[1] with regard to some wells for which bottom hole pressure data during well fracturing operations were available. It turned out that the observed data could not entirely be explained on the basis of the above theory, but that a finite tensile strength of the rock has to be allowed for.

It appears that the difficulty with the above model lies in the fact that fundamentally a two-dimensional stress state was introduced. In order to have a truly three-dimensional stress state, it seemed desirable to abandon the cylindrical geometry of the well and to regard the build-up of pressure in form of a spherical pressure center rather than of a pressure line.

The calculation of the stresses around a spherical cavity under pressure contained in an homogeneous elastic medium which is subject to

---

[1] Scheidegger, A. E.: Geol. u. Bauw. **27**, 45 (1962).

an arbitrary stress state at infinity can be built up from the stresses around a cavity if only a uniaxial tension (or compression) is present at infinity. The latter problem has been solved by SOUTHWELL and GOUGH[1]; the solution has also been reproduced by TIMOSHENKO and GOODIER[2].

Accordingly, the solution of the equations of elasticity theory in question can be represented as follows. Let us consider a Cartesian co-ordinate system x, y, z. The center of the spherical cavity is at the origin of the co-ordinate-system; the radius of the spherical cavity is denoted by a.

Table 10. *Underground Stresses as Calculated from Well Data*

Well	P_0 lb/sq in	P_f lb/sq in	P_c lb/sq in	Depth Feet	S_2 Horiz. lb/sq in	S_2 Vertical lb/sq in	S_3 Horiz. lb/sq in
Alabama A . . .	700	3200	3800	3600	−3200	−3600	−6501
Oklahoma C . .	1000	2400	3300	3900	−2400	−2900	−3890
Oklahoma D . .	1000	3200	4100	3300	−3200	−3300	−5783
Oklahoma F . .	200	2400	3100	2900	−2400	−2900	−5209
Venezuela . . .	990	1650	1820	1840	−1650	−1840	−2531

If we now assume that the solid with the cavity under consideration is under a tension S in the $\pm z$ direction (tension being counted positive), then it can be shown that the stresses induced can be represented as the superposition of three stress systems which are given by TIMOSHENKO and GOODIER[2].

If one now introduces a pressure p within the spherical cavity, then a fourth stress system must be superposed over the three already considered (cf. TIMOSHENKO and GOODIER[2]).

The equations mentioned above refer to a stress state where only one of the principal stresses of the geological stress tensor is different from zero, all others are zero. The formulas for a general tri-axial stress state can be obtained by superposing the above formulas for three different principal stresses.

The formulas mentioned above have been used by the writer[3] to calculate the geological stresses for the same wells as those mentioned above. The basic assumptions (fracture if tension occurs; vertical stress equals overburden pressure) are the same as those used earlier; the result of the calculations is shown in Table 10. It will be noted that the new

[1] SOUTHWELL, R. V., and H. J. GOUGH: Phil. Mag. **7**, No. 1, 71 (1926).

[2] TIMOSHENKO, S., and J. N. GOODIER: Theory of Elasticity, 2nd ed New York: McGraw-Hill Book Co. 1951.

[3] SCHEIDEGGER, A. E.: Geofis. Pura Appl. **46**, 66 (1960).

model does not require the introduction of a tensional strength of the rock.

In all cases that were investigated, one arrives at a geological stress state which corresponds to incipient wrench faulting. However as noted above, using the model of an open hole of infinite length it is not possible to account for the observed pressure data without the introduction of a finite tensile strength of the rock. Such a finite tensile strength, of course, could also be introduced in the new model (with a corresponding change in the calculated stresses), but this is not necessary. Therefore it appears that the second model is an improvement over the first one. In either case, it turns out that the geological stresses as calculated have about the same values. It is interesting to observe that in all cases that were investigated, the stresses correspond to a state of incipient transcurrent faulting. Whilst not too many inferences can be drawn from an analysis of only five wells, this is nevertheless remarkable.

2.44. Manifestations of Ground Stresses in Mine Shafts. Mines are holes in the Earth's crust which are not very different from wells, except that they are larger and generally more or less horizontal. It stands to reason, therefore, that one ought to be able to observe in mines, too, some effects which are caused by the presence of geological stresses.

Stresses in mines, naturally, have been the subject of many investigations. As with wells, the problem has generally been to determine the conditions under which a mine becomes unsafe (collapses) so that precautions can be taken to avoid such conditions. The problem, thus, is to determine the stresses around and in the vicinity of the variously shaped openings that are represented by mines. It is generally assumed that the rock behaves elastically up to a certain limit and plastically thereafter[1]. Also, elastic behavior of the rock may be presumed to be prevalent immediately during and after the excavation with some creeping readjustment taking place later.

Most workers on stresses in the vicinity of mines assume quite simple geologic stress states. A very common assumption is that the stress state is hydrostatic[2], the pressure corresponding to the weight of the overburden, or else that the stress is unidirectional in the vertical direction, with the horizontal stresses adjusting themselves in accordance with POISSON's number of the rock[3,4] assuming that there cannot

[1] FENNER, R.: Glückauf **74**, 681 (1938).

[2] ITERSON, F. K. T. VAN: Geol. en Mijnbouw **10**, 198 (1948).

[3] MORRISON, R., and D. F. COATES: Trans. Canad. Inst. Min. Met. **58**, 401 (1955).

[4] ROTTER, D.: Montanwiss. Lit. Ber. A, Beih. No. 2 (1960).

be any lateral expansion. In either of these cases, the tectonic stress field is ignored and, therefore, no inferences can be made, based upon these theories from observations in mines, with regard to the character of the tectonic stresses.

Doubts against the adequacy of assuming a simple geologic stress state have been uttered on a few occasions. The view that the undisturbed (by the mine) stress state should probably be regarded as tri-axial has been stated by CORTLETT and EMERY[1] in rather general terms. Similarly, GZOVSKII et al.[2] noted that the geological stress state present before any mines are built should be regarded as tri-axial. Neither of the papers mentioned above, however, proceeds to actually providing formulas where the effect of a tri-axial stress state upon the mine opening would be calculated and from which, conversely, it would be possible to calculate the geologic stress state from observations in mines.

A remarkable exception to the above pattern is contained in a paper by SONNTAG[3] who investigated a bi-axial stress state (the mine being considered as cylindrical; the principal stresses are orthogonal to the axis of the mine in the horizontal and vertical directions). Furthermore, SONNTAG assumes that the elastic modulus E also is different in different directions. Based upon solutions of the elasticity equations for the stress concentrations on the edge of holes given in SAWIN'S[4] book mentioned earlier, SONNTAG presented a variety of cases in which the stresses were calculated near the mine shaft.

Of particular interest, in connection with our problem, is a discussion by SONNTAG of the deformations near a circular mine shaft. If we assume that the principal stresses are σ_1 and σ_2, and that the corresponding moduli of elasticity are E_1 and E_2 (the principal directions of the modulus of elasticity and of the stresses coincide), then a circular shaft, as it is being driven into the rock, will be deformed. The radius r in the 1-direction will change by the amount Δr_1, the radius in the 2-direction by the amount Δr_2 so that

$$\frac{\Delta r_2}{\Delta r_1} = \left\{-\frac{\sigma_2}{\sigma_1}\frac{\sqrt{E_1}}{\sqrt{E_2}} + \left(1 - \frac{\sigma_2}{\sigma_1}\right)\right\} \Big/ \left\{-\frac{\sqrt{E_1}}{\sqrt{E_2}} + \left(1 - \frac{\sigma_2}{\sigma_1}\right)\right\}. \quad (2.44\text{–}1)$$

In all the above investigations, POISSON'S number is assumed as equal to zero.

[1] CORTLETT, A. V., and C. L. EMERY: Trans. Canad. Inst. Min. Met. **62**, 186 (1959).

[2] GZOVSKII, M. V. *et al.:* Freiberger Forschungsh. C **81**, 209 (1960).

[3] SONNTAG, A. G.: Bauingenieur **33**, No. 8, 287 (1958).

[4] SAWIN, G. N.: Spannungsverteilung am Rande von Löchern. Berlin: Verlag Technik 1956.

Formula (2.44–1) yields a direct means for calculating the ratio of the horizontal to vertical geological stresses from measurements on mine shafts. SONNTAG gives a nomograph for doing this, but the method does not seem to have been applied to any practical cases.

2.45. Surface Problems. The surface of the Earth represents a boundary condition for the geological stress field: The normal stress at the surface must be zero, and the shear stress must also vanish. Thus, a given regional tectonic stress system will adjust itself in such a fashion so as to satisfy the above conditions.

When a valley is cut into an area in which, as usual, a tectonic stress system is present, then the adjustment of the stresses to satisfy the above boundary conditions, may cause stress concentrations in certain areas. These stress concentrations may become so large that the rock fails as it becomes exposed.

Examples of rock failure owing to rock being exposed by geological forces, have been collected notably by KIESLINGER[1]. According to KIESLINGER, the result of such failure which is most often observed, is a splitting off of arched plates from the naked rock. A particularly notable instance of this happening—according to KIESLINGER—is in the Yosemite Valley in California. After this valley had been carved out by a glacier, the residual stresses forced whole plates to break off, parallel to the new surface, such as is seen in the *Royal Arches*.

The surface effects of the tectonic stresses have some significance with regard to engineering construction. In the latter context the problem has been discussed by AMPFERER[2,3] and by HORNINGER[4]. Accordingly, disregarding the effect of erosion, valleys have a general tendency to become narrower. The writer, however, is not aware of any calculations of the stresses involved.

2.46. Seismological Implications. A reasonable assumption is that all seismic shocks are related, in a general way, to present-day tectonic stresses of an area. Usually, the active seismic belts and the zones of recent mountain building run roughly parallel, and hence it may safely be assumed that the causes of orogenesis and of seismic activity have some relation to each other. Both mountain building and earthquakes are manifestations of the release of underground stresses. It stands to reason that the stress systems causing these two phenomena are the same and that it ought to be possible to verify this in some detail.

[1] KIESLINGER, A.: Geol. u. Bauw. **24**, 95 (1958).
[2] AMPFERER, O.: Sitzgsber. Akad. Wiss. Wien, Kl. I **149**, 51 (1940).
[3] AMPFERER, O.: Sitzgsber. Akad. Wiss. Wien, Kl. I **150**, 97 (1941).
[4] HORNINGER, G.: Geol. u. Bauw. **24**, 37 (1958).

Attempts at a correlation of seismic with tectonic observations have been reported on several occassions. As mentioned above, there is the well-known correlation between seismicity and mountain ranges (see Sec. 2.22).

It is natural to suspect that the earthquake faults determined by fault-plane solutions (see Sec. 2.24) should correlate with the pattern of tectonic faulting in any one area. However, attempts at searching

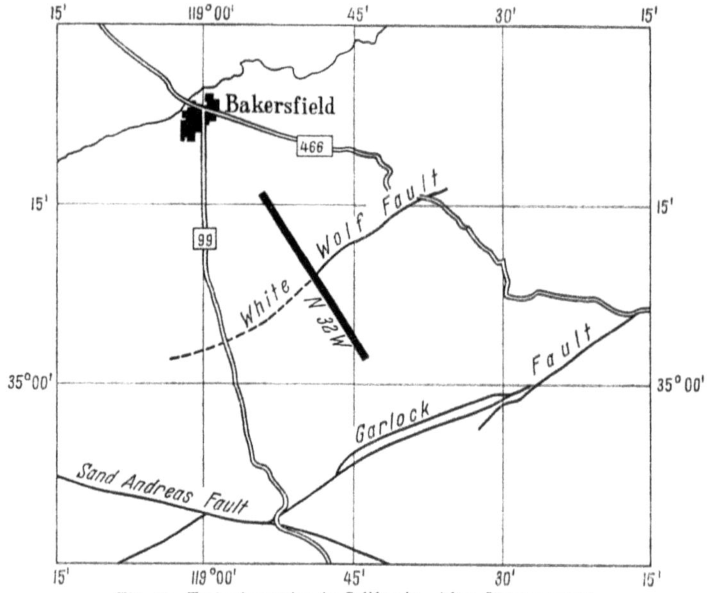

Fig. 43. Tectonic motion in California. After SCHEIDEGGER[1]

for such a correlation have generally ended in failure. It turned out that the actual faults evident in earthquakes are oriented in quite a random fashion and that there is no obvious correlation with the tectonic structure of the area in which the seismic shocks occurred. Only the *type* of fault that occurs in any one area seems to be characteristic of that area (cf. see 2.24).

In order to carry the above analysis further, it is necessary to take the fault plane solutions as they are published in the literature. As was noted earlier, the fault plane solutions themselves do not seem to have any regular pattern in any one area. Part of the reason for this may arise from the ambiguity which makes it impossible to know which is the fault plane and which is the auxiliary plane in any one fault plane solution. It is obvious, however, that there is one direction in any one fault plane solution which is entirely unambiguous; this is the direction of the null axis. The null axis is the intersection of the fault

[1] SCHEIDEGGER, A. E.: J. Geophys. Res. **64**, 1499 (1959).

plane with the auxiliary plane. It is the direction in which no motion occurs during an earthquake. It seems therefore reasonable to hypothesize that the maximum motion in any one area must be orthogonal to the null axes, or to the average direction of null axes of earthquakes in any one area. This gives one immediately a means to get at the tectonic stresses in an area.

Thus, we consider a series of earthquakes in a certain area; we assume we have fault plane solutions and consequently that we know

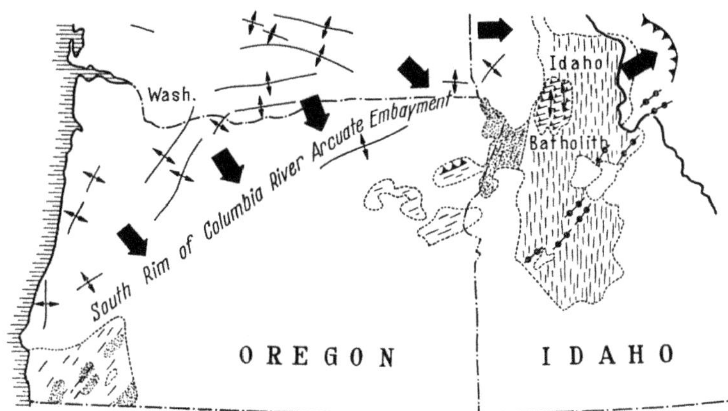

Fig. 44. Stresses in Nevada. After BADGLEY[1]

the directions of the null axes. We then determine that direction which is most orthogonal to these null axes: this, then, must be the tectonic motion direction of the area. The method to do this is by means of a least square analysis which is best carried out on a computer. As noted above, the reason for assuming the indicated connection between tectonic and earthquake motion is that the null axis is the direction in any one earthquake along which no motion occurs. This direction will tend to be orthogonal to the overall tectonic motion direction in the area.

The idea indicated above was applied to the northwest Pacific Ocean where rather consistent results were obtained[2]. Unfortunately, the number of fault plane solutions available in that area was rather small so that the analysis suffered somewhat from a lack of data. The theory was further tested by SCHEIDEGGER[3,4] in the case of Central Asia and California. The results were very promising inasmuch as it was possible to establish the tectonic motion patterns in the areas that were investigated. An example of the tectonic motion patterns obtained in California is shown in Fig. 43.

[1] BADGLEY, P. C.: A.I.M.E. Meeting, San Francisco, Paper No. 59—169 (1959).
[2] SCHEIDEGGER, A. E.: Bull. Seism. Soc. Amer. **48**, 369 (1958).
[3] SCHEIDEGGER, A. E.: Bull. Seism. Soc. Amer. **49**, 369 (1959).
[4] SCHEIDEGGER, A. E.: J. Geophys. Res. **64**, 1499 (1959).

2.47. Geological Manifestations. In Sec. 2.41, we have mentioned ANDERSON'S theory of faulting which predicts the position of faults in a given geological stress system. It stands to reason that the relationship between faults and stress systems postulated by ANDERSON'S theory can be used to infer the prevalent stresses from an analysis of faults.

A typical result obtained by BADGLEY[1] using essentially the above-indicated method, is shown in Fig. 44. The arrows indicate the direction of the main compression in the early Tertiary time.

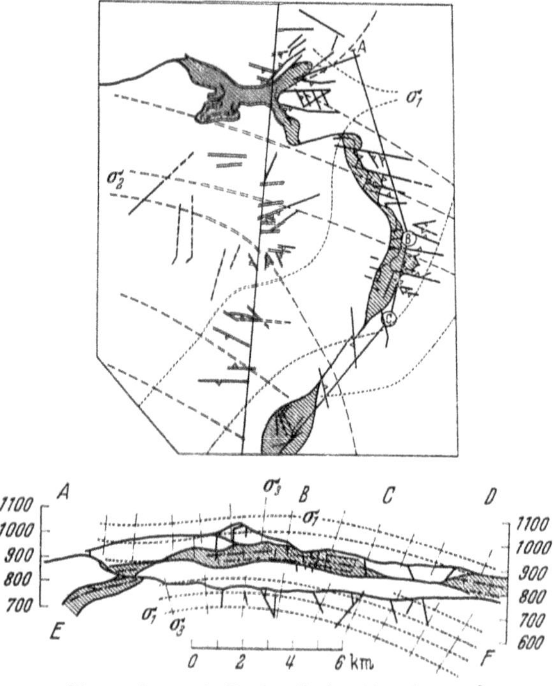

Fig. 45. Stresses in Donbass Basin. After GZOVSKII[2]

The above method has been applied in great detail by GZOVSKII et al.[2] to several areas of Russia. The reconstruction of the stress state for the southeastern depression of the saddle of the South Darbass Anticline (Donbass Basin) is shown in Fig. 45.

A somewhat more elaborate method for arriving at the tectonic stresses from a field analysis of faults has been proposed by LENSEN[3] who noted that the angle by which conjugate fault systems intersect each other, ought to be an indication of the relative compression in the

[1] BADGLEY, P. C.: A.I.M.E. Meeting, San Francisco, Paper No. 59—169 (1959).
[2] GZOVSKII, M. V. *et al.*: Freiberger Forschungsh. C **81**, 209 (1960).
[3] LENSEN, G. J.: New Zeal. J. Geol. Geophys. **1**, 565 (1958).

horizontal direction that is present. In a tectonic stress system which corresponds to the state of incipient transcurrent faulting, the faults will be at right angles to each other. In a stress system which corresponds to a state of incipient reverse faulting, the faults will be orthogonal to

Fig. 46. Isallo stress lines in New Zealand. After LENSEN[1]

the greatest compression and the angle between the strikes of conjugate systems will be zero (the dips will be orthogonal to each other). In a state of incipient normal faulting, the angle between the strikes of conjugate fault systems will also be zero, but the strikes will be parallel to the greatest horizontal compression. It is seen that the faults always lie symmetrical to the largest horizontal compression. The angle which is formed by the two conjugate systems and which contains the maximum horizontal pressure may be called "compressional angle", γ. In a state

[1] LENSEN, G. J.: New Zeal. J. Geol. Geophys. 1, 565 (1958).

of incipient transcurrent faulting, this angle γ is 90°, in incipient reversed faulting, 180°, and incipient normal faulting, 0°. From looking at lineaments, it is possible to determine the angle γ and to plot it on a map. Lines of equal γ are called "isallo stress lines". The character of a stress state can directly be read off an isallo stress map. LENSEN, generally, plots a different angle (which he calls ε) on his isallo stress maps, which is connected with the angle γ introduced above as shown in Table 11.

Table 11

γ (degrees)	ε (degrees)
0	−45.0
45	−22.5
90	0.0
135	+22.5
180	+45.0

In Fig. 46, we give one of LENSEN's isallo stress maps; however the lines do not refer to LENSEN's angle ε, but to the angle γ introduced above. It is interesting to note that LENSEN's isallo stress map correlates fairly well with a gravity map of the same area.

2.48. Stresses and Large-Scale Tectonics. An inspection of the large-scale tectonic features that are seen on the Earth's surface makes it evident that the stress state near the surface is not at all uniform. It is immediately obvious that many types of stresses must be present. If we look for instance at the Mid-Atlantic rift, there is very little doubt that an extensive tensional stress system must be present in the Atlantic Ocean. If we look at the fracture zones off the coast of North America in the Pacific Ocean, then it is quite clear that in that area an extensive transcurrent stress system must be present. Finally, if we look at a mountain range such as the Alps or the Rocky Mountains, then the overthrusts and nappes which are seen there cannot be explained unless an extensive compressional stress system is surmised. Thus, field work indicates that the stress systems within the surface of the Earth are very extensive and that they may be of very varied natures. There are areas in which tensional stresses are prevalent, areas in which transcurrent or shear stresses are prevalent and areas in which compressional stresses are prevalent.

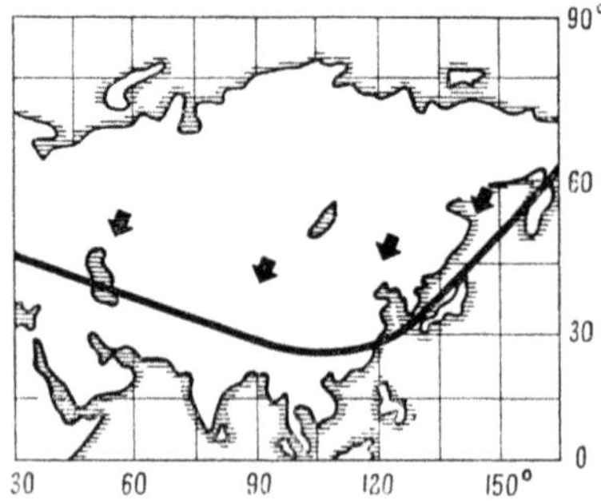

Fig. 47. Tectonic motion of Asia

All the various results that have been obtained by an analysis of the manifestations of the stresses are in agreement with the concept of a heterogeneous stress state within the Earth's crust. However, the stresses turn out to be homogeneous over fairly large areas, although they are different for different areas.

The manner in which the large-scale tectonic motions can be correlated with the tectonics of an area may be demonstrated as follows[1]. From earthquake fault plane solutions that have been obtained along the Pacific coast of Asia, it seems not unreasonable to assume that the Pacific is moving towards Asia. Such a movement has also been postulated by Japanese investigators on different grounds, so that the seismic results seem to be in accordance with those from other sources. If the results from all of Asia are plotted, one obtains the picture shown in Fig. 47. Making a generalization of the individual motions one might say that there is a relative motion of the land mass of Eurasia towards the south. Similar generalizations can be arrived at for other areas.

2.5. Data from Age Determinations

2.51. Principles. The geological time scale discussed in Sec. 1.22 begins with the Cambrian, owing to the absence of fossils in earlier strata. It is, however, inconceivable that the Earth would not have existed for a very long time before the advent of animals that could be fossilized. It is thus fortunate that the physics of radioactive decay has enabled one to extend the geological time scale backwards to almost 3 billion years ago.

Radioactive decay of fissionable material occurs spontaneously. The number of atoms decaying in any one (short) time interval is thus proportional to the total number of atoms present and to a decay constant λ. Thus the decay of e.g., rubidium can be written as follows:

$$-\frac{d}{dt}\,\mathrm{Rb}^{87} = \lambda\,\mathrm{Rb}^{87} \qquad (2.51\text{-}1)$$

where Rb^{87} indicates the number of atoms of rubidium-87 present in any quantity under consideration. Integrating, one obtains

$$\mathrm{Rb}^{87} = \mathrm{Rb}_0^{87}\, e^{-\lambda t}. \qquad (2.51\text{-}2)$$

Rubidium-87 decays into strontium-87; the number of radiogenic strontium-87 atoms contained in, say, a rubidium crystal must be equal to the number of rubidium atoms that have decayed (barring any losses) since the formation of that crystal. Thus

$$\mathrm{Sr}^{87} = \mathrm{Rb}_0^{87} - \mathrm{Rb}^{87} = \mathrm{Rb}_0^{87}\,(1 - e^{-\lambda t}) \qquad (2.51\text{-}3)$$

or

$$\mathrm{Sr}^{87} = \mathrm{Rb}^{87}\,(e^{\lambda t} - 1)\,. \qquad (2.51\text{-}4)$$

Thus, by measuring the ratio of $\mathrm{Sr}^{87}/\mathrm{Rb}^{87}$ it is possible to determine the time of formation of the crystal. This is the basis of radioactive age

[1] Scheidegger, A. E.: Publ. Dom. Obs. Ottawa **24**, 385 (1961).

determinations. The actual technique of carrying out the measurements is rather involved; the required isotope-ratios are obtained by means of mass-spectrometry. The materials, therefore, have to be changed chemically into such a state that they can be introduced into a mass spectrometer which, in itself, is not a simple piece of equipment.

The strontium decay scheme mentioned above is only one of several nuclear decay-reactions occurring in nature which are suitable for radioactive age determinations. For a naturally decaying material to be suitable for such age determinations it is necessary, firstly, that its half-life be comparable with the age of the Earth, and secondly, that the material be reasonably abundant in rocks. The decay schemes which are thus usable for the present purpose are shown in Table 12. A detailed discussion of the techniques involved in using these schemes may be found in an article by WILSON et al.[1].

Table 12. *Radioactive Decay Schemes Usable for Age Determinations*

Isotope	End product	Half life
K^{40}	Ca^{40} and A^{40}	1.3×10^{9} years
Rb^{87}	Sr^{87}	6.1×10^{10} years
Th^{232}	Pb^{208}	1.4×10^{10} years
U^{235}	Pb^{207}	7.1×10^{8} years
U^{238}	Pb^{206}	4.5×10^{9} years

2.52. An Extended Geological Time Scale. One of the outcomes of radioactive age determinations was the possibility of extending the geological time scale, as well as to assign more certain absolute ages to strata whose relative age had been determined earlier by means of index fossils. The scale thus found by WILSON is shown in Table 13[1].

Table 13. *Extended Geological Time Scale*

Events	Time (in billion years)	Rock types
Present day	0	Cenozoic group
	0.07	Mesozoic group
	0.2	Paleozoic group
First index fossil	0.5	
	1.0	Grenville-Huronian types
Change in orogenesis	2.0	Keewatin-Temiskaming types
Oldest known rocks	3.0	No rocks preserved
	4.0	
Origin of Earth	4.2	

[1] WILSON, J. T., R. D. RUSSELL u. R. M. FARQUHAR: Handbuch der Physik, Bd. 47, S. 288. Berlin-Göttingen-Heidelberg: Springer 1956.

A most interesting observation in considering this scale is that the so-called "Precambrian" time takes up about 5/6 of the Earth's history, the rest only 1/6. Thus, conventional geology usually is concerned with only the latter sixth of the life of the Earth. This has only recently been recognized.

Concurrently with the extension of the geological time scale, attempts have been made to elucidate the tectonic history of the Precambrian time. It is known that old shields have the appearance of worn-down mountain roots rather than that of homogeneous blocks. It may thus be conjectured that orogenetic activity has been going on for a long time before the appearance of the first fossilizable animals upon Earth.

Originally, the Precambrian areas had been divided into only two "ages", Proterozoic and Archean, the former being assumed as the younger. This division came about because of the presence of two types of Precambrian rocks, a sedimentary type resting upon a highly altered type. The former was simply called Proterozoic, the latter Archean. It was observed, however, soon after radioactive age determinations had been undertaken, that the classification into Proterozoic and Archean rocks is not a chronological one at all. Although in any one region Proterozoic rocks rest upon and are younger than Archean type of rocks, Proterozoic rocks are not at all of the same age and may be older in one area than the Archean type rocks in another.

In fact, the arrangement of ages in the Canadian Shield (which has been most thoroughly investigated) seems to suggest a phenomenon of *continental growth* (see Fig. 48). The Canadian Shield may be divided into "provinces", i.e. regions of roughly uniform ages. The oldest province is found around James Bay, younger ones progressively following on its side[1]. The above picture is due to WILSON et al.[1] and was established in 1956. Meanwhile, newer investigations[2] have fully confirmed the general structural character of the Canadian Shield; it is even possible to make a finer division into "provinces".

Accordingly, the oldest province in a shield would form a *continental nucleus*. There are indications that orogenesis in the continental nuclei occurred in a manner different from that occurring to-day: Mountain building processes taking place in continental nuclei produced many small sinuous belts characterized by poorly differentiated sediments and by a high proportion of basic volcanic materials. The change-over to present-day type of orogenesis (as described in Sec. 1.4) occurred about 2000 million years ago. An analysis of rock ages and tectonic patterns

[1] WILSON, J. T., R. D. RUSSELL and R. M. FARQUHAR: Canad. Min. Met. Bull. **49**, 550 (1956).

[2] LOWDON, J. A.: Geol. Surv. Can. Pap. 61–17 (1961).

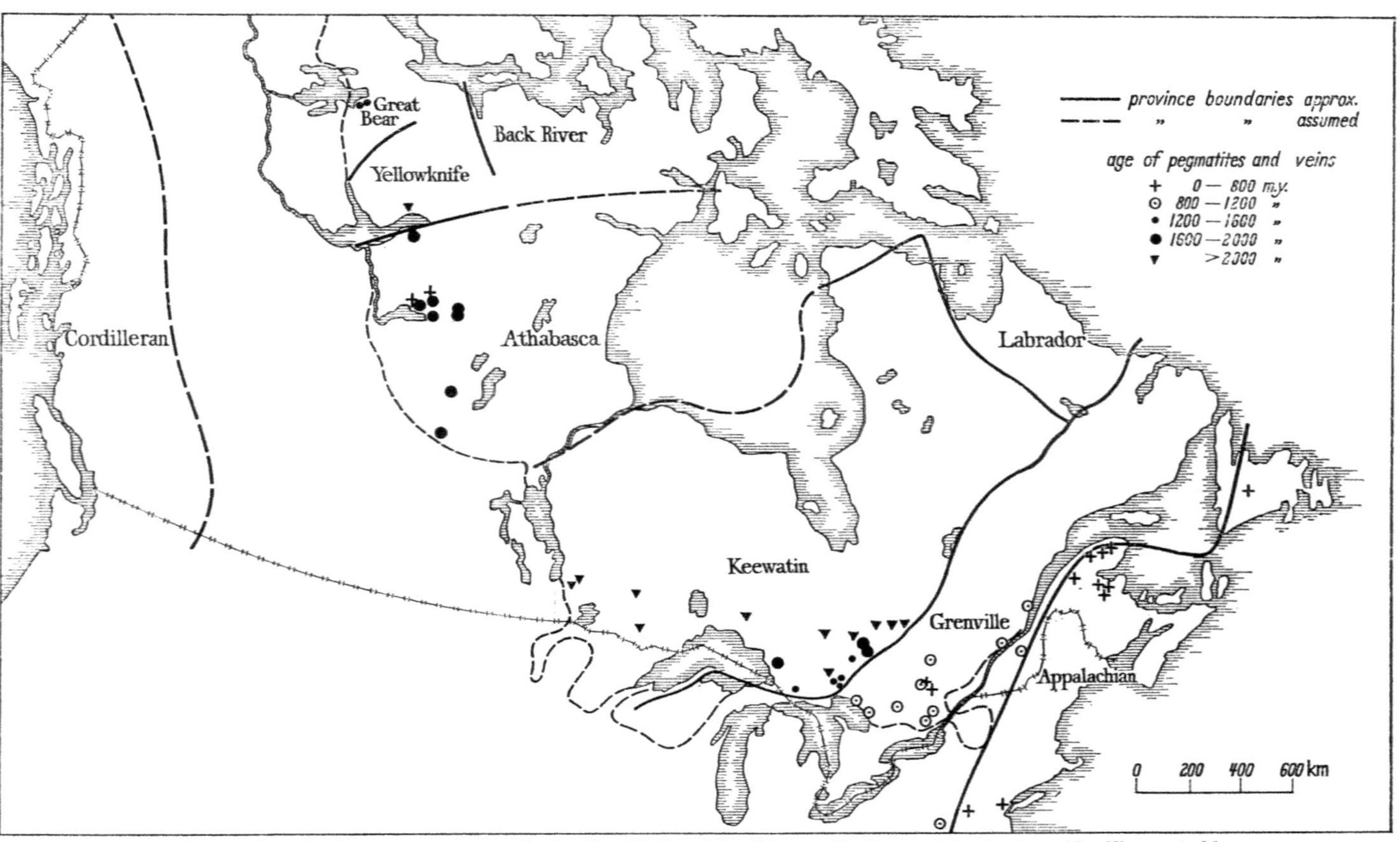

Fig. 48. Geological basement provinces in the Canadian Shield, as inferred from radioactive age determinations. After WILSON et al.[1]

[1] WILSON, J. T., R. D. RUSSELL u. R. M. FARQUHAR: Handbuch d. Physik, Bd. 47, S. 288. Berlin-Göttingen-Heidelberg: Springer 1956.

recognizable since that time yields that about 10 "modern-type" orogenetic cycles occurred up to the present (as mentioned already earlier; cf. Sec. 1.41).

The above age determinations refer to continental areas.

Rocks dredged from the oceans are all much younger than Precambrian. We have already mentioned that, at least on the ridges, no rocks seem to be older than Tertiary.

This general statement seems to be confirmed by the only radioactive age determination of an oceanic boulder that has come to the attention of the writer. This is the determination of the age of a boulder from the Mid-Atlantic Ridge by CARR and KULP[1]. The age turns out to be 30 ± 15 million years, the rather large error being due to the inherent unreliability of the method used.

2.6. Thermal Data

2.61. Surface Heat Flow Measurements. It has been postulated for some time that the Earth may essentially be a heat engine. This means that the energy causing geodynamic effects may stem from thermal phenomena. Unfortunately, the thermal history of the Earth is only very imperfectly known as it is closely linked with problems of the origin of the Earth and with the chemistry of the Earth's interior. These topics are beyond the scope of the present study, and we shall, therefore, only present here a survey of those investigations and speculations that are of importance with regard to problems of geodynamics.

Table 14. *Heat Flow Measurements.* (After ALLAN[2])

Region	No. of observations	Mean value
Africa	7	1.10×10^{-6} cal/cm² sec
Canada	8	0.96
United States .	9	1.46
Great Britain .	12	1.15
Pacific Ocean .	6	1.18
Atlantic Ocean.	5	0.98

The most tangible information regarding thermal properties of the Earth is obtained from surface heat flow measurements. Such measurements require the determination of the thermal gradient in a mine, bore hole or such like, and a determination of the thermal conductivity of the rock strata. The results from such measurements are remarkably uniform over the whole world. The average heat flow is found to be about

$$H = 1.2 \times 10^{-6} \text{ cal cm}^{-2} \text{ sec}^{-1} \qquad (2.61\text{–}1)$$

[1] CARR, D. R., and J. L. KULP: Bull. Geol. Soc. Amer. **64**, 253 (1953).

[2] ALLAN, D. W.: Endeavour **1954**, 89.

with very little variation from one place to another. The results have been summarized e.g. by ALLAN[1] and by BIRCH[2]. The data compiled by ALLAN are shown in Table 14.

Naturally, the value given in (2.61-1) for the heat flow is an *average* value; local patterns of "heat flow anomalies" do definitely occur.

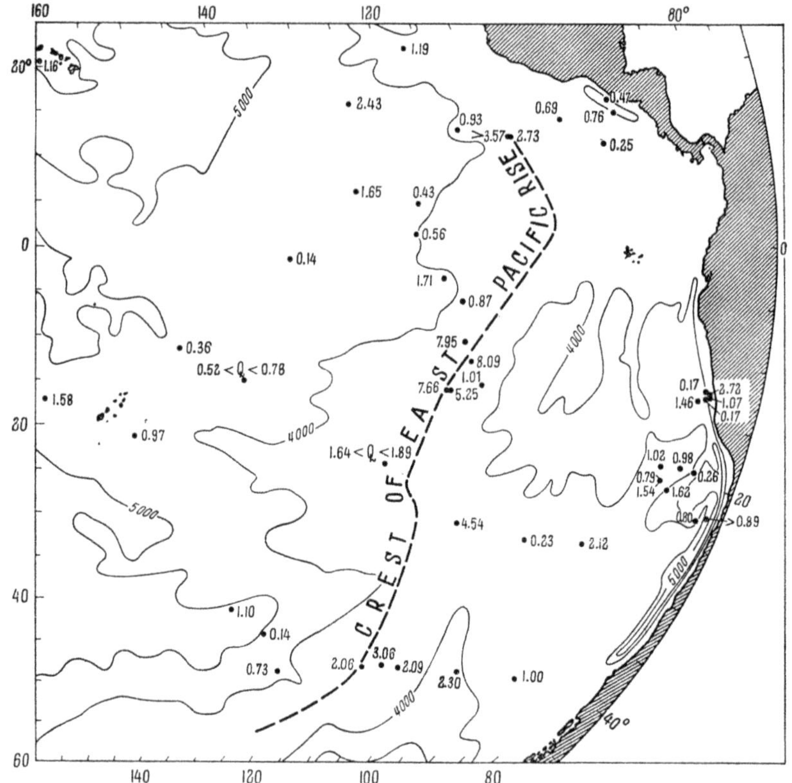

Fig. 49. Heat flow (in units of 10^{-6} cal cm^{-2} sec^{-1}) along the East Pacific Rise. After HERZEN[3]

Thus, HERZEN[3] found a distinctly anomalous pattern of heat flow along Pacific ocean ridges. In the latter case, the anomaly is an increase, as shown in Fig. 49. A similarly anomalously high heat flow value was found on the Mid-Atlantic Ridge[4].

Most significant is the observation that there is no difference between continental and oceanic heat flows. This suggests that each square kilometer of the Earth's surface is underlain by the same amount of

[1] ALLAN, D. W.: Endeavour **1954**, 89.

[2] BIRCH, F.: Geophysics **19**, 645 (1954).

[3] HERZEN, R. v.: Nature, Lond. **183**, 882 (1959).

[4] BULLARD, E. C., and A. DAY: Geophys. J. Roy. Astr. Soc. **4**, 282 (1961).

heat-generating radioactivity. This is very surprising in view of the fact that it is known that the continental rocks contain much more radioactive material than the oceanic rocks. Somehow, this difference must be compensated in the deeper layers of the Earth.

In view of the total heat flow taking place through the Earth's surface, the additional amount of heat released by volcanic activity is quite insignificant: it is probably only about 1/100 of the total in any one time interval[1, 2]. The energy lost to the Earth through heat flow is also very much greater than the energy lost by earthquake activity. However, in particularly seismic areas (such as Japan) the energy lost by earthquakes may perhaps equal that lost by heat flow[3].

2.62. Temperature in the Earth's Interior. In order to estimate the temperature distribution in the Earth's interior, it is necessary to make a hypothesis regarding the dependence of the volume coëfficient of thermal expansion α on the pressure p:

$$\frac{1}{\alpha} = \frac{1}{\alpha_0} + b p. \qquad (2.62\text{–}1)$$

This assumption is suggested by analogy with the compressibility-pressure relationship which is of the same form, and has been confirmed by results of the theory of solids[4]. The values of the constants $1/\alpha_0$ and b are as follows

$$\frac{1}{\alpha_0} = 2.4 \times 10^4\ {}^\circ\mathrm{K}, \qquad (2.62\text{–}2)$$

$$b = 6.2 \times 10^{-8}\ {}^\circ\mathrm{K\ cm^2\ dyne^{-1}}. \qquad (2.62\text{–}3)$$

It is now possible to make an estimate of the temperature distribution in the Earth[4, 5]. The adiabatic temperature gradient satisfies the equation

$$\frac{dT}{dp} = \frac{\alpha T}{\varrho C_p}. \qquad (2.62\text{–}4)$$

Here, T is the temperature, ϱ the density, and C_p is the specific heat at constant pressure. Inserting the hypothesis formulated in Eq. (2.62–1) into (2.62–4), one obtains

$$\frac{dT}{T} = \frac{dp}{\varrho C_p \left(\frac{1}{\alpha_0} + b p\right)}. \qquad (2.62\text{–}5)$$

---

[1] Verhoogen, J.: Amer. J. Sci. **244**, 744 (1946).

[2] Robson, G. R., and P. L. Willmore: Bull. Volc., Ser. II **17**, 13 (1955).

[3] Bullard, E. C.: In: The Earth as a Planet, ed. Kuiper, p. 120. 1954. — Gutenberg, B.: Quart. J. Geol. Soc. Lond. **112**, 1 (1956). — Tsuboi, C.: J. Phys. Earth **5**, 1 (1957).

[4] Jacobs, J. A.: Canad. J. Phys. **31**, 370 (1953).

[5] Jacobs, J. A.: Nature, Lond. **170**, 838 (1952); also Adv. in Geophysics **3**, 183 (1956).

If one further assumes that C_p is a constant, the above relationship can be integrated to yield:

$$\log \operatorname{nat} T = \frac{1}{C_p} \int \frac{dp}{\varrho \left(\frac{1}{\alpha_0} + b p \right)} . \tag{2.62–6}$$

This can be evaluated numerically, since the dependence of the density ϱ on depth (and hence on pressure) is approximately known (cf. Sec. 2.14). The result is the temperature distribution shown in Fig. 50 (after JACOBS)[1].

JACOBS' curve estimates the temperature in the Earth only below 1000 km depth. Information regarding the upper layers must be sought from different sources. Various authors have made estimates, mostly based upon specific assumptions, which have been collected by GUTENBERG[4]. A series of estimates is shown in Fig. 51.

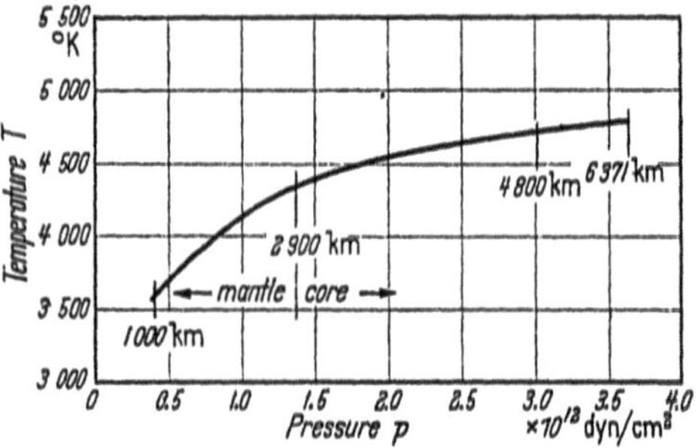

Fig. 50. Temperature distribution in the interior of the Earth. After JACOBS[2]

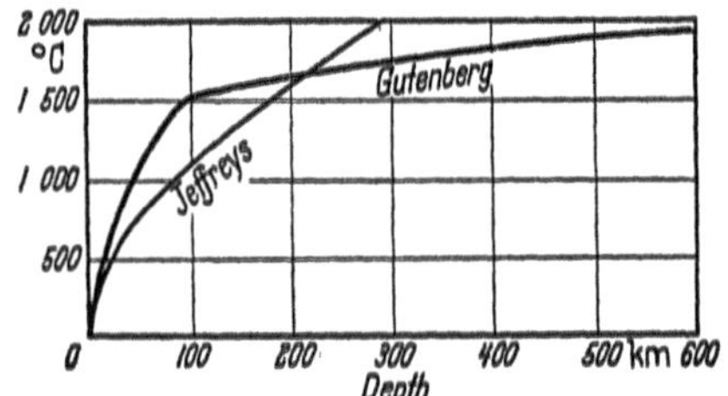

Fig. 51. Temperature distribution in the crust and upper mantle of the Earth; according to GUTENBERG and JEFFREYS[3]

Much of the above theory is obviously a great oversimplification. The conductivity of the mantle may be affected by radiative processes[5]; furthermore, complicated phase-transitions could take place. Much of the pertinent theory has been reviewed by GUTENBERG[3] so that there is no need to review it here.

2.63. Thermal History of the Earth. It is usually assumed that the Earth was at one time of its life a hot celestial body. The reason for this assumption will be explained more fully in Sec. 5.52; for the

---

[1] VERHOOGEN [VERHOOGEN, J.: Trans. Amer. Geophys. Un. **35**, 85 (1945)] disagrees with this as he assumes that the temperature at the core boundary fluctuates, in space as well as in time, between a lower limit of about 1500° C and an upper limit of 2500° C.

[2] JACOBS, J. A.: Canad. J. Phys. **31**, 370 (1953).

[3] GUTENBERG, B.: Physics of the Earth's Interior. New York: Academic Press 1959.

[4] In: Internal Constitution of the Earth (l.c.) p. 162.

[5] LYUBIMOVA, E. A.: Izv. Akad. Nauk SSSR., Ser. Geofiz. **1957**, No 5, 673 (1957).

present we shall accept it as a hypothesis which will enable us to follow the customary discussions of the Earth's thermal history as referring to geological time.

If the Earth was a hot body, it follows that its present condition was reached by cooling. The manner in which this occurred, however, is not at all clear. One of the most commonly accepted ideas is that if the Earth was first completely liquid, the heavy materials would have sunk to the center to form the core. The mantle would have solidified outwards from its base at the top of the liquid core and would have been cooling ever since. The statement that the solidification of the mantle would have started at the bottom follows from a comparison of the increase with depth of the melting point with the increase of the adiabatic temperature: ADAMS[1] has shown that of the two the melting point increases faster; hence solidification must have started from the bottom up.

The temperature at the Earth's surface is determined by the balance between solar radiation and heat loss into space; the heat supply from the Earth's interior is in this connection probably quite insignificant.

It has been only relatively recently realized that there is a significant source of heat within the Earth, viz. that represented by the radioactive materials contained in most common rocks. SLICHTER[2] analyzed the available data in the light of the existence of radioactivity, but unfortunately neglecting the change of the latter with time. Since the decay constants of most radioactive materials are at most comparable, and never much greater than the total life of the Earth, the change of the amount of radioactivity within the Earth must have been appreciable.

The problem can be tackled by trying to solve FOURIER's heat conductivity equation (in spherical co-ordinates)

$$\varrho C \frac{\partial T}{\partial t} = \frac{1}{r^2} \frac{\partial}{\partial r} \left(K r^2 \frac{\partial T}{\partial r} \right) + H(r, t) \qquad (2.63\text{–}1)$$

where $H(r, t)$ is the rate of production of heat by radioactivity per unit time and volume, C is the specific heat, K is the thermal conductivity and T is the temperature. The solution of this heat conductivity equation is quite difficult to achieve and can only be attempted numerically by means of high-speed electronic digital computers. This has been done by ALLAN[3] for two particular models which may be considered as reasonable. We present the results obtained by ALLAN in Table 15.

[1] ADAMS, L. H.: J. Wash. Acad. Sci. **14**, 459 (1924).

[2] SLICHTER, L. B.: Bull. Geol. Soc. Amer. **52**, 561 (1941).

[3] ALLAN: l.c.; also JACOBS, J. A.: Pub. Bur. Centr. Seism. Int. A **19**, 151 (1956) and Adv. in Geophysics **3**, 183.

Table 15. *Temperatures (°C) in* ALLAN's *Models*

Time (10^9 years)	Depth (km)					
	50	100	500	1000	2000	2900
			Model 1			
0	666	1220	2310	2856	3615	3958
0.25	905	1216	2339	2894	3656	3992
0.5	861	1166	2361	2930	3693	4024
1	763	1056	2382	2988	3753	4077
2	606	892	2358	3072	3844	4162
3	521	785	2308	3124	3913	4225
4	461	714	2248	3163	3970	4282
			Model 2			
0	666	1220	2310	2856	3615	3958
0.25	1447	1520	2330	2885	3643	3985
0.5	1424	1576	2346	2910	3667	4010
1	1267	1481	2369	2960	3705	4052
2	997	1245	2383	3017	3763	4117
3	825	1073	2347	3057	3808	4168
4	712	934	2298	3086	3841	4209

It may be pointed out that both of ALLAN's models have as characteristic features that the greater part of the mantle is and has been heating up during most of the geological time. This is in contrast to what had been believed until very recently and is of great consequence with regard to orogenetic thinking.

The basic assumption of the heat-conductivity equation as fundamental to the thermal history of the Earth makes no provision for the possible existence of thermal convection currents in the mantle. It is demonstrated, however, that the observed heat flows (which can be compared with those resulting from ALLAN's models) can well be explained by the assumption of conduction alone.

2.7. Electromagnetic Effects

2.71. The Earth's Magnetic Field. It has been well known for a long time that a magnetic field is associated with the Earth[1]. The main part of the field can be described as that of a magnetic dipole whose axis is somewhat offset with regard to the present axis of the Earth's rotation. Short term disturbances of the magnetic field are known to be caused by stray currents in the ionosphere, presumably induced by cosmic radiation. Other characteristic disturbances of the magnetic field are called secular variations. These appear to be regional phenomena; they also include the westward drift of non-axial characteristics[2,3].

[1] BALMER, H.: Beiträge zur Geschichte der Erkenntnis des Erdmagnetismus. Aarau: Sauerländer 1956.

[2] See e.g. RUNCORN, S. K.: Handbuch der Physik, Bd. 47, S. 498. Berlin-Göttingen-Heidelberg: Springer 1956. For many references.

[3] CHAPMAN, S., and J. BARTELS: Geomagnetism. Oxford: Univ. Press 1948.

The origin of the magnetic field of the Earth and of its characteristic disturbances are not properly understood. At present it is usually held that some magneto-hydrodynamic phenomenon in the core of the Earth is responsible for the magnetic field[1,2]. If this is true, then there is reason to believe that the axis of the dipole field approximating the Earth's magnetic field should always more or less coincide with the axis of rotation.

2.72. Paleomagnetism. It is observed that, in sediments containing iron oxide minerals, the magnetized grains are oriented in a definite direction. It must therefore be postulated that an orienting magnetic field was present during the deposition of the sediments, and the contention[3] is that, under sufficiently quiet conditions, the Earth's magnetic field is strong enough to effect this. It is therefore reasonable to expect that an analysis of the "remnant" magnetization of rocks would yield information regarding the magnetic field of the Earth in times past.

If suitable rocks from various parts of the world are analyzed regarding their remnant magnetization, it may be hoped that the results can be interpreted in terms of a path of wandering of the Earth's magnetic pole. By implication this, then, would also indicate the path of the pole of rotation.

The above idea makes sense only if rocks are sufficiently stable so that they do not change their magnetization during the millions of years of burial. This question has been discussed by various writers[4-8] and has by no means been completely solved[9]. However, if certain conditions are fulfilled, it stands to reason that the magnetism of the rocks may be preserved for a long time.

It is found that the various ancient positions of the pole can be joined by a fairly smooth curve. This, however, is possible only if no distincion is made between North and South Pole. It must be assumed that either the magnetic field of the Earth can undergo a spontaneous reversal, or that a self-reversal of the remnant magnetization of the rocks can occur during consolidation. Only if this is assumed can one define a coherent path of "polar wandering". The polar path determined in this fashion

[1] ELSASSER, W. M.: Rev. Mod. Phys. **22**, 1 (1950); **28**, 135 (1956).
[2] HERZENBERG, A.: Phil. Trans. Roy. Soc. Lond. A **250**, 543 (1958).
[3] RUNCORN, S. K.: Handbuch der Physik, Bd. 47, S. 370. Berlin-Göttingen-Heidelberg: Springer 1957.
[4] FRÖHLICH, F.: Z. Geophys. **24**, 228 (1958).
[5] KOBAYASHI, K.: J. Geomagn. Geoelectr. **10**, No. 3, 99 (1959).
[6] MEIKLEJOHN, W. H., and R. E. CARPENTER: J. Appl. Phys. Suppl. **31**, No. 5, 164 S (1960).
[7] HALL, J. M., and R. N. NEALE: Nature, Lond. **188**, 805 (1960).
[8] EVERITT, C. W. F.: Phil. Mag. **6**, 713 (1961).
[9] QUIRING, H.: Forsch. u. Fortschr. **31**, No. 7, 193 (1957).

turns out to be fairly reasonable and more or less in agreement with that inferred from paleoclimatic evidence. The path shown in Fig. 52 has been determined by RUNCORN[1] from the analysis of North American rocks.

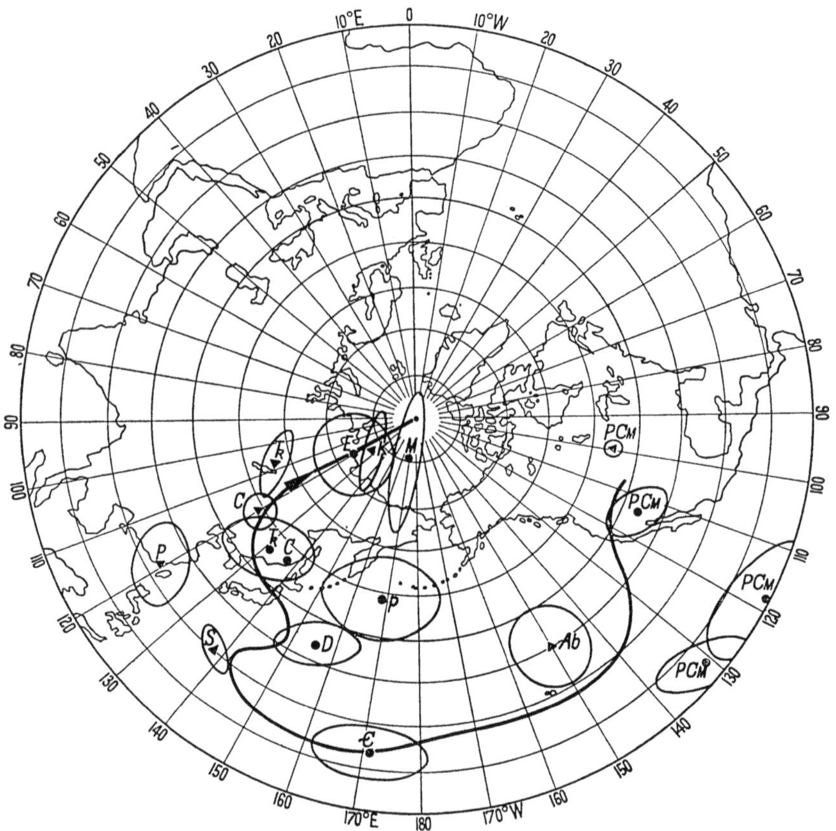

Fig. 52. Migration of the North Pole as indicated by paleomagnetic evidence. After RUNCORN[1]. *Triangles:* Pole position obtained from rocks in North America; *Circles:* Pole positions obtained from rocks in Europe. *M*—Miocene, *E*—Eocene, *K*—Cretaceons, *Ꞧ*—Triassic, *P*—Permian, *C*—Carboniferons, *D*—Devonian, *S*—Silurian, *Ꞓ*—Cambrian, *Ab*—Algonkian, PC_M—Precambrian

The paleoclimatic evidence suggests not only that a shift of the poles took place, but also that large continental drifts occurred, particularly in view of the simultaneous glaciation of several southern continents which are now widely separated. It is interesting to observe that similar conclusions are suggested by paleomagnetic work. CLEGG et al.[2] made a very careful study of this question, using rock samples collected in India. They arrived at a suggested position of the North

[1] RUNCORN, S. K.: Bull. Geol. Soc. Amer. **67**, 301 (1956).

[2] CLEGG, J. A., E. R. DEUTSCH and D. H. GRIFFITHS: Phil. Mag. **1**, 419 (1956).

Pole for the Eocene at 28° N, 85° W. This is widely different from the position of 75° N, 120° E which is given in Fig. 52 for that same epoch. The discrepancy can be resolved if a drift of India is assumed with regard to North America. The postulated drift of India would be northward, about 6000 km in 60 million years. In other areas, similar results have been suggested. The whole problem of the interpretation of paleomagnetic data has recently been reviewed by Cox and Doell[1] and by Irving[2].

Paleomagnetic data have also been collated with other phenomena. Opdyke and Runcorn[3] investigated ancient wind directions and found agreement with paleomagnetic results. Evison[4] found evidence in paleomagnetic data for the theory of continental growth.

2.73. Magnetic Anomalies. The (relatively) steady magnetic field of the Earth may show anomalies in two ways: First, short term variations may occur with time (these will be discussed in the next Section) and second, small steady local deviations from the over-all field of the Earth may be found. These local deviations are the true "magnetic anomalies".

The local magnetic anomalies seem to be caused by bodies of magnetic material fairly close to the surface. Thus, the survey of magnetic anomalies has mostly been used for prospecting purposes for ores; a famous such case is known from the Kursk region in the Soviet Union[5].

In connection with geophysical investigations of a fundamental importance, the survey of magnetic anomalies yielded the result that large displacements took place on the fracture zones in the Pacific. Thus, Vacquier[6] quotes a right lateral displacement of 160 km across the Murray Fracture Zone and corresponding displacements across other fracture zones

2.74. Telluric Currents. It has been found that electric currents in the Earth are associated with the time-changes of the magnetic field. These currents are termed "telluric currents". If one measures the magnetic and electric field variations simultaneously, then it is possible to obtain an idea of the distribution of the electric conductivity at depth. The theory of such measurements was first given by Cagniard[7]; a good review has later been published by Porstendorfer[8].

---

[1] Cox, A., and R. R. Doell: Bull. Geol. Soc. Amer. **71**, 645 (1960).

[2] Irving, E.: Geophys. J. Roy. Astr. Soc. **1**, 224 (1958); **2**, 51 (1959); **3**, 96, 444 (1960).

[3] Opdyke, N. D., and S. K. Runcorn: Endeavour **18**, 26 (1959).

[4] Evison, F. F.: Geophys. J. Roy. Astr. Soc. **4**, 320 (1961).

[5] Oelsner, C.: Bergakademie **12**, 771 (1961).

[6] Vacquier, V.: Nature, Lond. **183**, 452 (1959).

[7] Cagniard, L.: Geophysics **18**, 605 (1953).

[8] Porstendorfer, G.: Tellurik. Freiberger Forschungsh. C **107** (1961).

Unfortunately, not too many surveys of the above type have been made. The area analyzed best is undoubtedly Japan where RIKITAKE[1] has made many investigations. The latter author came to the conclusion that a hypothetical dipole is located beneath the central part of Japan whose explanation is not yet entirely certain[2]. His investigations also led RIKITAKE[1,2] to postulating a model of the Earth's crust in the vicinity of Japan which is based on the assumption that there is a wedge-shaped intrusion of low-conductivity material that lies ordinarily near the surface, into great depths (to 700 km). This fits well together with the idea that there is some sort of a geosyncline where an island arc is located; the deepest deep-focus earthquakes would occur at the lower boundary of the wedge. A schematic cross-section of RIKITAKE's model is shown in Fig. 53.

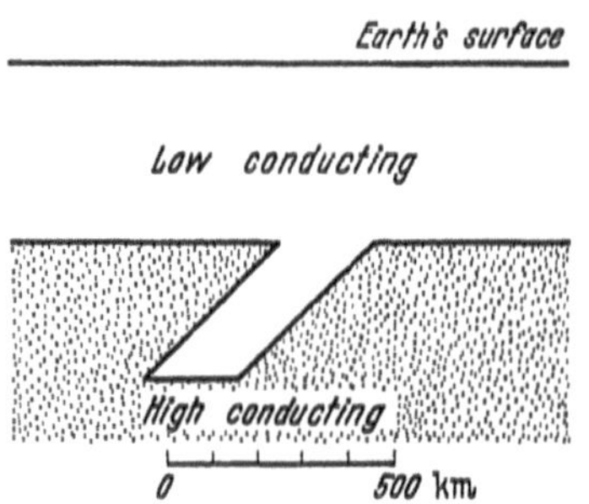

Fig. 53. Schematic cross-section of the electrical conductivity beneath Japan. After RIKITAKE[3]

2.8. Geochemical Data[4]

2.81. Geochemistry of the Interior of the Earth. A further body of information regarding the Earth may be obtained from chemical investigations. From the spectroscopic analysis of celestial bodies it has been inferred that the chemical composition of the universe is quite uniform. The latter consists throughout of the same elements and their relative abundances are everywhere more or less identical. Such differences as exist can usually be explained by the prevailing local conditions such as high temperatures which permit certain elements to be used up in thermonuclear reactions[5].

If the knowledge of cosmic abundances of elements is carried over to the Earth, then it is possible to make some inferences regarding the chemical composition of its interior. Combining this information with the knowledge of density, pressure and seismic velocity distribution with depth leads to the following conclusions.

[1] RIKITAKE, T.: Bull. Earthquake Res. Inst. **34**, 291 (1956); **36**, 1 (1958); **37**, 545 (1959).

[2] RIKITAKE, T.: Geophys. J. Roy. Astr. Soc. **2**, 276 (1959).

[3] RIKITAKE, T.: Bull. Earthquake Res. Inst. **34**, 291 (1956).

[4] The fundamental treatise on this subject is by K. RANKAMA and T. G. SAHAMA: Geochemistry. Chicago: Chicago Univ. Press 1950.

[5] MASON, B.: Principles of Geochemistry. New York: J. Wiley & Sons 1952

The *mantle*, which reaches from below the Mohorovičić discontinuity to a depth of 2900 km probably consists of dunite, peridotite or eclogite, as these are the only materials that have elastic constants of the right order of magnitude to yield the high seismic velocities observed at that depth. Within the mantle, at a depth of about 900 km, there is perhaps a further discontinuity (Birch discontinuity), involving a change of phase, or a chemical change or both[1]. This follows from the observation that the variation of the observed compressibility with depth cannot be accounted for by the compression of a homogeneous material between 200 and 800 km depth (BULLEN's *C*-layer), but that compression alone can account very well for the observed compressibility between 900 and 2900 km depth. Hence, there may be a discontinuity at between 800 and 900 km depth. Thus, the "transition layer" (BULLEN's *C*-layer) poses certain problems, and a number of conjectural calculations have been made[2, 3] (cf. also p. 105).

The *core* which reaches from 2900 km to the center of the Earth, presumably consists of a mixture of iron and nickel. The latter hypothesis is supported by the abundance of iron in meteoritic material; meteorites being thought of as débris from a planet of similar constitution as the Earth which for some reason disintegrated. However, all that is *really* known about the core of the Earth is that it must have a high density and that it probably has a high electrical conductivity. The latter follows from the customary explanation of the Earth's magnetic field in terms of magneto-hydrodynamic convection currents. Both these properties conceivably could also be exhibited by a metallic phase of magnesium-iron silicates (RAMSEY's hypothesis[4]) or hydrogen (KUHN-RITTMANN's hypothesis[5]) which would be stable at the high pressures prevailing at the depth in question. It is not quite clear, though, whether such a hypothesis can be maintained in the light of quantum mechanical estimates of the density to be expected in such a phase. It seems, therefore, that the assumption of a core consisting of an iron-nickel alloy is still the most satisfactory one in the light of present knowledge.

2.82. Geochemistry of the Crust. The crust is the only part of our globe which is directly accessible. Because of the strict differentiation of the material of the Earth into several layers, it is possible to treat the crust as a separate chemical unit.

---

[1] BIRCH, F.: Trans. Amer. Geophys. Un. **32**, 533 (1951).
[2] SHIMAZU, Y.: J. Earth Sci., Nagoya **6**, 12, 31 (1958).
[3] KNOPOFF, L., and R. J. UFFEN: J. Geophys. Res. **59**, 471 (1954).
[4] RAMSEY, W. H.: Month. Not. Roy. Astron. Soc. **108**, 404 (1948).
[5] KUHN, W., u. A. RITTMANN: Geol. Rdsch. **32**, 215 (1941).

The materials found in the crust show an extreme variety. However, there seem to be only two basic types which have been called *andesitic* and *basaltic*.

Most of the continental material can be classified as andesitic. Andesite is a volcanic lava containing about 60% silica, the remainder being Al_2O_3, CaO and other constituents. When it becomes worn-down, it causes various types of sedimentary rocks to be formed; finally due to

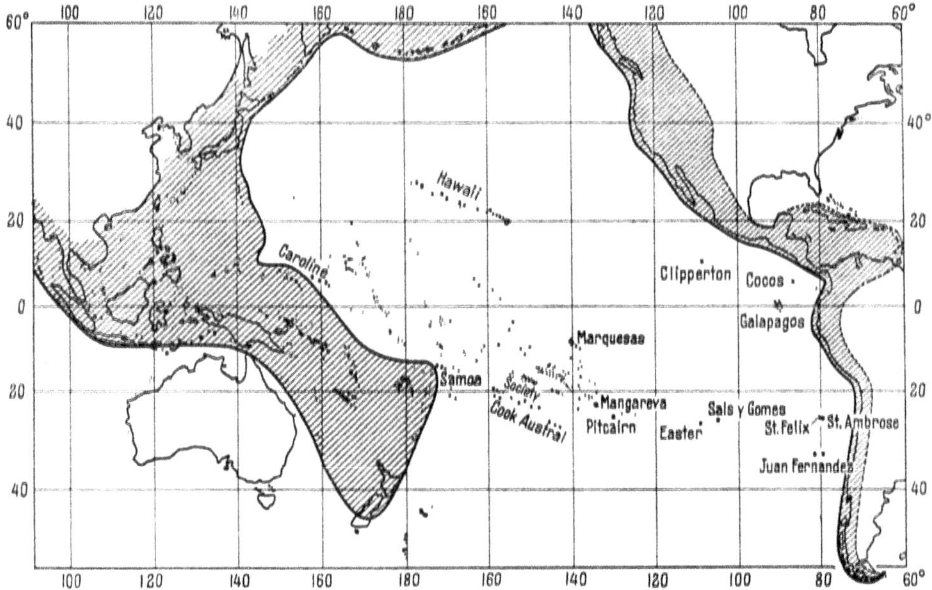

Fig. 54. The Andesite Line, separating andesitic (shaded) from non-andesitic areas in the Pacific. After CHUBB[1]

subsequent metamorphism it becomes granite and grano-diorite. Thus, the evolution of continental rocks can be considered as a geochemical cycle. Starting from andesitic lava, one has first crystallization, then weathering and transportation which produce sediments. The sediments become consolidated to form sedimentary rocks; thence metamorphism takes over and produces metamorphic rocks[2, 3], and finally anatexis (remelting) may take place and recreate the lava. The passage of any particular chemical element through its "geochemical cycle"[4] can be followed and it is found that the picture created in this fashion is essentially consistent.

[1] CHUBB, L. J.: Geol. Mag. **71**, 289 (1934).

[2] FYFE, W. S. *et al.*: Metamorphic Reactions and Metamorphic Facies. Geol. Soc. Amer. Memoir No. 73 (1958).

[3] READ, H. H.: The Granite Controversy. London: Murby 1957.

[4] ENGELHARDT, W. v.: Nova Acta Leopold. **21**, No. 143, 85 (1959).

In contrast to the andesitic rocks of the continents, there are the basaltic ones of the ocean bottoms. The basalt-cycle is similar to the andesite cycle. Starting out with basaltic lava, one has crystallization and sedimentation. However, upon metamorphosis one obtains gabbro and not granite. If anatexis takes place, one obtains again the lava.

As mentioned above, basalt seem to be the material of the ocean bottom. Basaltic type rocks are mostly found in oceanic islands. Volcanoes at the margins of continents seems to spew out mostly andesitic lava, whereas the lava found on mid-oceanic islands is mostly basaltic. The reason for the difference in chemical composition of the continental crust and the oceanic crust is not clear; it seems to run parallel with the marked differences in physiographic and geophysical properties. The boundary between basaltic and andesitic rocks and lavas in the Pacific ocean has been called the "*Andesite Line*" (see Fig. 54).

It is a great puzzle why the material of the crust (i.e. basalt and andesite) should be of such rare abundance in the Earth as a whole; the mantle, according to Sec. 2.81, is assumed to consist of dunite, peridotite and eclogite. In order to get around this difficulty, it has been suggested by LEES[1] and others[2,3] that the Mohorovičić discontinuity is a phase-discontinuity and not a chemical discontinuity at all. Accordingly, the mantle would consist of a high-pressure phase of basalt and andesite. Unfortunately, the whole mechanism of phase-transitions is not yet sufficiently well known to come to any definite conclusion about this point. It may be expected that the hole to be drilled in to the Earth's mantle (mohole) will settle this question in the near future.

[1] LEES, G. M.: Quart. J. Geol. Soc. Lond. **109**, 217 (1953).
[2] LOVERING, J. F. L.: Trans. Amer. Geophys. Un. **39**, 947 (1958).
[3] KENNEDY. G. C.: Trans. Amer, Geophys. Un. **41**, 283 (1960). — Amer. Sci. **47**, 491 (1959).

III. The Mechanics of Deformation

3.1. Finite Strain in Rheological Bodies

3.11. The Physics of Deformation. The basic problem in the study of the Earth's crust is to understand its deformations. Unfortunately, the general physics of deformations is not as well understood as one might desire.

Deformations can occur in two fundamentally different ways: viz. continuously or discontinuously. Under continuous deformation we understand a condition where neighboring points in a material always remain neighboring points, whereas in discontinuous displacements this is not the case.

Several branches of the theory of continuous deformation have been very intensively developed. It is well known that the theory of elasticity has been carried to a high degree of refinement; the same is true for the hydrodynamics of viscous fluids. Unfortunately, the materials of which the Earth's crust is composed, are very unlikely to fit either of these theories. Of those branches of the continuous displacement theory that are more or less well developed, the theory of plasticity has the most bearing upon the displacements observed in the Earth's crust. However, the theory of plasticity has been developed for the description of the behavior of a metal during cold working, and it cannot be expected, therefore, that its application to the Earth's crust will lead to entirely satisfactory results. The material in the Earth's interior shows a very complicated behavior, possibly of the types discussed in the various theories of "rheology". However, these theories all appear to be more or less heuristic and therefore incomplete.

Very important is the discussion of discontinuous displacements. There are many instances where ruptures, fissures, fractures and such like occur in the Earth's crust. Although humans have been breaking things since the inception of civilization, it is an unfortunate fact that the whole subject of fracture is only very incompletely understood. There are quite a number of rule-of-thumb criteria of fracture or better: of when a structure is supposed to be safe so as *not* to fracture,—but the basic problem of describing the progress of a fracture surface in a given body under given external stresses, has not yet been solved.

We shall, in the following sections, consider the various aspects of the theory of deformations one by one.

3.12. The Structure of a Finite Strain Theory. In order to obtain a description of the dynamics of continuous media, there are various steps that must be observed[1-3]. In the first place, one must decide upon a description of the deformation. Once this has been achieved, one must express various physical laws: the condition of continuity, the law of motion, and boundary conditions. We shall discuss these steps one by one.

a) Measure of Displacement. Let us assume that a certain volume W (which may be infinite) of space is filled with matter. The volume W, and the way in which it is filled, changes with time t.

The points of space may be specified by giving three Cartesian co-ordinates x_i such that the line element is defined as follows:

$$ds^2 = dx_i \, dx_i. \tag{3.12-1}$$

In this formula, the summation convention has been used, which stipulates that one has to sum over all indices that occur twice.

The above scheme characterizes the geometrical space ("co-ordinate space") occupied or potentially occupied by the continuous medium. The next task is to characterize the medium. It is well known[2] that this can be done by introducing three parameters ξ_α. The whole medium is characterized if the parameters run through all points of a volume Ψ in "parameter space", i.e. the space of the ξ's.

We shall assume that the space Ψ of the parameters is endowed with a Cartesian metric, so that a line element $d\sigma$ can be defined as follows:

$$d\sigma^2 = d\xi_\alpha \, d\xi_\alpha. \tag{3.12-2}$$

It is always possible to make such a parameter transformation that the parameters are equal to the co-ordinates of all the particles at a particular time, say t_0. It is often convenient to do this.

With the above characterization of a continuous medium and the geometrical space occupied by it, one can proceed to describe motions. The complete motion is obviously given if the geometrical co-ordinates of each particle of the medium are known for all times:

$$x_i = x_i(\xi_\alpha, t). \tag{3.12-3}$$

Thus, the specification of three functions of the three parameters plus time determines the motion. This type of description is often called the *material* form of the description of motion.

---

[1] The following is after A. E. Scheidegger: Canad. J. Phys. **34**, 498 (1956).

[2] Truesdell, C.: J. Rat. Mech. Anal. **1**, 125 (1952).

[3] Prager, W.: Introduction to Mechanics of Continua. Boston: Ginn & Co. 1961.

One also could have provided a description of the motion in another way, viz. by solving Eqs. (3.12–3) for the ξ's:

$$\xi_\alpha = \xi_\alpha(x_i, t). \tag{3.12–4}$$

Physically, this means that one states which "particle" is at a given time at any given spot. This is called the *spatial* form of the description of motion.

Although the specification of the functions in Eqs. (3.12–3) completely describes the motion, it is often convenient to introduce various other quantities. This is so because it is cumbersome to express the equations of motion directly in terms of the quantities introduced heretofore.

An important kinematical notion is the concept of *strain*. Strain, in the finite theory, is defined as half the difference of the squared distance between neighboring points in two states, one of which is arbitrarily called "state of zero strain". The element of distance between neighboring particles is:

$$d s^2 = \frac{\partial x_i}{\partial \xi_\alpha} \frac{\partial x_i}{\partial \xi_\beta} d\xi_\alpha d\xi_\beta = \varkappa_{\alpha\beta} d\xi_\alpha d\xi_\beta \tag{3.12–5}$$

where

$$\varkappa_{\alpha\beta} = \frac{\partial x_i}{\partial \xi_\alpha} \frac{\partial x_i}{\partial \xi_\beta}. \tag{3.12–6}$$

Thus, the element of distance between two particles defines a symmetric tensor $\varkappa_{\alpha\beta}$ in parameter space. If we denote the (time independent) tensor of the state of zero strain by $\zeta_{\alpha\beta}$, we can define the "material strains" as follows:

$$\varepsilon_{\alpha\beta} = \frac{1}{2}(\varkappa_{\alpha\beta} - \zeta_{\alpha\beta}) = \frac{1}{2}\left(\frac{\partial x_i}{\partial \xi_\alpha} \frac{\partial x_i}{\partial \xi_\beta} - \zeta_{\alpha\beta}\right). \tag{3.12–7}$$

Eq. (3.12–5) permits a different interpretation of strain from that given above. For, this Eq. (3.12–5) can be taken as a fundamental metric form in a certain space $\Sigma(t)$ which has been called the "material strain space". In this instance, one should note that the $d\xi$'s are contravariant vectors and could be expressed by using superscripts instead of subscripts (employing the notation of Riemannian geometry):

$$d s^2 = \varkappa_{\alpha\beta} d\xi^\alpha d\xi^\beta. \tag{3.12–8}$$

The elements (points) of this strain space are the parameters; the line element is ds, which is the line element of the co-ordinates. The metric in strain space is a function of time. In the strain space $\Sigma(t)$, the tensor $\varkappa_{\alpha\beta}$ can be used to raise and lower indices, if the contravariant metric tensor is defined as follows:

$$\varkappa_{\alpha\beta}\varkappa^{\beta\gamma} = \delta_\alpha^\gamma \tag{3.12–9}$$

where δ_α^γ signifies the Kronecker symbol.

Because of Eq. (3.12–5), the metric in strain space must be flat. This means that the contracted Riemann-Christoffel curvature tensor in Σ must be zero. This imposes six conditions upon the $\varkappa$'s and hence upon the strains. These conditions are very well known in the theory of elasticity where they are called "compatibility conditions".

Henceforth, we shall consider $\varkappa_{\alpha\beta}$ (and therewith $\varepsilon_{\alpha\beta}$) as a tensor in parameter space, and not as a metric. The problem, therefore, will be to determine the components of a tensor "field" $\varepsilon_{\alpha\beta}$ as a function of time.

As a final kinematical notion, one can introduce the concept of density. Postulating the medium as homogeneously of density ϱ_0 in the unstrained state, we define, in accordance with the principle of continuity:

$$\left.\begin{aligned}\varrho(\xi_\alpha, t) &= \varrho_0 \sqrt{\det\zeta} / \sqrt{\det\varkappa} \\ &= \varrho_0 \sqrt{\det\zeta_{\alpha\beta}} / \sqrt{\det(2\varepsilon_{\alpha\beta} + \zeta_{\alpha\beta})}\,.\end{aligned}\right\} \tag{3.12–10}$$

The above definition of strain was termed "material" because it gives the line element ds at time t in terms of the parameters. It is customary in hydrodynamics to term all description of motion in terms of the parameters (which are identifiable with the co-ordinates at t_0) as "material" and this terminology is being retained here.

Since parameter and co-ordinate space are entirely homologous, it is obviously possible to reverse the rôles played by the two[1, 2]. Thus, let the line element $d\sigma$ of parameter space (i.e. the line element of the material at time $t = t_0$) be expressed in terms of the co-ordinates at the time t. One obtains:

$$d\sigma^2 = \frac{\partial\xi_\alpha}{\partial x_i}\frac{\partial\xi_\alpha}{\partial x_j}\,dx_i\,dx_j = k_{ij}\,dx_i\,dx_j \tag{3.12–11}$$

where

$$k_{ij} = \frac{\partial\xi_\alpha}{\partial x_i}\frac{\partial\xi_\alpha}{\partial x_j}\,. \tag{3.12–12}$$

This means that the distance $d\sigma$, which was taken up at t_0 by two points of the medium differing at time t by dx_i, is specified by a symmetric tensor k_{ij} in co-ordinate space.

It is fairly easy to calculate the connection between spatial and material (as defined above) distances. One obtains:

$$k_{ij} = \varkappa_{\alpha\beta}\frac{\partial\xi_\alpha}{\partial x_m}\frac{\partial\xi_m}{\partial x_i}\frac{\partial\xi_\beta}{\partial x_r}\frac{\partial\xi_r}{\partial x_j}\,. \tag{3.12–13}$$

In particular, the state of zero strain is characterized by the following tensor:

$$z_{ij} = \zeta_{\alpha\beta}\frac{\partial\xi_\alpha}{\partial x_m}\frac{\partial\xi_m}{\partial x_i}\frac{\partial\xi_\beta}{\partial x_r}\frac{\partial\xi_r}{\partial x_j}\,. \tag{3.12–14}$$

[1] DEUKER, E. A.: Dtsch. Math. 5, 546 (1941).

[2] ECKART, C.: Phys. Rev. 73, 373 (1948).

Thus, one can define the "spatial strain" as follows:

$$e_{ij} = \frac{1}{2}(k_{ij} - z_{ij}) = \frac{1}{2}\left(\frac{\partial \xi_\alpha}{\partial x_i}\frac{\partial \xi_\alpha}{\partial x_j} - z_{ij}\right). \qquad (3.12\text{–}15)$$

In the above formulas, it should be noted that the tensor z_{ij} characterizing the state of zero strain is no longer time-independent. The spatial density becomes:

$$\left.\begin{aligned} \varrho(x_i, t) &= \varrho_0\left(\sqrt{\det z}/\sqrt{\det k}\right) \\ &= \varrho_0\sqrt{\det z_{ij}}/\sqrt{\det(2e_{ij} + z_{ij})} \end{aligned}\right\} \qquad (3.12\text{–}16)$$

which is the same as the material density. The latter statement is easy to check as the expression for z as well as k can be written as the product of the matrix ζ or $\varkappa$, respectively, with the same matrices. Upon the formation of the determinants, the determinants of these matrices can be factorized out and cancel as they are the same in numerator and denominator.

It will be noted that the tensor k_{ij} also, if taken as a metric tensor, describes a flat metric. Hence it must satisfy the condition that the Riemann-Christoffel curvature tensor, formed with k_{ij}, must be the zero tensor. This leads to six compatibility relations for k_{ij}, and hence for e_{ij}, as was the case for the material strains.

Finally, one may make a few remarks regarding "convected co-ordinates". We have seen above that the tensors $\varkappa_{\alpha\beta}$ and k_{ij}, being symmetric tensors, can be thought of as metric tensors in certain spaces. This is the approach to finite displacement rheology which has been taken by OLDROYD[1]. The tensor $\varkappa_{\alpha\beta}$, in fact, can be regarded as describing the metric in a "convected" co-ordinate system, viz. in a system whose co-ordinate lines are given at $t = t_0$ as Cartesian co-ordinate lines, and are moving along with the medium. If the motion is expressed in terms of such "convected co-ordinates", one has to introduce the whole formalism of Riemannian geometry, which is actually quite unnecessary in view of the fact that, after all, the medium is moving in an ordinary Euclidean space. "Convected" co-ordinates, therefore, appear as rather clumsy means of describing the dynamics of continuous matter.

b) Continuity Condition. In order to expand the theory further, it is necessary to define time derivatives of the various quantities introduced above.

Owing to the occurrence of parameters and of co-ordinates, time differentiation of any function can be performed either with the parameters held constant, or with the co-ordinates held constant. Although the complete reciprocity between parameter space and co-ordinate space

[1] OLDROYD, J. G.: Proc. Roy. Soc. Lond., Ser. A 200, 523 (1950).

has already been demonstrated by DEUKER[1], this fact is not commonly borne to light in the customary presentation of rheological theories.

We shall denote the time derivative of a scalar function of the parameters and co-ordinates, for constant parameters, by D/Dt:

$$\frac{D}{Dt} f(\xi, t) = \frac{\partial f}{\partial t}\bigg|_{\text{constant } \xi} \tag{3.12–17}$$

and the time derivative with the co-ordinates held constant, by

$$\frac{\Delta}{\Delta t} f(x, t) = \frac{\partial f}{\partial t}\bigg|_{\text{constant } x} . \tag{3.12–18}$$

With the definition of time derivatives, one is now in a position to formulate the continuity condition. It is

$$\frac{\Delta \varrho}{\Delta t} + \frac{\partial}{\partial x_i}\left(\varrho \frac{D x_i}{D t}\right) = 0 . \tag{3.12–19}$$

In this presentation of the continuity equation, care has been taken to indicate the various types of time derivatives, according to our notation. It becomes then apparent that the usual form of the continuity equation is somewhat cumbersome since it contains functions which have x as well as ξ as arguments.

c) The Equations of Motion. A similar situation occurs in the equations of motion which are usually written as follows:

$$\frac{\partial \tau_{ik}}{\partial x_k} + \varrho\left(f_i - \frac{D^2 x_i}{D t^2}\right) = 0 . \tag{3.12–20}$$

The equations of motion contain several important concepts. The quantity τ_{ik} is called the stress tensor, f_i the specific volume force; ϱ is as usual the density of the material. It is customary to define the stress tensor as a function of x_i. The stress tensor τ_{ik} is postulated in such a fashion that, upon any imagined closed surface within a body, there exists a distribution of stress vectors (tractions) $p_{(n)i}$ whose resultant and moment are equivalent to those of the actual forces of material cohesion exerted by the material outside upon that inside[2], and that these stress vectors can be written as follows

$$p_{(n)i} = \tau_{ij} n_j \tag{3.12–21}$$

where n_j is the normal unit vector to any surface element under consideration. It is customary to represent the normal component of the stress vector upon a given surface by σ, the tangential one by τ, and to call them tension and shear, respectively.

[1] DEUKER, E. A.: Dtsch. Math. 5, 546 (1941); see also A. E. GREEN and W. ZERNA: Phil. Mag. (7) 41, 313 (1950).

[2] TRUESDELL, C.: J. Rat. Mech. Anal. 1, 125 (1952).

A much used representation of the stress tensor has been devised by MOHR[1]. In a two-dimensional stress state, one can represent the stress tensor by the locus of all the corresponding points in a $\sigma-\tau$

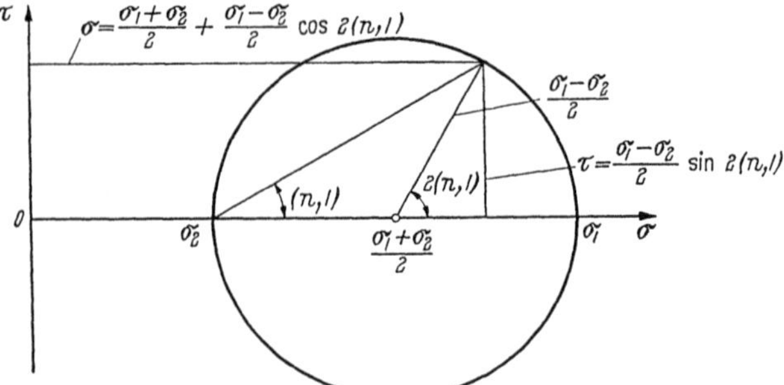

Fig. 55. Mohr circle for a two-dimensional stress state

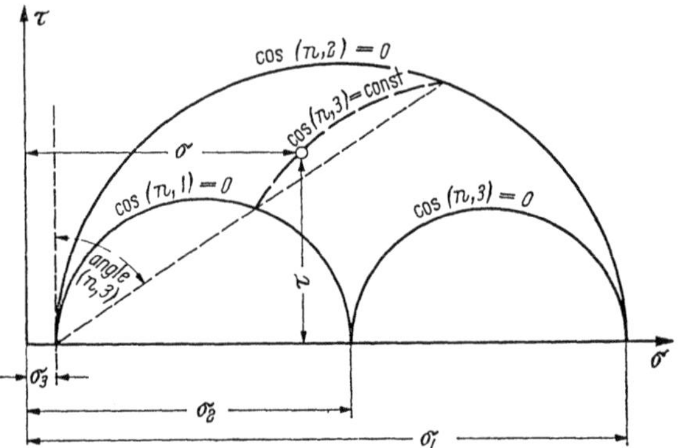

Fig. 56. Mohr diagram for a three-dimensional stress state

diagram. This locus turns out to be a circle (MOHR's circle, see Fig. 55), for one has:

$$\sigma = \frac{\sigma_1 + \sigma_2}{2} + \frac{\sigma_1 - \sigma_2}{2} \cos 2(\boldsymbol{n}, 1), \tag{3.12–22}$$

$$\tau = (-)\frac{\sigma_1 - \sigma_2}{2} \sin 2(\boldsymbol{n}, 1) \tag{3.12–23}$$

where σ_1 and σ_2 denote the principal[2] stresses. In the three-dimensional case, one can accordingly represent a stress tensor by three limiting circles, each corresponding to the plane defined by each pair of dimensions.

[1] MOHR, O.: Abhandlungen aus dem Gebiete der technischen Mechanik, 3. Aufl. Berlin: Wilh. Ernst & Sohn 1928.

[2] i.e. the tractions for those orientations of the surface in the point under consideration for which the shear is zero.

Each point in the area between these three circles defines a possible orientation of a surface element, and the corresponding values for σ, τ read from the diagram are then the normal and tangential components of the stress vector acting upon that surface, respectively (cf. Fig. 56).

d) Rheological Condition. The relations noted heretofore are not sufficient to determine the behavior of continuous matter. What is needed is a connection between the kinematical quantities and the dynamical quantities. Such relations are commonly termed "rheological conditions", often also called "constitutive equations".

The rheological condition, being an equation of state, must be independent of the motion of the medium as a whole. It, therefore, appears as natural to use as kinematical variables the strains $\varepsilon_{\alpha\beta}$ or the components of the tensor $\varkappa_{\alpha\beta}$ describing the state of the medium in parameter space.

Many types of rheological equations have been postulated (see Sec. 3.3), but it is to the credit of OLDROYD[1] to have made a satisfactory enumeration of possible variables and the form in which they may enter the rheological condition. OLDROYD states that, in its general form, the rheological condition can be written as a set of integrodifferential equations in parameter space, of which six are independent, relating the stresses τ_{ik}, the finite strains $\varepsilon_{\alpha\beta}$ (or else the components of the tensor $\varkappa_{\alpha\beta}$), the temperature $T(\xi, t)$ and such physical constants as may pertain to the medium.

The quantities τ_{ik}, $\varepsilon_{\alpha\beta}$, and τ are functions of time. The fact that it is permissible for the rheological condition to contain integrations of time (with fixed ξ) allows for the possibility of occurrence of effects which depend on the total previous strain-history at a certain point of the medium. In simple cases, the integrodifferential equations can be reduced to differential equations by the processes of rearrangement and differentiation.

3.13. The Possible Schemes of Dynamics. In order to solve a concrete dynamical problem, it is necessary to express all the equations of the last section in one type of independent variable. The necessary equations, as outlined above, are (i) the continuity equation, (ii) the equations of motion, (iii) the rheological condition, and (iv) boundary and initial conditions. The functions sought after will be either $x_i(\xi_\alpha, t)$ or $\xi_\alpha(x_i, t)$. According to which set of the unknowns is chosen, one obtains two different schemes of dynamics that are possible.

It is observed that the equations of motion and the boundary conditions (in most cases) can be more easily expressed in co-ordinate space than in parameter space. For the rheological condition, and in some

[1] OLDROYD, J. G.: Proc. Roy. Soc. Lond., Ser A **200**, 523 (1950).

cases for the boundary conditions (if there are free surfaces), the reverse is true. Before proceeding to a solution of a certain problem, one will therefore have to decide at this point whether it will be preferable to work in parameter space or in co-ordinate space.

The transformation schemes to pass from one set of variables to the other are rather involved[1]. Fortunately, in many applications it is not required to deal with them in full generality and it is therefore not necessary to reproduce them here in detail. For most practical applications the passage from one set of unknowns to the other can be accomplished easily owing to the special assumptions which are introduced in the respective theories.

3.14. Additional Stress and Strain. It is evident from the above exposition that the problem of determining the displacement from a set of boundary conditions and from the knowledge of the appropriate rheological condition is, in its full generality, a very difficult one. It is therefore often necessary to make certain simplifications.

A device which is often used is the restriction that all that one attempts to calculate, is the deviation from a certain "standard" stress- (and therefore also strain-) state. Thus, assuming a certain stress state, one aims at determining the "additional" stress and the corresponding "additional" strains. It is in this instance often possible to assume that the "additional strains" are small.

A justification of the above-indicated procedure lies in the fact that the tensor of zero strain is, in reality, a badly defined quantity;—at least in the manner as it was introduced in Eq. (3.12–7). One can therefore argue that one could have just as well taken any other tensor and called it "zero strain tensor",—thus, for instance, the tensor of the state from which one wants to calculate the deviation. It should be mentioned, however, that there is really a physical way conceivable in which the zero strain state can be defined. This is as follows: Imagine that one cuts a small volume element surrounding the point in which the zero strain state is to be defined. Assume further that all the tractions across its surface be removed. The material, then, presumably, will reach (at least after a long time) an equilibrium state; the latter may be taken as the zero strain state. It is evident that the zero strain state defined in this manner may not be integrable; i.e. it is quite possible that no position of the body exists in which all the strains (and therewith the stresses after infinite time) are removed.

The above physical definition of a zero strain state introduces certain complications. First of all, with the loss of integrability, one also loses the ordinary compatibility relations for the strains;—i.e. the latter have

[1] See SCHEIDEGGER, A. E.: Canad. J. Phys. 34, 498 (1956).

to be replaced by some different relations expressing that the *additional* displacements starting from a *real* state must take place in Euclidean space. It is therefore logically much more satisfactory to admit as zero strain states only actually possible states of the body although the latter may not be zero strain states physically.

Under the above conditions it seems logical to *define* the chosen zero strain state also as a zero stress state;—i.e. one concerns oneself only with the stresses *additional* to those present in the zero strain state. However, if this be done, it is evident that the rheological condition becomes, in general, a function of the zero strain state. The only instance where the rheological condition does not depend on the zero strain state is when the latter is a linear relationship between stress and strain: for only then is the additional displacement the same function of the additional stress, regardless of the amount of pre-strain. This is the case only in infinitesimal elasticity theory (see Sec. 3.21 *infra*).

The device of using additional stress and strain is very common in geodynamics. The "standard" state is the hydrostatic state of the geoid; all stresses and strains of interest are deviations therefrom.

3.15. Finite Strain in Tectonic Deformations[1]. It is extremely difficult to apply the general formalism of finite strain theory to geodynamic investigations. For the sake of simplicity, it will be necessary to assume that the functions $x_i(\xi_\alpha)$ introduced above are linear. Thus, we shall now write

$$x_i = a_{i\alpha}\xi_\alpha + x_i^{(0)} \tag{3.15-1}$$

where the $a_{i\alpha}$ and $x_i^{(0)}$ represent schemes of coëfficients that may be functions of time. The inverse relations are then given by

$$\xi_\alpha = A_{\alpha i}x_i + \xi_\alpha^{(0)} \tag{3.15-2}$$

where the $a_{i\alpha}$ and $A_{\alpha i}$ represent matrices such that

$$\alpha_{i\alpha}A_{\alpha j} = \delta_{ij} \tag{3.15-3}$$

and where

$$x_i^{(0)} = -a_{i\alpha}\xi_\alpha^{(0)} \quad \text{or} \quad A_{\beta i}x_i^{(0)} = -\xi_\beta^{(0)}. \tag{3.15-4}$$

The components of the tensors $\varkappa_{\alpha\beta}$ and k_{ij} introduced above become therefore

$$\varkappa_{\alpha\beta} = (\partial x_i/\partial\xi_\alpha)(\partial x_i/\partial\xi_\beta) = a_{i\alpha}a_{i\beta}, \tag{3.15-5a}$$

$$k_{ij} = (\partial\xi_\alpha/\partial x_i)(\partial\xi_\alpha/\partial x_j) = A_{\alpha i}A_{\alpha j}. \tag{3.15-5b}$$

The tensor $\varepsilon_{\alpha\beta}$ becomes

$$\varepsilon_{\alpha\beta} = \tfrac{1}{2}(a_{i\alpha}a_{i\beta} - \delta_{\alpha\beta}). \tag{3.15-6}$$

[1] See SCHEIDEGGER, A. E.: Canad. J. Phys. **40**, 761 (1962).

It will be noted that Eqs. (3.15–1), (3.15–2) represent transformations of the whole of space into itself, representing, in our case, the deformation of rock strata. In general, such deformations are subject to certain conditions. First of all, in most geological deformations it is usually permissible to assume that the total volume of the rocks is conserved. This requires that the Jacobian determinant of the transformation be 1:

$$|a_{i\alpha}| = 1. \tag{3.15–7}$$

For the determinant of $\varkappa$ this yields

$$|\varkappa| = |a_{i\alpha} a_{i\beta}| = |a_{i\alpha}|\,|a_{i\beta}| = 1. \tag{3.15–8}$$

Hence also

$$|k| = 1. \tag{3.15–9}$$

The problem now arises to ascertain as much information about the tensor k_{ij} as is possible from field observations. Two possibilities come to one's mind in order to accomplish this, viz. (i) the observation of the deformation of pebbles and (ii) the observation of faults.

a) Deformation of pebbles. If one finds pebbles embedded in a formation, then it may be possible to ascertain the principal directions of k directly. If these pebbles have (a more or less) ellipsoidal shape, and if one knows that the pebbles were originally spherical, then the principal axes of the ellipsoid immediately represent the principal directions of the tensor k. The shape of an originally spherical object is given by a quadratic equation. In the system of principal axes this equation becomes

$$K_{11} x_1^2 + K_{22} x_2^2 + K_{33} x_3^2 = 1 \tag{3.15–10}$$

which is the central equation of an ellipsoid with the principal half-axes L_i

$$L_i = K_{ii}^{-\frac{1}{2}}. \tag{3.15–11}$$

The K_{ii} are then the eigenvalues of the tensor k.

b) Faults. A fault is, in abstraction, thought of as a discontinuous deformation in which, along a "fault plane", two surfaces slide over each other, the displacement itself being represented by a "displacement vector" lying in the fault plane.

In order to find the principal axes of k in the vicinity of a fault, we regard the latter as a limiting case of a continuous displacement. In case an original sphere overlaps the two sides of a fault, it becomes clear that the sphere will be very much stretched in the direction of the displacement vector. The direction of the displacement vector, thus, should represent one of the sought-after principal axes of the tensor k.

The second principal axis of k will be orthogonal to the displacement vector and orthogonal to the fault plane: in this direction, one has maximum shortening.

The third principal axis is normal to the displacement vector, and lies in the fault plane. Along it, there is no deformation. This axis, therefore, represents the direction of the intermediate axis of the tensor ellipsoid.

In geological field work, three types of faults are commonly discerned (see Sec. 1.61). The first type, called *normal* fault, has a displacement vector normal to the strike of the fault, the sides moving away from each other. The second type (*reversed* fault) is the same as a normal fault, but with a reversed direction of motion. In the third type (*transcurrent* fault), the displacement vector is parallel to the strike. It is seen that in all these types of faults, the strike direction, the dip direction and the normal to these two directions are principal directions of k. In mixed type faults, further complications might occur. It should be noted that the intermediate principal axis is given by the strike direction in normal and reversed faults, and by the dip direction in transcurrent faults.

3.2. Elasticity and Plasticity

3.21. Infinitesimal Elasticity Theory. Infinitesimal elasticity theory is obtained from the general scheme of rheological dynamics if it is assumed that

(a) the displacements are always small

(b) the rheological condition is as expressed by HOOKE's law.

These two assumptions completely define a mathematical theory of deformation.

Assumption (a) permits one to express the co-ordinates as follows [cf. (3.12–3)]

$$x_i(\xi_\alpha, t) = x_i(\xi_\alpha, 0) + u_i(\xi_\alpha, t) \tag{3.21–1}$$

where u_i is called displacement. As indicated above, the displacements are assumed as small so that their squares can be neglected. It is also customary to identify the parameters with the co-ordinates at time $t=0$ so that

$$x_i(t=0) = \xi_i. \tag{3.21–2}$$

Under these assumptions, the expression for the (material) strains (3.12–7) reduces to

$$\varepsilon_{ik} = \frac{1}{2}\left(\frac{\partial u_i}{\partial x_k} + \frac{\partial u_k}{\partial x_i}\right). \tag{3.21–3}$$

Consequently, the compatibility relations for the strains (i.e. the condition that the Riemann-Christoffel curvature tensor of the metric

$\varepsilon_{ik} + \delta_{ik}$ is zero) turns out to be

$$\frac{\partial^2 \varepsilon_{(i)(i)}}{\partial x_j^2} + \frac{\partial^2 \varepsilon_{(j)(j)}}{\partial x_i^2} = 2 \frac{\partial^2 \varepsilon_{ij}}{\partial x_i \partial x_j}, \tag{3.21-4a}$$

$$\frac{\partial^2 \varepsilon_{(i)(i)}}{\partial x_j \partial x_k} = \frac{\partial}{\partial x_{(i)}} \left(- \frac{\partial \varepsilon_{jk}}{\partial x_i} + \frac{\partial \varepsilon_{ik}}{\partial x_j} + \frac{\partial \varepsilon_{ij}}{\partial x_k} \right). \tag{3.21-4b}$$

In the above scheme the strain ε_{ik} as defined in (3.21-3) represents actually a tensor as is indicated by the notation. Care should be taken in referring to the literature since the definition of "strain" is not always that given here. Notably it will often be found that the "shear" strains are taken as twice the corresponding components as defined here. This, however, effects that the strain is no longer representable as a tensor.

Assumption (b) (*i*.e. HOOKE's law) can be stated as follows (see e.g. JEFFREYS[1])

$$\tau_{ik} = \lambda \delta_{ik} \varepsilon_{jj} + 2\mu \varepsilon_{ik} \tag{3.21-5}$$

which expresses that there is proportionality of the isotropic and deviatoric components of stress and strain tensors separately. The constants λ and μ are called "LAMÉ's constants". The quantity μ is also often referred to as "rigidity". The rheological Eq. (3.21-5) refers to isotropic bodies only. Modifications have to be made for crystals.

The elasticity constants introduced in (3.21-5) are not the only ones that are possible. In principle, any two constants that are linearly independent functions of λ and μ could be used. Thus, the following constants have been employed (cf. LOVE[2]):

YOUNG's modulus E

$$E = \frac{\mu(3\lambda + 2\mu)}{\lambda + \mu} \tag{3.21-6a}$$

POISSON's ratio m

$$m = \frac{\lambda}{2(\lambda + \mu)} \tag{3.21-6b}$$

bulk modulus k (incompressibility)

$$k = \lambda + \tfrac{2}{3}\mu. \tag{3.21-6c}$$

There are many good books on elasticity to which the reader is referred for further details[2-5].

---

[1] JEFFREYS, H.: Cartesian Tensors. London: Cambridge University Press 1931.

[2] LOVE, A. E. H.: A Treatise on the Mathematical Theory of Elasticity, 4th ed. London: Cambridge University Press 1927.

[3] e.g. TIMOSHENKO, S., and J. N. GOODIER: Theory of Elasticity, 2nd ed. New York: McGraw-Hill 1951.

[4] MUSKHELISHVILI, N. I.: Some Basic Problems of the Mathematical Theory of Elasticity (transl. from Russian). Groningen: Noordhoff 1953.

[5] GREEN, A. E., and W. ZERNA: Theoretical Elasticity. Oxford: Clarendon Press 1954.

In the present study, we shall be chiefly concerned with equilibrium problems. In this connection, the introduction of stress functions (Airy functions) has been proven to be very convenient. This is particularly true for two limiting cases: the plane strain state and the plane stress state[1].

In the *plane strain state* one assumes that

$$\left.\begin{aligned} \varepsilon_{33} = \varepsilon_{23} = \varepsilon_{31} = 0 \\ \varepsilon_{11}, \varepsilon_{22}, \varepsilon_{12} \neq 0. \end{aligned}\right\} \tag{3.21–7}$$

It can easily be verified that the conditions of equilibrium are satisfied if the following assumption is made

$$\tau_{11} = \partial^2 \varphi / \partial x_2^2, \tag{3.21–8a}$$

$$\tau_{22} = \partial^2 \varphi / \partial x_1^2, \tag{3.21–8b}$$

$$\tau_{12} = - \partial^2 \varphi / \partial x_1 \partial x_2 \tag{3.21–8c}$$

where φ is a stress function. It must satisfy the following differential equation (the Laplacians to be taken in two dimensions)

$$\text{lap lap } \varphi = 0. \tag{3.21–9}$$

Any solution of (3.21–9) generates a solution of the plane strain equilibrium problem.

A similar situation holds for the *plane stress state*. We suppose

$$\tau_{33} = 0; \quad \tau_{23} = \tau_{31} = 0 \tag{3.21–10}$$

whereupon it can be shown that a solution of the problem is again obtained if one sets

$$\tau_{11} = \partial^2 \varphi / \partial x_2^2, \tag{3.21–11a}$$

$$\tau_{22} = \partial^2 \varphi / \partial x_1^2, \tag{3.21–11b}$$

$$\tau_{12} = - \partial^2 \varphi / \partial x_1 \partial x_2 \tag{3.21–11c}$$

where φ is again a stress function that must satisfy the following differential equation

$$\text{lap lap } \varphi = 0. \tag{3.21–12}$$

An important special case of elasticity theory concerns the existence of body waves. We note that the equations of motion can be written in terms of the displacements as follows, provided external forces are absent:

$$\varrho \frac{\partial^2 u_i}{\partial t^2} = (\lambda + \mu) \frac{\partial [\partial u_j / \partial x_j]}{\partial x_i} + \mu \text{ lap } u_i. \tag{3.21–13}$$

---

[1] Cf. e.g. JAEGER, J. C.: Elasticity, Fracture and Flow. London: Methuen & Co. Ltd. 1956.

If the divergence is formed of both sides of (3.21–13), one obtains

$$\varrho \frac{\partial^2 \Theta}{\partial t^2} = (\lambda + 2\mu) \operatorname{lap} \Theta \tag{3.21–14}$$

with

$$\Theta = \partial u_j / \partial x_j, \tag{3.21–15}$$

and, similarly, if the curl of (3.21–13) is formed, one obtains

$$\varrho \frac{\partial^2}{\partial t^2} [\operatorname{curl}(u_i)] = \mu \operatorname{lap} [\operatorname{curl}(u_i)]. \tag{3.21–16}$$

Eqs. (3.21–14) and (3.21–16) have the form of wave equations. They imply that a dilatational distrubance Θ may be transmitted with a phase velocity c_p of

$$c_p = \sqrt{(\lambda + 2\mu)/\varrho} \tag{3.21–17}$$

whereas a rotational disturbance may be transmitted with a phase velocity c_s of

$$c_s = \sqrt{\mu/\varrho}. \tag{3.21–18}$$

Both types of body waves are found in the Earth (cf. Sec. 2.11) where they have been called P and S waves, respectively. Dispersion is usually absent so that the above phase velocities are equal to the speed with which a disturbance travels through the Earth.

The above discussion can be extended and it can then be shown that, in addition to body waves, *surface waves* may also exist. The surface waves are usually subject to dispersion so that a distinction has to be made between group velocity and phase velocity. However, we shall not concern ourselves here with the detailed discussion of elastic waves which belongs into a treatise on elasticity theory. In that connection, it is standard text-book material[1–3].

A further interesting application of the equations of elasticity theory is INGLIS'[4] determination of the stresses in a plate which has an elliptic hole. INGLIS employed curvilinear co-ordinates α, β which are connected with a Cartesian system as follows:

$$\left.\begin{aligned} x &= c \cosh \alpha \cos \beta \\ y &= c \sinh \alpha \sin \beta . \end{aligned}\right\} \tag{3.21–19}$$

The stress determination of INGLIS using the methods of general elasticity theory is quite straightforward although it is somewhat tedious. INGLIS considered several cases to which we shall have to refer later.

[1] Cf. e.g. EWING, M., W. S. JARDETZKY and F. PRESS: Elastic Waves in Layered Media. New York: McGraw-Hill Book Co. 1957.

[2] BREKOVSKIKH, L. M.: Waves in Layered Media (transl. from Russian). New York: Academic Press 1960.

[3] MIKLOWITZ, J.: Appl. Mech. Rev. 13, 865 (1960) gives a good list of references.

[4] INGLIS, C. E.: Trans. Inst. Naval Architects 55, Part 1, 219 (1913).

In detail, the Inglis solution proceeds as follows. Introducing the above curvilinear co-ordinates, we denote the displacements at any point by u_α, u_β. The corresponding strains are $\varepsilon_{\alpha\alpha}, \varepsilon_{\alpha\beta}$ and $\varepsilon_{\beta\beta}$. The kinematic conditions for the strains are

$$\left.\begin{aligned} \varepsilon_{\alpha\alpha} &= h_1 \frac{\partial u_\alpha}{\partial\alpha} + h_1 h_2 u_\beta \frac{\partial}{\partial\beta}\left(\frac{1}{h_1}\right), \\ \varepsilon_{\beta\beta} &= h_2 \frac{\partial u_\beta}{\partial\beta} + h_1 h_2 u_\alpha \frac{\partial}{\partial\alpha}\left(\frac{1}{h_2}\right), \\ 2\varepsilon_{\alpha\beta} &= \frac{h_1}{h_2}\frac{\partial}{\partial\alpha}(h_2 u_\beta) + \frac{h_2}{h_1}\frac{\partial}{\partial\beta}(h_1 u_\alpha), \end{aligned}\right\} \tag{3.21-20}$$

where

$$\left.\begin{aligned} h_1^2 &= (\partial\alpha/\partial x)^2 + (\partial\alpha/\partial y)^2, \\ h_2^2 &= (\partial\beta/\partial x)^2 + (\partial\beta/\partial y)^2. \end{aligned}\right\} \tag{3.21-21}$$

The volume dilatation is given by

$$\Theta = h_1 h_2 \left[\frac{\partial}{\partial\alpha}\left(\frac{u_\alpha}{h_2}\right) + \frac{\partial}{\partial\beta}\left(\frac{u_\beta}{h_1}\right)\right] \tag{3.21-22}$$

and the rotation by

$$2\omega = h_1 h_2 \left[\frac{\partial}{\partial\alpha}\left(\frac{u_\beta}{h_2}\right) - \frac{\partial}{\partial\beta}\left(\frac{u_\alpha}{h_1}\right)\right]. \tag{3.21-23}$$

In our curvilinear co-ordinates, the stress-strain relations take the following form:

$$\left.\begin{aligned} (1-m)\frac{\partial\Theta}{\partial\alpha} - (1-2m)\frac{\partial\omega}{\partial\beta} &= 0, \\ (1-m)\frac{\partial\Theta}{\partial\beta} + (1-2m)\frac{\partial\omega}{\partial\alpha} &= 0 \end{aligned}\right\} \tag{3.21-24}$$

where m is, as usual, POISSON's ratio. The above two equations state that $(1-m)\,\Theta + i(1-2m)\,\omega$ is a function of $\alpha + i\beta$. Taking for instance

$$(1-m)\,\Theta + i(1-2m)\,\omega = \text{const}\,\frac{e^{-n(\alpha+i\beta)}}{\sinh(\alpha+i\beta)} \tag{3.21-25}$$

one obtains

$$\left.\begin{aligned} &F(1-m)\,\Theta \\ &\quad = \text{const}\,\{-e^{-(n-1)\alpha}\cos(n+1)\,\beta + e^{-(n+1)\alpha}\cos(n-1)\,\beta\}, \end{aligned}\right\} \tag{3.21-26}$$

$$\left.\begin{aligned} &F(1-2m)\,\omega \\ &\quad = \text{const}\,\{e^{-(n-1)\alpha}\sin(n+1)\,\beta - e^{-(n+1)\alpha}\sin(n-1)\,\beta\} \end{aligned}\right\} \tag{3.21-27}$$

with

$$u = \alpha\, u_\alpha/h; \qquad v = \beta\, u_\beta/h \tag{3.21-28}$$

and

$$F \equiv \cosh 2\alpha - \cos 2\beta; \qquad h^2 = 2/[F c^2]. \tag{3.21-29}$$

Now, u and v can be determined from the two partial differential equations. One obtains

$$\left.\begin{aligned} u &= A_n[(n+Q)\, e^{-(n-1)\alpha} \cos(n+1)\beta \\ &\quad + (n-Q)\, e^{-(n+1)\alpha} \cos(n-1)\beta] + \varphi, \\ v &= A_n[(n-Q)\, e^{-(n-1)\alpha} \sin(n+1)\beta \\ &\quad + (n-Q)\, e^{-(n+1)\alpha} \sin(n-1)\beta] + \psi \end{aligned}\right\} \quad (3.21\text{–}30)$$

where Q signifies $3-4m$ and φ, ψ are conjugate functions satisfying the Laplace equation; they may be chosen as equal to

$$\left.\begin{aligned} \varphi &= \text{const}\, e^{-n\alpha} \cos n\beta, \\ \psi &= \text{const}\, e^{-n\alpha} \sin n\beta. \end{aligned}\right\} \quad (3.21\text{–}31)$$

The solutions are valid for any integer n; the coëfficients A_n, B_n must be determined from boundary conditions. The values of the strains can be deduced from (3.21–31) by means of Eq. (3.21–24); the stresses, then, are obtained from the stress-strain relations

$$\left.\begin{aligned} \tau_{\alpha\alpha} &= \frac{E}{1+m}\left(\varepsilon_{\alpha\alpha} + \frac{m}{1-2m}\Theta\right), \\ \tau_{\beta\beta} &= \frac{E}{1+m}\left(\varepsilon_{\beta\beta} + \frac{m}{1-2m}\Theta\right), \\ \tau_{\alpha\beta} &= \frac{E}{1+m}\varepsilon_{\alpha\beta}. \end{aligned}\right\} \quad (3.21\text{–}32)$$

The final result is

$$\left.\begin{aligned} F^2\tau_{\alpha\alpha} = A_n\{&(n+1)\, e^{-(n-1)\alpha} \cos(n+3)\beta + (n-1)\, e^{-(n+1)\alpha} \cos(n-3)\beta \\ &- [4e^{-(n+1)\alpha} + (n+3)\, e^{-(n-3)\alpha}] \cos(n+1)\beta \\ &+ [4e^{-(n-1)\alpha} - (n-3)\, e^{-(n+3)\alpha}] \cos(n-1)\beta\} \\ + B_n\{&n\, e^{-(n+1)\alpha} \cos(n+3)\beta + (n+2)\, e^{-(n+1)\alpha} \cos(n-1)\beta \\ &- [(n+2)\, e^{-(n-1)\alpha} + n\, e^{-(n+3)\alpha}] \cos(n+1)\beta\}, \end{aligned}\right\} \quad (3.21\text{–}33\,\text{a})$$

$$\left.\begin{aligned} F^2\tau_{\beta\beta} = A_n\{&-(n-3)\, e^{-(n-1)\alpha} \cos(n+3)\beta - (n+3)\, e^{-(n+1)\alpha} \cos(n-3)\beta \\ &+ [(n-1)\, e^{-(n-3)\alpha} - 4e^{-(n+1)\alpha}] \cos(n+1)\beta \\ &+ [(n+1)\, e^{-(n+3)\alpha} + 4e^{-(n-1)\alpha}] \cos(n-1)\beta\} \\ - B_n\{&n\, e^{-(n+1)\alpha} \cos(n+3)\beta + (n+2)\, e^{-(n+1)\alpha} \cos(n-1)\beta \\ &- [(n+2)\, e^{-(n-1)\alpha} + n\, e^{-(n+3)\alpha}] \cos(n+1)\beta\}, \end{aligned}\right\} \quad (3.21\text{–}33\,\text{b})$$

$$\left.\begin{aligned} F^2\tau_{\alpha\beta} = A_n\{&(n-1)\, e^{-(n-1)\alpha} \sin(n+3)\beta + (n+1)\, e^{-(n+1)\alpha} \sin(n-3)\beta \\ &- (n+1)\, e^{-(n-3)\alpha} \sin(n+1)\beta - (n-1)\, e^{-(n+3)\alpha} \sin(n-1)\beta\} \\ + B_n\{&n\, e^{-(n+1)\alpha} \sin(n+3)\beta + (n+2)\, e^{-(n+1)\alpha} \sin(n-1)\beta \\ &- [(n+2)\, e^{-(n-1)\alpha} + n\, e^{-(n+3)\alpha}] \sin(n+1)\beta\}. \end{aligned}\right\} \quad (3.21\text{–}33\,\text{c})$$

As indicated above, these solutions (3.21–33) are valid for any integer n; the most general solution of the problem is therefore obtained by making a superposition for *all* possible n. Then, the values for the coëfficients A_n, B_n have to be chosen in such a manner as to satisfy the prescribed boundary conditions. In many instances, this is a very simple matter; INGLIS, in fact, discussed a variety of solutions. The cases which will be of specific interest to us in connection with geodynamics will be discussed later in their proper context (cf. Secs. 3.53 and 7.52).

Finally, we may note that the basic equations of elasticity allow solutions to exist which are unstable. This means that, in addition to the "straightforward" solution, there also exist solutions in which the deformations exceed all bounds. Such occurrences are referred to as *buckling*. A simple case where this takes place may be visualized by imagining a thin rod being compressed. The "straightforward" solution simply represents a shortening of the rod lengthwise, but this becomes unstable as soon as the compression exceeds a certain value. The rod then buckles sideways.

3.22. Dislocations. The infinitesimal theory of elasticity leads to the notion of *dislocation* by the observation that the displacement u_i corresponding to a given strain ε_{ik} in a multiply-connected body may not be single-valued. In order to restore one-valuedness of the displacement, one has to introduce surfaces of discontinuity which make the body once again singly-connected. Such discontinuities are called dislocations. It is thus evident that dislocations require the existence of non-evanescible circuits in the body. In the limit, the multiple connection of the body can be achieved by the assumption of singular lines upon which the strains are not continuous. These singular lines must either be closed in themselves or else begin and end on the external surface of the body. They are the rims of the surfaces of discontinuity restoring the single-connectedness referred to above.

Let the strain field ε_{ik} be given and calculate the displacement u_i at a point x_i^1 proceeding from a point x_i^0. The displacement is given by the line integral[1]

$$u_i(x_i) = \int_{x^0}^{x^1} \frac{\partial u_i}{\partial x_k} dx_k \tag{3.22–1}$$

taken along any path from x^0 to x^1. It can then be shown that the value $u_i(x_i)$ is not independent of the path.

Owing to the definition of strain, one has in general

$$\frac{\partial u_i}{\partial x_k} = \varepsilon_{ik} - \omega_{ik} \tag{3.22–2}$$

[1] See e.g. LOVE: l.c., p. 221; also: STEKETEE, J. A.: Canad. J. Phys. 36, 192 (1958).

with

$$2\omega_{ik} = -\frac{\partial u_i}{\partial x_k} + \frac{\partial u_k}{\partial x_i}. \tag{3.22-3}$$

Hence

$$u_i^1 - u_i^0 = \int_{x^0}^{x^1} \varepsilon_{ik}\, dx_k - \int_{x^0}^{x^1} \omega_{ik}\, dx_k. \tag{3.22-4}$$

The second integral can be written as follows (omitting the x in the limits of the integral, and writing only its superscript)

$$\left.\begin{aligned} -\int_0^1 \omega_{ik}\, dx_k &= -\int_0^1 \omega_{ik}\, d(x_k - x_k^1) \\ &= +\,\omega_{ik}^0 (x_k^0 - x_k^1) + \int_0^1 (x_k - x_k^1)\, d\omega_{ik} \end{aligned}\right\} \tag{3.22-5}$$

where

$$d\omega_{ik} = \frac{\partial \omega_{ik}}{\partial x_l}\, dx_l. \tag{3.22-6}$$

We now have the identity

$$\frac{\partial \omega_{ik}}{\partial x_j} = \frac{\partial \varepsilon_{jk}}{\partial x_i} - \frac{\partial \varepsilon_{ij}}{\partial x_k} \tag{3.22-7}$$

as one may easily verify by differentiation. Hence we have

$$u_i^1 = u_i^0 + \omega_{ik}^0 (x_k^0 - x_k^1) + \int_0^1 \Lambda_{ik}\, dx_k \tag{3.22-8}$$

with

$$\Lambda_{ik} = \varepsilon_{ik} + (x_l - x_l^1)\left\{\frac{\partial \varepsilon_{kl}}{\partial x_i} - \frac{\partial \varepsilon_{ik}}{\partial x_l}\right\}. \tag{3.22-9}$$

It is observed that the Λ_{ik} satisfy the following differential equations

$$\frac{\partial \Lambda_{ik}}{\partial x_l} = \frac{\partial \Lambda_{il}}{\partial x_k} \tag{3.22-10}$$

so that the integral in (3.22–8) is the same for all reconcileable paths. Thus, in any singly-connected body, the displacements must be single-valued, because along any circuit that returns to any point under consideration, Λ_{ik} satisfies the expression (3.22–8), and because all circuits are "evanescible", i.e. can be contracted into a point. By the same token, it is obvious that Λ_{ik} in a multiply connected body can be chosen such that the line integral in (3.22–8) around a non-evanescible circuit is not zero. This proves the existence of dislocations.

3.23. Plasticity. Mathematical elasticity theory finds a natural extension in what is called the mathematical theory of plasticity[1]. For, is has been noted that bodies strained to a certain point (called the "elastic limit") often show a behavior which can be described with

[1] Hill, R.: The Mathematical Theory of Plasticity. Oxford: Clarendon Press 1940. 356 pp.

good success fairly simply without having to go into the intrinsecacies of general finite-strain rheology.

Thus, most metals and many other substances begin to yield in a very special way when they are strained beyond the elastic limit. The best known "yield criterion" is that of MISES[1]; it can be written as follows:

$$(\tau_{11}-\tau_{22})^2+(\tau_{22}-\tau_{33})^2+(\tau_{33}-\tau_{11})^2+6(\tau_{12}^2+\tau_{23}^2+\tau_{31}^2)=6k^2. \quad (3.23\text{–}1)$$

Here, k is a parameter which depends on the amount of pre-strain. Thus, if a body of the type considered here be stressed beyond its elastic limit, then the stress state is such in each point of the "plastic" region, that Eq. (3.23–1) is satisfied. The assumed variability of the coëfficient k automatically takes into account the experimentally-observed phenomenon of "strain-hardening": the coëfficient is found to increase with the amount of work W_p that has been put into the *plastic* deformation:

$$6k^2 = f(W_p). \quad (3.23\text{–}2)$$

The increment of plastic work dW_p in a plastic-elastic body may be written

$$dW_p = \tau_{ij}\left(d\varepsilon_{ij} - \frac{\partial\tau_{ij}}{2\mu}\right) = \tau'_{ij}\left(d\varepsilon_{ij} - \frac{d\tau_{ij}}{2\mu}\right) \quad (3.23\text{–}3)$$

where the dash (′) indicates the deviatoric component of a tensor:

$$a'_{ij} = a_{ij} - \tfrac{1}{3}a_{nn}\delta_{ij} \quad (3.23\text{–}4)$$

and where use has been made of the (experimentally indicated) assumption that all volume changes during plastic deformation are elastic. A remark should perhaps be made concerning the strain increment $d\varepsilon$ occurring in the last formulas. In Eq. (3.23–3), the strain increment indicates the increment of the total material strain, the latter as defined by (3.12–7). However, since the plasticity formulas only refer to the strain *increment*, it is customary to think of the plastic strains in terms of an infinitesimal theory.

It is thus seen that the phenomenon of plasticity can best be described by saying that there are two "regions" of behavior of the body: an elastic one and a plastic one. In the elastic region, HOOKE's law is satisfied, and in the plastic region, the stresses satisfy the yield condition (3.23–1). The yield condition itself depends on the strain history of the body.

Within the plastic region, the strain increment $d\varepsilon_{ij}$ can be split at any instant into an elastic component $d\varepsilon_{ij}^e$ and into a plastic component $d\varepsilon_{ij}^p$:

$$d\varepsilon_{ij} = d\varepsilon_{ij}^e + d\varepsilon_{ij}^p. \quad (3.23\text{–}5)$$

[1] MISES, R. v.: Göttinger Nachr., math.-phys. Kl. **1913**, 582.

The rheological equation for the elastic part of the strain increment can be obtained from HOOKE's law, but for the plastic part one has, so far, only the yield condition, which is not sufficient to make the displacement determined. An additional assumption is therefore necessary for which REUSS[1] has proposed:

$$d\varepsilon_{ij}^{p} = \tau_{ij}' \, d\lambda \tag{3.23-6}$$

where $d\lambda$ is a scalar factor of proportionality which has to be determined experimentally. It expresses the amount of strain-hardening that the body exhibits. Eq. (3.23-6) signifies that the principal axes of stress and strain always coincide.

Exact solutions of the problem of finding the displacement pattern in a plastic-elastic material in its full generality are very difficult to achieve. Only few such solutions are available.

It is somewhat easier to obtain solutions if the elasticity of the material is disregarded, i.e. if one concerns oneself with a plastic-rigid material. Naturally, this can constitute only an approximation to reality. A further simplification is reached if one confines oneself to conditions of *plane strain*. This implies (a) that the (plastic) flow is everywhere parallel to a given plane (e.g. the x, y plane) and (b) that the motion is independent of z, the direction orthogonal to that plane.

In this case, the conditions for the determination of a problem reduce to

(a) the yield condition

$$\tfrac{1}{4}(\sigma_x - \sigma_y)^2 + \tau_{xy}^2 = k^2 \tag{3.23-7}$$

(b) the equilibrium conditions (cf. 3.12-20)

$$\partial\sigma_x/\partial x + \partial\tau_{xy}/\partial y = 0, \tag{3.23-8a}$$

$$\partial\tau_{xy}/\partial x + \partial\sigma_y/\partial y = 0 \tag{3.23-8b}$$

(c) the condition of zero volume change

$$\partial u_x/\partial x + \partial v_y/\partial y = 0 \tag{3.23-9}$$

(d) the rheological (stress-strain) equation (from Eq. 3.23-6)

$$\frac{2\tau_{xy}}{\sigma_x - \sigma_y} = \left\{\frac{\partial u_x}{\partial y} + \frac{\partial v_y}{\partial x}\right\} \Big/ \left\{\frac{\partial u_x}{\partial x} - \frac{\partial v_y}{\partial y}\right\}. \tag{3.23-10}$$

Here, u_x and v_y are the velocity-components along the x and y axes. The Eqs. (3.23-7/10) are sufficient to determine the unknowns σ_x, σ_y, τ_{xy}, u_x, v_y; since they are homogeneous in the velocities, they do, however, not really involve the time-element. The "velocities", therefore, may be replaced by any monotonous functions of displacement.

[1] REUSS, A.: Z. angew. Math. Mech. **10**, 266 (1930).

It turns out that the above set of Eqs. (3.23–7/10) is hyperbolic. The characteristics are called "slip lines"[1]. It is possible to derive several theorems regarding the geometry of such slip lines. This permits one to calculate the slip lines for a variety of boundary conditions. Of the many cases where slip line fields have been determined, we show here two examples. In Fig. 57 we show the slip lines resulting from a symmetrical stress state around a point as calculated by NADAI[2]. The region of plastic deformation is separated by a circle from that which remains rigid. Second, in Fig. 58 we show the plastic slip lines underneath a cylindrical stamp as determined by HENCKY[3].

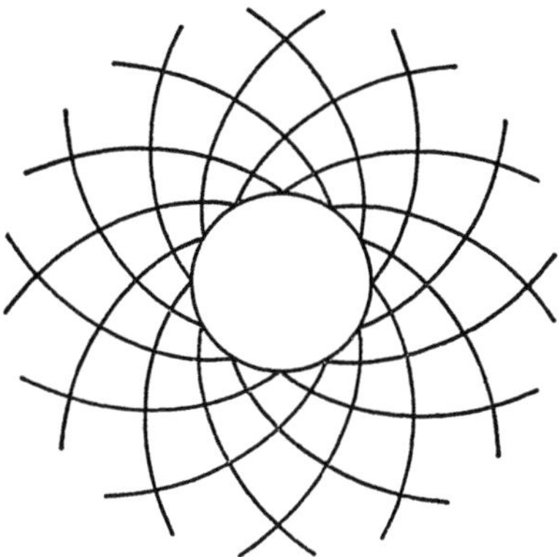

Fig. 57. Slip lines in the form of logarithmic spirals in a symmetrical stress state. After NADAI[2]

The slip lines are not just mathematical inventions, but do have physical reality. Characteristics allow for certain differential quotients to be discontinuous across them which permits actual physical discontinuities to exist.

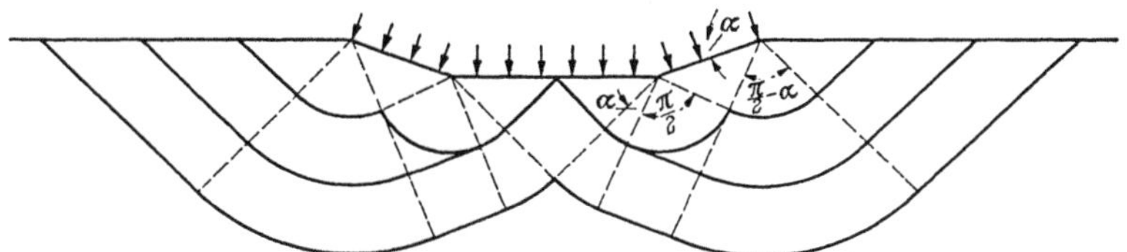

Fig. 58. Plastic slip lines underneath a cylindrical stamp. After HENCKY[3]

Another type of simplification of the general plasticity theory is obtained if the principles of plasticity are applied to the failure of thin steel plates under tension. This has been done by BIJLAARD[4] who made the assumption that, in a thin plate, plastic flow by preference must occur in bands in whose length-direction the dimensions of the material do not change during flow. This assumption immediately yields that at any point in an elastic-plastic plate there are two directions along which the above condition is fulfilled. Thus, in a homogeneous stress state, BIJLAARD expects two sets of plastic bands to develop; the angle which

[1] On this subject, see the excellent review by SOBOTKA, Z.: Appl. Mech. Rev. **14**, 753 (1961).

[2] NADAI, A.: Z. Physik **30**, 106 (1924).

[3] HENCKY, H.: Z. angew. Math. Mech. **3**, 241 (1923).

[4] BIJLAARD, P. P.: Rapport Ass. Gén. Ass. Géol. U.G.G.I. Edinburgh 1936. 23 pp.

they enclose is of the order of 110°. Correspondingly, the angle of the bands with the principal direction of stress (minimum pressure) is 55°. Confirmation of BIJLAARD's theory has been obtained by experiments with thin steel plates under tension.

An extension of the above concepts immediately leads to a visualization of the possibility of failure of a material by *plastic necking:* Imagine a rod under tension. In any cross-section that is slightly smaller than the rest of the rod (owing to ever-present small inhomogeneities) there is a stress-concentration. This will have the effect that plastic flow will first occur in this particular cross-section which will thereby become even narrower. This, in turn, increases the stress concentration, a neck will develop and the rod will finally fail.

3.3. Hydrodynamics of Viscous Fluids[1]

3.31. Fluids Kinematics. It seems useful to review first some of the basic facts about fluid kinematics. The theorems ensuing from that theory hold for any kind of flow.

If the *velocity vector* is drawn at every point of a moving fluid, a vector field is obtained. If this vector field is independent of time, the flow is "steady". If the curl of the velocity field is formed, a second vector field is obtained, the *vorticity field*. Finally, one can define a *vortex tube* by drawing all the vortex lines through every point of a small closed curve. The *strength* of such a vortex tube is obtained by integrating the scalar product of the vorticity vector with the unit vector normal to and of the magnitude of the element of area over any cross section of the tube. It is a fundamental theorem of fluid kinematics that this strength is constant all along the tube. Because of this property, vortex tubes (and therefore also vortex lines) cannot begin or end in the fluid: they must be closed curves, or extend to the boundaries of the fluid.

A further important concept is that of *circulation*. Mathematically, the circulation C around a closed circuit lying entirely within the fluid is defined as follows

$$C = \oint v_s \, ds \tag{3.31-1}$$

where v_s is the component of the fluid velocity tangent to the element ds of the circuit. The integral is to be taken once completely around the circuit. In virtue of STOKES' theorem, the circulation is equal to the total strength of all the vortex tubes going through the circuit. The

[1] Many good monographs exist on this subject to which the reader is referred for details; e.g. LAMB, H.: Hydrodynamics. London: Cambridge Univ. Press 1932; PAI, S. I.: Viscous Flow Theory (2 Vols.), New York: D. van Nostrand 1957; GOLDSTEIN, J.: Modern Developments in Fluid Dynamics (2 Vols.), Oxford: University Press 1938, etc.

circulation around any given vortex tube is therefore also constant all along the tube.

The definitions of vorticity, vortex lines, vortex tubes and circulation, together with the theorems concerning the strength of a vortex tube and the connection of circulation therewith, are purely kinematic or geometrical matters, completely independent of the presence or absence of stress. They hold, therefore, for *any* kind of flow.

3.32. Dynamics of Viscous Fluids. The rheological condition for a viscous fluid (with η a constant of the medium called its viscosity), is for the deviatoric components

$$\tau_{ik} = 2\eta \dot{\varepsilon}_{ik} \quad (i \neq k) \tag{3.32-1}$$

and for the isotropic components (provided there is no bulk viscosity)

$$\partial \varrho / \partial p = \varrho \beta_f \tag{3.32-2}$$

where p is the pressure ($p = -\frac{1}{3}\tau_{ii}$), ϱ the density of the fluid and β_f another constant—the compressibility. The above rheological conditions, together with the continuity conditions etc. applying in any *general* medium, completely define the dynamics of viscous (Newtonian) fluids.

The set of conditions outlined above can be combined to yield various differential equations which are applicable under various conditions. The best known of these equations is that of NAVIER and STOKES[1]. It is applicable to incompressible fluids. Because of its fundamental importance it is re-stated here:

$$\boldsymbol{v} \operatorname{grad} \boldsymbol{v} + \partial \boldsymbol{v}/\partial t = \boldsymbol{F} - (1/\varrho) \operatorname{grad} p - (\eta/\varrho) \operatorname{curl} \operatorname{curl} \boldsymbol{v}. \tag{3.32-3}$$

Here, $\boldsymbol{v}$ is the local velocity vector of a point of the fluid, t is time, and $\boldsymbol{F}$ the volume force per unit mass.

Theory and experiment show that for high flow velocities the flow pattern becomes transient although the boundary conditions remain steady: eddies are formed which proceed into the fluid at intervals. For any one system, there seems to be a "transition point" below which steady flow is stable. Above the "transition point" the steady flow becomes unstable and forms eddies. The steady flow is often termed "laminar", the unstable flow "turbulent".

Although the transition point has been calculated from the Navier-Stokes equation for certain simple systems, it is obvious that such a calculation is a very difficult undertaking. One has therefore to take recourse to experiments to determine when turbulence will set in. If some systems can be shown to be dynamically similar, then the transition point in one system will have a corresponding point in the dynamically similar system which can be calculated.

[1] See LAMB, H.: Hydrodynamics. London: Cambridge University Press 1932.

It has been shown by REYNOLDS that flow system which are geometrically similar, are also dynamically similar if the following "Reynolds number" (denoted by Re) is the same in both systems

$$Re = \varrho v d/\eta \tag{3.32–4}$$

where all the constants have the same meaning as before and d is a characteristic diameter of the system.

It has been observed that in straight circular tubes (which are naturally all geometrically similar), turbulence will set in if $Re = 2000$. This value, however, is tied up with the assumption that what is under consideration, is a straight tube. In other systems (for instance curved tubes), the "critical" Reynolds number (at which turbulence sets in) may be quite different.

A further important case is that of a viscous fluid flowing in laminar, parallel flow past a sphere of radius a. The resistance R offered by the sphere has been calculated by STOKES; it is

$$R = 6\pi a \eta v \tag{3.32–5}$$

where v is the flow-velocity at infinity.

3.33. Thermohydrodynamics of Viscous Fluids. We shall turn now our attention to the problem of the free thermal convection currents in various materials between two surfaces of different temperatures, subject to a gravitational field. Heat transfer by liquids of the Newtonian type[1] has been investigated because of the importance of heat transmission in engineering problems. Some results are thus available in textbooks[2]. These results have been obtained with particular reference to water and air for which the viscosity is very small

Dimensional analysis shows that the motion depends only on two dimensionless numbers: on the Reynolds number mentioned above and on the product of Grashoff and Prandtl numbers (sometimes called Rayleigh number). Let c denote the specific heat per unit mass of the material, D a characteristic diameter, g the gravity acceleration, k the thermal conductivity, ΔT the temperature difference, v a characteristic velocity of the fluid in motion, β the coëfficient of thermal expansion (per °C), η the viscosity and ϱ the density, then the Grashoff number G is given by the equation

$$G = D^3 \varrho^2 g \beta \, \Delta T/\eta^2 \tag{3.33–1}$$

and the Prandtl number P by

$$P = c\eta/k. \tag{3.33–2}$$

[1] For a summary see e.g. SCHEIDEGGER, A. E.: Bull. Geol. Soc. Amer. **64**, 127 (1953).

[2] See e.g. MCADAMS, W. H.: Heat Transmission, 2nd ed. New York: McGraw-Hill 1942.

For laminar motion, the product λ of Grashoff and Prandtl numbers (Rayleigh number) governs the thermal convection. JEFFREYS[1, 2] has deduced theoretically that no stable thermal convection should occur unless λ is at least 1709. In his papers, the direction of the thermal gradient is supposed to coïncide with the gradient of gravity; the two surfaces are horizontal. The convective motion between the two surfaces is pictured as occurring in the form of convection cells[3]. A convection cell contains a vortex tube which is closed within the cell. The picture is thus in agreement with the kinematical properties of vorticity. An example is shown in Fig. 59.

This theory has been tested experimentally. SCHMIDT and others[4, 5] have employed an optical method to study the mechanism of heat transfer by natural convection above a horizontal plate. The photographs do indeed show a cell-like pattern under certain conditions. When λ is above 2000, alternate portions of the fluid circulate upward and downward in streams of substantial width. As the characteristic product increases, the rate of fluid circulation increases, until finally turbulence ensues.

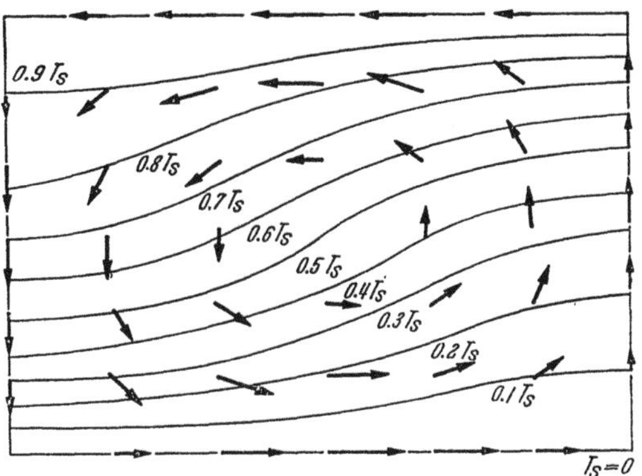

Fig. 59. Temperature and velocity distribution within one half of a convection cell. Temperature gradient is 10% in excess of critical gradient. After OKAI[6]

JEFFREYS' results have been reviewed and extended by LOW[7]. PELLEW and SOUTHWELL[8] extended the results of JEFFREYS and LOW, and made investigations under different sets of boundary conditions. Qualitatively, the previous results are confirmed, but it is shown that any *oscillatory* convective motion must of necessity decay.

---

[1] JEFFREYS, H.: Phil. Mag. (7) **2**, 833 (1926).

[2] JEFFREYS, H.: Proc. Roy. Soc. Lond., Ser. A **118**, 195 (1928).

[3] This has originally been postulated by BÉNARD, H.: Ann. chim. et phys. **23**, 62 (1901).

[4] SCHMIDT, R. J., and S. W. MILVERTON: Proc. Roy. Soc. Lond., Ser. A **152**, 586 (1935).

[5] SCHMIDT, R., J. and O. A. SAUNDERS: Proc. Roy. Soc. Lond., Ser. A **165**, 216 (1938).

[6] OKAI, B.: J. Phys. Earth **7**, 1 (1959).

[7] LOW, A. R.: Proc. Roy. Soc. Lond., Ser. A **125**, 180 (1929).

[8] PELLEW, A., and R. V. SOUTHWELL: Proc. Roy. Soc. Lond., Ser. A **176**, 312 (1940).

The calculations mentioned above make some rather restrictive assumptions: the surface of the fluid is fixed, the temperature gradient is constant etc. Therefore, approximation methods have been developed which allow one to abandon some of these restrictions[1-4]. In addition, SOBERMAN[5] investigated the effect of transient heating conditions.

3.4. Other Types of Rheological Behavior

3.41. Principles. The "ideal bodies" discussed so far (excluding the plastic body) constitute the system of "classical" bodies. According to the general remarks made earlier, it is obvious that it cannot be hoped that these classical bodies will provide anything but a crude classification of all rheological behavior that can be envisaged. The most general rheological condition that can be thought of is a very complicated affair. In order to obtain a physical picture of some of the possibilites, the system of classical bodies must be enlarged.

An extension of the classical bodies to approximate some of the more commonly found natural bodies is usually done by, first of all, restricting oneself to infinitesimal strains. This, of course, is a severe limitation of generality and, in fact, usually quite an inconsistent procedure as, if any type of true rheological effects occur at all, the displacements become large, at least after the elapse of sufficient time. Nevertheless, a very approximate idea of the physical possibilities may be obtained by assuming the strains as infinitesimal. At any rate, it is customary to do this in most engineering literature, whether it be justified or not. A convenient summary of this type of treatment has been given by REINER[6]. We shall present here a discussion of the more important cases.

3.42. Maxwell Liquid. In terms of infinitesimal strain terminology, the part of the rheological equation referring to the deviatoric components for the Hooke (elastic) solid can be written as follows

$$\tau_{ik} = 2\mu\,\varepsilon_{ik} \quad (i \neq k) \tag{3.42–1}$$

and for the viscous (Newtonian) liquid as follows

$$\tau_{ik} = 2\eta\,\dot{\varepsilon}_{ik} \quad (i \neq k)\,. \tag{3.42–2}$$

[1] REID, W. H., and D. L. HARRIS: Phys. Fluids **1**, 102 (1958).
[2] MALKUS, W. V. R., and G. VERONIS: J. Fluid Mech. **4**, 225 (1958).
[3] OKAI, B.: J. Phys. Earth **7**, 1 (1959).
[4] YIH, C. S.: Quart. J. Appl. Math. **17**, 25 (1959).
[5] SOBERMAN, R. K.: Phys. Fluids **2**, 131 (1959).
[6] REINER, M.: Twelve Lectures on Theoretical Rheology. Amsterdam: North-Holland Publishing Co. 1949.

Combination of both gives (after REINER[1])

$$\dot{\varepsilon}_{ik} = \dot{\tau}_{ik}/(2\mu_M) + \tau_{ik}/(2\eta_M) \quad (i \neq k). \tag{3.42–3}$$

A similar combination with appropriate constants can be set up for the isotropic components. The quantities μ_M and η_M are constants of the body; they are often referred to as (Maxwell-) rigidity and viscosity. However, it should be noted that the names "rigidity" and "viscosity" are not very good ones. The ordinary viscous fluid is obtained if, in (3.42–3) $\mu_M \to \infty$ in which case η_M becomes the fluid viscosity. Similarly, for $\eta_M \to \infty$ one obtains an elastic solid with rigidity μ_M. It is for this reason that the two constants μ_M and η_M have been called "rigidity" and "viscosity". However, "Maxwell constants" would be a better name.

The rheological Eq. (3.42–3) describes a liquid which shows stress-relaxation. That the latter occurs can be seen immediately if it is assumed that the deformation is kept constant (i.e. $\dot{\varepsilon} = 0$). Then the stress diminishes exponentially with a time constant τ

$$\tau = \eta_M/\mu_M. \tag{3.42–4}$$

Furthermore, if a constant stress is applied, it is seen that deformation occurs at a constant *rate*. This phenomenon is called *creep*.

Stress relaxation phenomena have been studied extensively by MAXWELL[2]. Hence the name "Maxwell liquid" for the material at present under consideration.

3.43. Kelvin Solid. The equations of the Hooke solid and of the viscous liquid in their infinitesimal-strain form as displayed in (3.42–1/2) can also be combined in a different manner, viz.[3,4]

$$\tau_{ik} = 2\mu_K \varepsilon_{ik} + 2\eta_K \dot{\varepsilon}_{ik} \quad (i \neq k) \tag{3.43–1}$$

where again a similar combination with appropriate constants could be written down for the diagonal components ($i = k$). Again, the quantities μ_K and η_K are constants of the body which are again often referred to as "rigidity" and "viscosity", respectively. Again, these names are not very good. One notes that a "Kelvin solid" becomes an elastic solid if $\eta_K \to 0$ and a viscous fluid if $\mu_K \to 0$. These are the opposite limits to those required to yield the same classical bodies in the case of Maxwell liquids. The use of the terms "rigidity" and "viscosity" indiscriminately in the case of Maxwell liquids and Kelvin bodies has given

[1] REINER, M.: loc. cit.

[2] MAXWELL, J. C.: Collected Works **2**, 26 (1866).

[3] MEYER, O. E.: J. reine angew. Math. **78**, 130 (1874). — Ann. Physik **1**, 108 (1874).

[4] KELVIN, Lord: Collected Works **3**, 1 (1878).

rise to much confusion. Correspondingly, we should prefer to call μ_K and η_K "Kelvin constants".

The Kelvin solid is characterized by an elastic after-effect: if a stress change is performed, the body will eventually reach that state which would correspond to HOOKE's law,—but only exponentially. The time constant is again η_K/μ_K. A loading-unloading diagram has therefore the characteristics shown in Fig. 60.

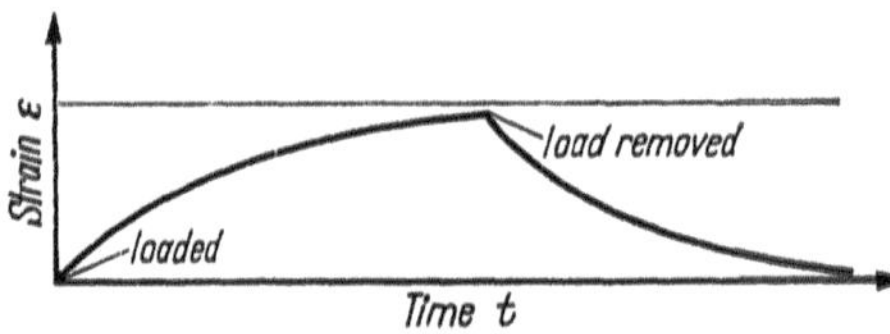

Fig. 60. Loading-unloading diagram of a Kelvin body. After REINER[1]

The elastic after-effect characteristic of a Kelvin body also causes free oscillations to be damped. The damping constant can be calculated simply as follows. The equation of motion of a Kelvin body is

$$\mu_K x + \eta_K \dot{x} = -c\ddot{x} \tag{3.43-2}$$

where c is some constant of the system. Trying for a solution

$$x = \exp[(\alpha + i\omega)\, t] \tag{3.43-3}$$

one obtains

$$\mu + \eta(\alpha + i\omega) = -c(\alpha^2 - \omega^2 + 2\alpha i\omega). \tag{3.43-4}$$

Equating real and imaginary parts yields

$$\mu + \eta\alpha = -c(\alpha^2 - \omega^2), \tag{3.43-5}$$

$$\omega\eta = -c\, 2\alpha\omega \tag{3.43-6}$$

and hence one obtains

$$\alpha = -\eta/(2c), \tag{3.43-7}$$

$$\omega = \sqrt{\frac{\mu}{c} - \frac{\eta^2}{4c^2}}. \tag{3.43-8}$$

Finally, eliminating the constant c yields

$$\alpha^2 + \frac{2\mu\alpha}{\eta} + \omega^2 = 0. \tag{3.43-9}$$

The last equation can be used to determine μ/η if α and ω are measured in a given system.

Finally, it may be remarked that more generalized bodies are obtained if one introduces a yield stress ϑ_{ik} into Eq. (3.43-1). Setting $\mu = 0$ to avoid further complications, one then has

$$\tau_{ik} = \vartheta_{ik} + 2\eta\,\dot{\varepsilon}_{ik} \quad (i \neq k) \tag{3.43-10}$$

[1] REINER, M.: loc. cit.

which is the constitutive equation of a "Bingham solid". Here, η is called the "plastic viscosity". Further complications can be achieved by including in the constitutive equations more and more terms. However, the basic shortcomings outlined in Sec. 3.41 of this type of theory are thereby not overcome, and it may be necessary to allow for finite, nonlinear deformations. Indications that this is so have been presented by KNOPOFF and MACDONALD[1] who found that the observed attenuation even of small amplitude stress waves in the Earth cannot be accounted for by a linear rheological law.

3.44. Heat Convection in General Rheology. It must be expected that convection currents, as discussed in Sec. 3.33 can also occur in "rheological" materials other than viscous fluids[2]. Unfortunately, very little has yet been done in solving specific problems which might have a bearing upon heat convection in general bodies. REINER[3] calculated the plastic flow between two rotating cylinders. If the inner cylinder rotates and the outer one is fixed, then that gives a very rough picture for part of a convection tube.

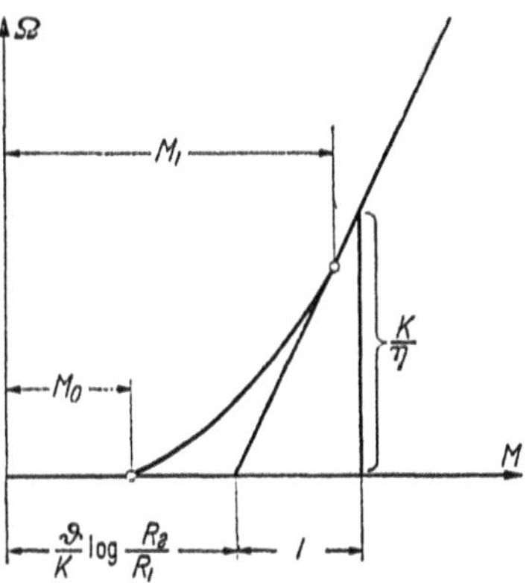

Fig. 61. Connection between torque and angular momentum in REINER's[3] rotation viscosimeter

REINER made his calculations under the assumption of the Bingham law of motion, for a fixed inner and rotating outer cylinder (a rotation viscosimeter). The material between the two cylinders will not flow if the torque M acting between the cylinders is smaller than a certain value M_0. If M exceeds the value M_0, the material will start to flow where the stress is largest, i.e. at the surface of the inner cylinder. As long, however, as the stress on the inner surface of the outer cylinder is below the yield stress ϑ, there will be a shell of solid material adjacent to the outer cylinder. When the torque becomes greater than another definite value M_1, all the material between the cylinders will flow. The connection between the torque M transmitted between the two cylinders and the relative angular velocity Ω is shown in Fig. 61. K is a certain geometrical constant. The velocity distribution between the cylinders was also calculated. It takes the form of a rather complicated expression with little bearing upon the problems here under consideration.

[1] KNOPOFF, L., and G. J. F. MACDONALD: Revs. Mod. Phys. **30**, 1178 (1958).
[2] SCHEIDEGGER, A. E.: Bull. Geol. Soc. Amer. **64**, 127 (1953).
[3] REINER, M.: J. Rheol. **1**, 5 (1930).

In order to obtain a better idea about the mechanics of plastic flow, more qualitative arguments have to be used. An interesting study of possible analogies between plastic and viscous flow has been made by OLDROYD[1]. The plasticity equations are solved "in the large" for cylinders of various shapes moving through each other. The plastic lines of flow are compared with the viscous lines of flow (see Fig. 62). The difference is not very great, apart from the fact that in plastic flow one finds a solid kernel in those regions where in viscous flow there is a small velocity gradient.

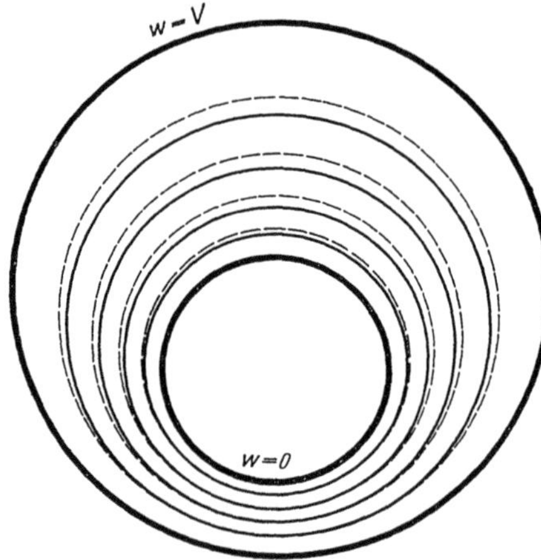

Fig. 62. Comparison of plastic and viscous flow lines in a special case. After OLDROYD[1]

It is now our task to obtain an idea of what may happen when a plastic material is heated from below. For vanishing yield stress, the Bingham body is a Newtonian liquid. The presence of a yield stress will affect the formation of convection cells to such an extent that it will need greater temperature differences to start the motion than that corresponding to a characteristic product of Grashoff and Prandtl numbers equal to 1709. Moreover, in view of OLDROYD'S results, it is very likely that the center of the vortex tube in the cell will rotate as a solid. The velocity distribution and the transmission of torques in the plastic region would be similar to that in REINER'S rotation viscosimeter.

If the yield stress is assumed to be rather large, however, the above arguments are questionable. A new phenomenon may take place whose nature may not be revealed by simple analogies. For large velocity gradients, a phenomenon might be found in the study of plastic flow which would be analogous to turbulence in a Newtonian liquid: the flow should become unstable instead of being steady and laminar. Phenomena of this sort are important in the pipe-line transportation of certain materials and therefore attempts have been made to find criteria for the onset of turbulence in plastic flow. HERSEY[2] used the method of dimensional analysis to find such criteria. His investigations were made for a general law of motion of REINER and his equations contain many unknown functions. Even if his equations are reduced to BINGHAM'S

[1] OLDROYD, J. G.: Proc. Cambr. Phil. **53**, 396, 521 (1947).
[2] HERSEY, M. D.: J. Rheol. **3**, 23 (1932).

law, some unknown functions are still left so that HERSEY's criteria are of little practical use, except that for vanishing yield stress his equations reduce to REYNOLD's criterion which is thereby confirmed.

Thus, it must be expected that, under very great temperature differences, vortices would be formed at the boundaries which would dissipate into the plastic medium, as this was the case with viscous liquids. The flow is then no longer steady.

3.5. Discontinuous Displacements

3.51. The Physics of Fracture. As outlined earlier, the analysis of discontinuous displacements has a very important bearing upon the physics of the Earth's crust, since fissures, faults and related phenomena are very common occurrences.

The basic problem of describing the dynamics of fracture processes has not yet been solved. Although man has been very much concerned with fracture and related phenomena, most of his interest has been directed towards establishing criteria for the safety of structures rather than towards a detailed analysis of how a collapse occurs. One is therefore faced with the situation that there is a multitude of fracture criteria which in conjunction with assumed "safety factors", can be used to build safe structures, but that there are not many investigations into the detailed mechanism of fracture that could be used to explain observed conditions in the Earth's crust.

The existing investigations into the theory of fracture can be separated into various sections. Firstly, there are the heuristic descriptions, encompassing all the engineering theories and fracture criteria; secondly, there are the "microscopic theories" which try to explain the observed values of "strength" (i.e. the resistance to fracture) of materials in terms of their known molecular forces, and thirdly, there are a few attempts to give a proper analytical expression to the heuristic investigations mentioned above;—in such a fashion that the stresses are indeed components of a tensor and the displacements are defined as vector fields. It is this last group of investigations which has the most direct bearing upon the physics of the Earth's crust.

However, since the last group of investigations has not been carried to a high degree of refinement, it is necessary to extract as much information as possible from the engineering theories. These are naturally based on experiments[1, 2]. We shall therefore discuss all the aspects of fracture theories that might prove relevant in connection with a study of the Earth's crust.

[1] ANDERSON, A. L. *et al.*: Norw. Geotech. Inst. Publ. No. 21 (1957).

[2] VINOGRADOV, S. D., i. K. I. KUZNETSOVA: Izv. Akad. Nauk SSSR., Ser. Geofiz. **1959**, No. 5, 756 (1959).

3.52. Phenomenological Theories. A great number of heuristic investigations into fracture and fracture criteria have been made;—as outlined above, principally in order to establish when a structure is safe.

When discussing fracture criteria, one must take care not to confound with "fracture" other phenomena of failure. For the engineer it usually matters little whether a structure collapses because of concentrated plastic deformation by necking, or whether it collapses because of fracture of some of its members: the end result is, for him, the same. The engineering literature, therefore, often simply speaks of "failure" without specifying whether a fracture mechanism or some other process is meant. We have dealt with failure by (elastic) buckling and by plastic instability (necking) earlier, and we shall concern ourselves here only with fracture in the proper sense.

A survey of the types of fracture that can occur in a material has been made by OROWAN[1] who distinguishes between the following cases: (a) brittle fracture, (b) fibrous fracture, (c) shear fracture, (d) fatigue fracture, (e) integranular fracture, and (f) fracture by molecular sliding. We shall discuss these cases in their proper order.

a) Brittle Fracture. Brittle fracture is the only type of fracture that occurs in completely brittle substances. It is that type of fracture which is theoretically best understood. It is characterized by a high velocity of propagation, producing a bright, smooth fracture surface.

Under an uniaxial stress state, brittle fracture occurs in an isotropic medium if the tension reaches a (for the material) critical value (called brittle strength of the material); the fracture surface is normal to the direction of the tensile stress.

In a triaxial stress state, the condition for brittle fracture is not so simple and thus, several criteria have been proposed. Mostly used is still the time-honored hypothesis of COULOMB, later modified by MOHR[2] which states: "For an isotropic medium fracturing under the action of three unequal principal stresses, the surface of fracture is parallel to the direction of the intermediate principal stress and inclined at an angle $\varphi \leqq 45°$ (30° is a good average) toward the maximum principal pressure." The originators of this statement arrived at it by modifying the hypothesis that the maximum shear surface would be the fracture surface, until reasonable agreement with observations was obtained.

The value of the stress at which fracture occurs, e.g. in terms of the "tensile strength" (as defined above) of the material must be inferred from microscopic considerations and will be discussed later. However, it may be said that corresponding to the idea that the maximum shear

[1] OROWAN, E.: Rep. Progr. Phys. 12, 186 (1949).

[2] MOHR, O.: Abhandlungen aus dem Gebiete der technischen Mechanik. 3. Aufl. Berlin: Wilh. Ernst & Sohn 1928.

surface would be the fracture surface, it also had been thought originally that the fracture condition would be one of critical shear stress. This was not found to correspond to observation and therefore was also modified by MOHR[1] to yield his famous heuristic fracture criterion. Each possible stress state is represented in the $\sigma - \tau$ diagram by a family of circles (MOHR's diagram, cf. Fig. 56). Those stress states, one of whose circles touches an empirically determined envelope, are thought to produce fracture. When referring to MOHR's criterion, it should be noted that it has been applied to any type of failure, and not only to brittle fracture.

The high velocity of the spread of a crack in brittle fracture can be explained by noting that the only work required for the latter is that necessary to overcome the cohesion between the atoms on either side of an existing crack. This work is so small that it can be supplied by the elastic energy stored in the material just prior to its disintegration.

b) Ductile Fracture. Ductile fracture is characterized by a very slow propagation of existing cracks. The fracture process can be stopped at any instant simply by stopping the continuation of external deformations. It thus appears that ductile fracture is a non-elastic type of fracture in which plastic deformation and related phenomena play an essential rôle. In ductile fracture, the energy expended against the cohesive forces of the material is negligible compared with the energy of plastic deformation that has to be expended in order to extend an existing crack. The crack can therefore spread only if the external forces continue to do work.

The typical fracture of cylindrical specimens under tension is the cup-and-cone pattern. The bottom of the "cup" appears velvety to the naked eye, and deeply jagged if it is viewed through a microscope (cf. OROWAN[2]); this has been called the "fibrous" type of fracture. The fracture "surface" (if it can be called such) is roughly perpendicular to the greatest tension. The cup-and-cone fracture always begins with the fibrous fracture at the bottom of the cup, in the center of the specimen.

The sides of the cup-and-cone fracture are usually very smooth; they represent the "shear" type of fracture. This fracture follows a surface of maximum shear strain.

Since the cup-and-cone fracture always starts at the bottom of the cup, the criterion for its occurrence must be one for the occurrence of fibrous fracture. The fact that the fibrous fracture starts first has often been thought due to the fact that the axial tension is highest in the center of the specimen. This suggests that a criterion assuming a critical tensile strength would be appropriate. However, matters are not quite so simple.

[1] MOHR, O.: Abhandlungen aus dem Gebiete der technischen Mechanik. 3. Aufl. Berlin: Wilh. Ernst & Sohn 1928.

[2] OROWAN, E.: Rep. Progr. Phys. **12**, 186 (1949).

Ductile materials, under certain conditions (especially around notches) often break in a fashion that is very typically brittle (notch brittleness). Ductile fracture, thus, cannot be described separately from notch brittle fracture. In order to describe properly the fracture possibilities of a ductile material, one has therefore to assume two fracture conditions. The most recent version of this idea is due to OROWAN[1]. In a cylindrical specimen under tension, in the absence of a notch or crack, the plastic strain is given by the yield stress curve $Y(\varepsilon)$. In the presence of a notch, however, the axial stress Y_n rises from the value Y of the yield stress at the tip of the notch to a maximum in the center of the specimen. The amount R by which it rises depends on the notch. The fracture condition can then be

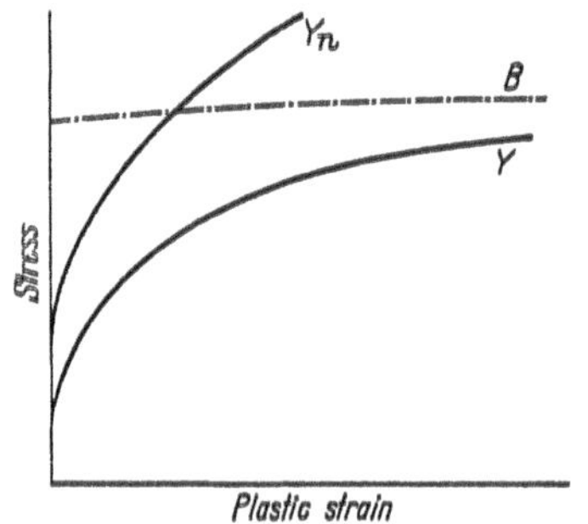

Fig. 63. Diagram representing the condition for notch brittle fracture. After OROWAN[2]

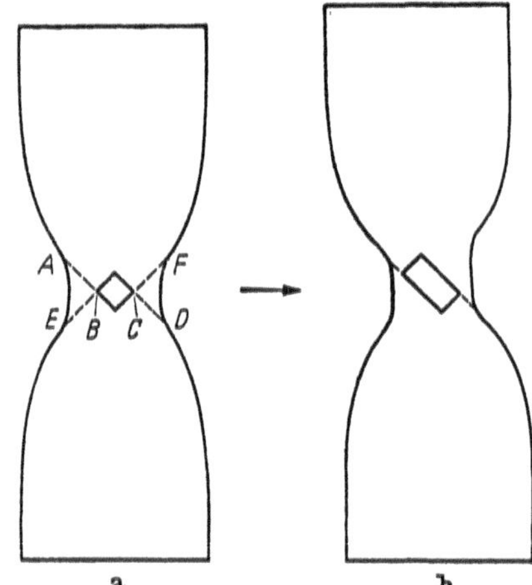

Fig. 64a and b. Possible mechanism of crack propagation in the neck of a tensile specimen under ductile fracture. After OROWAN[3]

respresented in graphical form as shown in Fig. 63. In connection with this figure, OROWAN[2] then states the fracture condition as follows: "In the absence of a notch or crack, the Y_n curves end by ductile fracture in points the locus of which is not a curve. With a sufficiently sharp crack, the Y_n curve may intersect the curve B and then brittle fracture occurs." The shape and position of the critical curve B, of course, must be found empirically.

It thus appears that fibrous fracture represents a crack propagation process which is largely governed by the laws of plastic deformation in the neighborhood of the crack. The fact that molecular cohesion is being overcome plays, in the pattern of the deformation, only an incidental rôle. This general statement can, for instance, be illustrated by the consideration of the possible special mechanism shown in Fig. 64[3]. Herein,

[1] OROWAN, E.: Trans. Inst. Eng. Shipbuild. Scotl. **1945**, 165 (1945).
[2] OROWAN, E.: Rep. Progr. Phys. **12**, 209 (1949).
[3] OROWAN, E.: Rep. Progr. Phys. **12**, 220 (1949).

the dotted lines indicate slip planes, the center square indicates a small quadrangular crack extending across the neck. Slipping along the slip lines (by plastic deformation) produces an elongation of the crack.

c) Fatigue. A peculiar type of fracture that may occur under cyclic stressing in a material, especially in metals, is fatigue fracture. Although it does not appear that fatigue fractures play any major rôle in processes taking place in the Earth's crust, a brief summary thereof will be given here for the sake of completeness.

The most important facts about fatigue have been stated by OROWAN[1] as follows: (i) fatigue fracture may occur after a certain critical number of applications of a given stress cycle is exceeded; (ii) sometimes there exists a critical amplitude of an alternating stress cycle below which the material can undergo any number of applications of the stress without breaking, but above which fatigue fracture occurs after a certain number of applications.

It appears that fatigue fracture is due to ductile fracture in weak inclusions in an otherwise elastic material, i.e. in plastic intrusions in an elastic matrix. During the repeated application of the stress cycle, the plastic intrusions get worked plastically until a ductile crack may appear therein. Once this has happened, stress concentrations at the tip of the crack may induce brittle fracture (or possibly notch brittle fracture) in the rest of the material.

With this model, OROWAN[1] was also able to explain the curious fact, mentioned above, of the existence, in certain cases, of a "safe" stress amplitude below which no fatigue fracture occurs in any number of stress cycles. A modern review of the problem has also been given by THURSTON[2].

d) Intergranular Fracture. Materials consisting of polycrystalline aggregates contain[2] two different components, *viz.* the crystal grains and the grain boundaries. The same is true for any type of heterogeneous agglomerate. The grains and the boundaries exhibit entirely different mechanical behavior. The grain boundaries show characteristics of viscosity. If the temperature is less than a certain critical value, they are rigid; but if the critical temperature is exceeded, they behave like a viscous liquid[3]. They have no definite yield stress, so that the rate of sliding would presumably be proportional to the prevailing shear stress if the grain boundaries were smooth enough to allow uninhibited sliding. Owing to the intermittent interlocking of the grains, sliding, if the stresses are continuously increased, also occurs intermittently. Once the

[1] OROWAN, E.: Rep. Progr. Phys. **12**, 221 (1949).
[2] THURSTON, R. C. A.: Trans. Canad. Inst. Min. Met. **60**, 390 (1957).
[3] OROWAN, E.: Rep. Progr. Phys. **12**, 228 (1949).

stresses are large enough to overcome the largest geometrical non-conformity at the grain boundaries, slow continuous sliding occurs during which cavities open up between the grains and, eventually, fracture may occur. This type of fracture has also been called "creep fracture".

A similar mechanism may occur in a somewhat related fashion in any material that contains two or more structural elements of different mechanical properties. Assume that a "surface of weakness" (e.g. an old fracture surface) representing one type of structural element, exists in an otherwise homogeneous material representing the second type. The surface may be very "bumpy" and therefore geometrically interlocked. If an external strain is applied at a constant time rate, the material may give along the surface of weakness whenever the stresses have become big enough to overcome the largest geometrical nonconformity by deforming the material itself in certain spots. Once this has happened, sliding may occur along the surface (by some viscous slip-mechanism) until a new interlocking occurs.

3.53. Microscopic Theories. *a) The Problem.* The fact that materials fracture under sufficiently high stresses must have its ultimate explanation in molecular considerations. In order to obtain such explanations, microscopic theories of fracture have been devised. In particular, the aim is to explain the experimentally observed *strength* of materials.

It has been recognized very early that the strength of a material as calculated from the physics of molecular cohesion in a homogeneous material is much higher than the actually observed fracture strength of common materials. The order of magnitude of molecular cohesion, or molecular strength, can be estimated as follows[1]. If a solid is strained uniformly, (brittle) fracture must occur if the elastic energy in the solid can provide the surface energy necessary to produce a crack. The latter is $2S$ for a specimen of unit thickness if S is the specific surface energy; we shall consider only the two dimensional case. Thus, a large portion (let us assume: one-half) of the energy $2S$ must be present just prior to the instant of fracture in the molecules in the immediate neighborhood of the (future) surface of fracture. On the other hand, if σ_m denotes the fracture stress, the elastic energy density prior to fracture is $\sigma_m^2/2E$ (E being YOUNG's modulus), and the strain energy e (per unit thickness) present between the two atomic planes which will be separated by the ensuing fracture process, is

$$e = a\sigma_m^2/2E, \tag{3.53-1}$$

if the planes are originally at the atomic distance a from each other.

[1] OROWAN, E.: Rep. Progr. Phys. **12**, 192 (1949).

Thus, we obtain:

$$\sigma_m = \sqrt{\frac{2SE}{a}}. \qquad (3.53\text{–}2)$$

Using typical values for most common brittle materials, it turns out that the "molecular" fracture strength σ_m is 10 to 1000 times greater than that stress at which fracture actually occurs.

b) The Griffith Theory of Brittle Fracture. To overcome this discrepancy, GRIFFITH[1–4] made the assumption that the observed values of strength were due to the presence of very small cracks inherent in every material, at which there is a stress concentration sufficient to overcome the molecular strength. Thus, during fracture, it would be actually the molecular cohesion which would be overcome; the stress causing this, however, would not be the overall-average stress in the material, but the local stress at the tips of existing cracks.

In order to calculate the magnitude of the stress concentration at the end of a very small elliptical crack, GRIFFITH made use of the solution by INGLIS (cf. Sec. 3.21) for the stresses at the ends of an elliptical cavity in a two-dimensional case under a given tension σ. INGLIS obtained for the strain energy e_σ (per unit thickness) in this case

$$e_\sigma = -\pi c \sigma^2/E \qquad (3.53\text{–}3)$$

where $2c$ is the length of the crack (i.e. c is the major axis of the ellipse) and E is, as before, YOUNG's modulus. The corresponding surface energy e_S represented by the crack, according to earlier remarks, is

$$e_S = 4cS \qquad (3.53\text{–}4)$$

where S is again the specific surface energy. The crack will spread if, for increasing length of the crack, the decrease in strain energy is greater than the increase in surface energy[4]. Thus the equilibrium size of the crack is given by

$$\frac{\partial}{\partial c}(4cS - \pi c^2\sigma^2/E) = 0 \qquad (3.53\text{–}5)$$

or

$$\sigma = \sqrt{\frac{2ES}{\pi c}}. \qquad (3.53\text{–}6)$$

In view of the above remarks, it is obvious that the crack can expand only if the stress given by (3.53–6) is exceeded; hence this stress is the fracture stress in the presence of the crack.

[1] GRIFFITH, A. A.: Phil. Trans. Roy. Soc. Lond. A **221**, 163 (1920).
[2] GRIFFITH, A. A.: 1st Intern. Conf. Appl. Mech. Delft A **55** (1924).
[3] OROWAN, E.: Rep. Progr. Phys. **12**, 192 (1949).
[4] PETCH, N. J.: Progr. Metal Phys. **5**, 1 (1954).

The solution given above for the fracture stress can be recalculated for a penny-shaped crack. The resulting expression differs from that above only by a numerical factor; it is

$$\sigma = \sqrt{\frac{2ES}{\pi(1-m^2)c}} \tag{3.53-7}$$

where m is POISSON's ratio.

Recent reviews and generalizations of the Griffith fracture theory have been published on several occasions in the literature[1-3]. An interesting application of the theory to fracture in rocks has been given by BRACE[4] who took the friction of the sides into account. STEKETEE[5] has shown that Griffith cracks can also be regarded as dislocations.

c) Griffith Fracture Criterion. The above theory can be extended to yield a fracture criterion for brittle material. Assuming that isotropic materials contain cracks of all orientations, GRIFFITH[6, 7] used the solution by INGLIS[8] for obtaining the maximum tensile stress at the tip of the crack of the most dangerous orientation, again in the two-dimensional case. Assuming that σ_1 and σ_2 are the principal stresses, with $\sigma_1 > \sigma_2$, GRIFFITH states his criterion as follows: Fracture occurs

$$\left.\begin{array}{lll} \text{(i) if} & 3\sigma_1+\sigma_2>0 & \text{when} \quad \sigma_1=K \\ \text{(ii) if} & 3\sigma_1+\sigma_2<0 & \text{when} \quad (\sigma_1-\sigma_2)^2+8K(\sigma_1+\sigma_2)=0 \end{array}\right\} \tag{3.53-8}$$

where K is the tensile strength for uniaxial stressing.

One of the consequences of the Griffith criterion is that, in uniaxial compression, the most dangerous cracks would be those at 45° to the stress, and they should propagate in their own planes. However, OROWAN[7] states that, "although fracture surfaces at about 45° are common in compressive tests, failure by cracking *parallel* to the direction of compression is almost equally often observed with glasses and stones". On the other hand, OROWAN[9] found that recent high pressure measurements can be correlated satisfactorily by GRIFFITH's criterion.

d) The Crack Propagation Velocity. The Griffith theory gives a criterion for the stress at which a crack begins, i.e. is nucleated. However, it has been noted by GILMAN[10] that the brittle fracture of a material

[1] SANDERS, J. L.: J. Appl. Mech. **27**, 352 (1960).
[2] SACK, R. A.: Proc. Phys. Soc. Lond. **58**, 729 (1946).
[3] CLAUSING, D. P.: Quart. Colo. School Mines **54**, No. 3, 285 (1959).
[4] BRACE, W. F.: J. Geophys. Res. **65**, 3477 (1960).
[5] STEKETEE, J. A.: Canad. J. Phys. **36**, 1168 (1958).
[6] GRIFFITH, A. A.: Ist Intern. Conf. Appl. Mech. Delft A **55**, 1924).
[7] OROWAN, E.: Rep. Progr. Phys. **12**, 192 (1949).
[8] See Sec. 3.21.
[9] OROWAN, E.: Rep. Progr. Phys. **12**, 200 (1949).
[10] GILMAN, J. J.: J. Appl. Phys. **27**, 1262 (1956).

takes place in two stages: nucleation at the atomic level, and then propagation through the material in which the crack had been nucleated. (A third stage is the spread to neighboring pieces of material if the latter is inhomogeneous.)

The Griffith theory deals only with the nucleation of a crack. In addition, it ignores the fact that at the tip of the crack where stress-concentration takes place, the principal stresses are not equal so that high shear stresses are present which should cause plastic flow. This plastic flow should prevent the crack from propagating because it would relieve the stress concentration. The difficulty can be resolved by noting a remark of MOTT[1] stating that it may be necessary for stresses to persist for some time in order to cause plastic flow. Thus, if a crack propagates fast enough, plastic flow would not have time to become established and hence the crack could indeed propagate.

The crack propagation velocity in an elastic body is determined by three factors. The driving force is the elastic energy H_E which is released during the propagation. This is balanced by the surface energy H_S necessary to separate the two sides of the crack and by the kinetic energy H_K associated with the rapid sideways motion of the material during crack formation. The latter can be written as follows:

$$H_K = \frac{v_c^2}{v_0^2} \frac{H_E}{B^2} \qquad (3.53\text{-}9)$$

where B is a constant, v_c is the crack velocity, and v_0 is the velocity of sound in the material. Solving this for the crack velocity[2] yields

$$v_c = B v_0 \sqrt{1 - \frac{H_S}{H_E}}\,. \qquad (3.53\text{-}10)$$

This relation implies that the crack cannot propagate unless the Griffith criterion ($H_E > H_S$) is satisfied, and that the velocity of crack propagation can never be larger than the velocity of sound in the material. ROBERTS and WELLS[3] calculated the constant B for an internal crack in a plate; they found B approximately equal to 0.38. Comparison of observed elastic crack velocities with theory shows that the experimental values approach the theoretical ones quite closely. This indicates that the propagation of cracks in elastic materials is quite satisfactorily understood.

For materials that are not ideally elastic, one must assume that the stress difference at the top of the crack will cause plastic flow. Thus, the driving energy H_E is not only balanced by the sum of kinetic and

[1] MOTT, N. F.: Proc. Roy. Soc. Lond., Ser. A **220**, 1 (1953).

[2] GILMAN, J. J.: J. Appl. Phys. **27**, 1262 (1956).

[3] ROBERTS, D. K., and A. A. WELLS: Engineering **178**, 820 (1954).

surface energy required in producing the crack, but in addition there is energy dissipation. This dissipation results from the plastic deformation. The energy balance equation thus reads as follows

$$H_E = H_P + H_K + H_S \tag{3.53–11}$$

where H_P is the energy dissipated in plastic deformation. GILMAN[2] has shown that the work of plastic deformation can be written as follows

$$H_p = W/v_c \tag{3.53–12}$$

where W is a constant that depends on the stress-strain rate relation and on the size of the material. The equation for the crack velocity is thus modified to read as follows:

$$v_c = B v_0 \sqrt{1 - \frac{W}{H_E} \frac{1}{v_c} - \frac{H_S}{H_E}}. \tag{3.53–13}$$

This equation demonstrates that a crack must have a certain critical velocity $\tilde{v}_c$

$$\tilde{v}_c = W/H_E \tag{3.53–14}$$

before it can propagate spontaneously in a brittle fashion. This is the result anticipated by MOTT. If the intrinsic crack velocity is less than the critical velocity, then the crack can not spread by itself at all and it can only open up in accordance with work performed by external forces.

Discussions of the kinematics of cracks which are slightly different from that given above, have also been published by BERRY[1], BRUECKNER[2], and by McCLINTOCK and SUKATME[3].

3.54. Analytical Attempts. As mentioned above, not many actually analytical attempts to describe fracture are available. The geometry of a fracture can be represented by the introduction of discontinuous functions into the general pattern of finite strain theories. Describing the displacements again in terms of co-ordinates and parameters as in Sec. 3.12:

$$x_i = x_i(\xi_\alpha, t), \tag{3.54–1}$$

one can represent a discontinuity in the body if the x_i are discontinuous functions of the ξ_α. It appears from physical experience that discontinuities are distributed in sheets through the material. Thus, let

$$F(\xi_1', \xi_2' . \xi_3', t) = 0 \tag{3.54–2}$$

describe a surface of discontinuity S. Eq. (3.54–2) implies that the surface S may change its position with the passing of time. As indicated

[1] BERRY, J. P.: J. Mech. Phys. Solids **8**, 194 (1960).

[2] BRUECKNER, H. F.: Trans. Amer. Soc. Mech. Engrs. **80**, 1225 (1958).

[3] McCLINTOCK, F. A., and S. P. SUKATME: J. Mech. Phys. Solids **8**, 187 (1960).

in Eq. (3.54–2), an arbitrary point on the surface S is denoted by ξ'_α. The rim of the surface is a line that must obviously be closed in itself or begin and end on the boundary of the body. The discontinuity at the surface S can be described by giving the vector X_i of "jump" in x_i if one passes from one side to the other. The discontinuity is thus given as a vector field, defined on the surface S:

$$X_i = X_i(\xi'_\alpha, t). \qquad (3.54\text{–}3)$$

The vector of jump can be calculated if a circuit around the edge of the discontinuity is drawn, as follows:

$$X_i = \oint\limits_{\substack{\xi'_\alpha \to \xi'_\alpha \\ \text{encircling rim}}} \frac{\partial x_i}{\partial \xi_\alpha} d\xi_\alpha. \qquad (3.54\text{–}4)$$

For physical reasons, only such displacement fields are allowed where this integral does not depend on the circuit (but it will naturally depend on ξ'_α). It is seen that a proper choice of the discontinuity will physically represent a fracture sheet.

A formulation of the dynamics of the above-introduced discontinuities has not yet been achieved. Such attempts as have been made introduce *dislocations* (as the simplest type of discontinuity) to describe fractures. One has thus a sheet of dislocations, the rim of the sheet representing the edge of the fracture, and the jump vector at every point describing the relative slip. If the fracture extends, the rim sweeps further through the body and the jump vectors change their size with time. Of particular interest is the case where one has a fracture with fixed rim, and where at a certain instant some of the jump vectors increase suddenly, i.e. some of the dislocations "snap". This can be thought of as a process sweeping at a finite velocity across the whole fracture sheet. However, such "laws" as have been postulated to describe the process, are entirely heuristic. The above mechanism has been advanced as a possible description of earthquake phenomena and will be referred to again in the appropriate Section (7.25).

3.6. Rheology of the Earth: The Basic Problem of Geodynamics[1]

3.61. General Considerations. The chief problem of the science of geodynamics is to determine deformations within the Earth and upon its surface. On the surface, the present-day deformations are known and the task is to explain the latter in terms of stresses that could be considered as reasonable. In any properly defined deformation theory, it is a matter of mathematical analysis to determine the stresses from the

[1] The following in after SCHEIDEGGER, A. E.: Canad. J. Phys. **34**, 383 (1957).

boundary conditions. Thus, if the "deformation theory" applying to the Earth were properly defined, it would be, in principle, a straightforward (though not necessarily easy) matter to calculate the stresses from the (known) strains and then to look for causes of the latter. Unfortunately, it is a fact that the "deformation theory" applying to the Earth is not known, thus leaving the matter of finding the causes of its present-day appearance wide open to speculation.

It is thus evident that the basic problem of geodynamics is to determine the proper rheological conditions in the Earth. The explanation of the present-day physiography would then follow more or less automatically.

The difficulties inherent in finding the proper rheological conditions applicable to the Earth arise from two causes. First, the state of the material in all but the uppermost few kilometers of the Earth's crust is not easy to envisage. Pressures and temperatures are such that it is unlikely that they can be duplicated in the laboratory in the near future, leaving only theoretical guessing to determine the behavior of the material under consideration. Second, the time elements involved are for the greater part such that, even if experiments involving the correct temperatures and pressures could be performed, the human life span would be millions of times too short to obtain the desired answers. This, again, forces one to speculate.

The time element in the rheological equations is a concept which, in general, has not been sufficiently noted. Terms like "rigid" and "fluid" have a meaning only if it is stated what the time intervals are in which the stresses in question are applied. In order to survey what *can* be said about the rheology of the material within the Earth, it is therefore necessary to make a distinction between various typical duration-intervals for the stresses. It is convenient to classify stresses according to whether their duration is "short", "intermediate", or "long". We shall consider stresses as short if they have a typical duration of the order of 3 seconds, as intermediate if their typical duration is of the order of 3 years, and as long if their typical duration is of the order of 100 million years. It will be noted that the ratio of subsequent typical stress durations is approximately 3×10^7. As limits for the interval for which stresses may be considered as short, intermediate or long, one may take the geometric mean between the "typical" durations (i.e. about 4 hours and 15000 years).

The ratio of subsequent "typical" durations, being 3×10^7, is very large. It can therefore not be expected that the rheological behavior deduced for one typical duration can be extrapolated for another. The various typical durations have to be studied separately.

3.62. Stresses of Short Duration. According to the above, we consider stresses as of short duration if the typical times involved are of the order of 3 seconds;—or, if the upper limit of their duration is about 4 hours. As might well be expected, the rheological behavior of the various layers of the Earth is best known for this time range. It is found that (infinitesimal) elasticity theory gives a good description of the phenomena for small displacements and it is therefore natural to classify the findings as to whether they apply to the material being stressed *below* or *beyond* the elastic limit. The vast majority of experimental investigations that have been made, concern stresses of short duration.

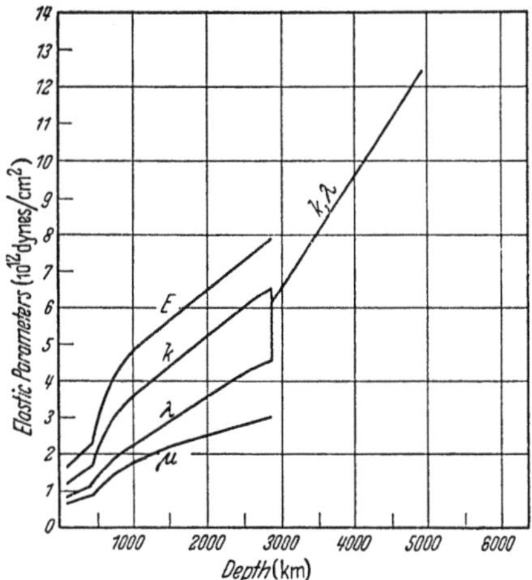

Fig. 65. Variation of elastic parameters in the Earth's interior. After BULLEN[1]

a) Stresses Below the Elastic Limit. Stresses of short duration (in the above sense) below the elastic limit occur in the Earth during the passing of earthquake waves. Seismology, therefore, yields important information bearing upon the question under consideration. It is found that the upper layers of the Earth, to a depth of 2900 km, behave as an elastic solid, below as an elastic fluid.

Seismology, directly, yields only the variation of seismic velocity with depth (cf. Fig. 31). In order to determine the elastic parameters therefrom, one has to make certain assumptions regarding the density variation within the Earth. BULLEN[1] studied various possibilities and found that they all yield a remarkably uniform result to a depth of about 413 km. The corresponding values for the elasticity constants are therefore quite definite down to that depth. Below 413 km, the values obtained are somewhat more doubtful; BULLEN's estimates are given in Table 16; these estimates have been plotted in Fig. 65.

BULLEN's table starts at a depth designated by "M" which indicates the layer just below the Mohorovičić discontinuity. Above this discontinuity the crust is very heterogeneous. The elastic constants of the various type of rocks for short-duration stresses can be measured in the

[1] BULLEN, K. E.: An Introduction to the Theory of Seismology, 2nd ed. London: Cambridge: University Press 1959.

laboratory[1-3]. Typical values are YOUNG's modulus $E=4\times10^{11}$ c.g.s., and POISSON's ratio $m=0.25$.

Table 16. BULLEN's *Values for the Elasticity Constants*
(Units for λ, μ, k, E in 10^{12} dynes/cm^2; m is a number)

Depth km	λ	μ	k	E	m = POISSON's ratio
M	0.74	0.63	1.16	1.60	0.269
100	0.80	0.67	1.24	1.70	0.272
200	0.90	0.74	1.38	1.89	0.275
300	1.01	0.81	1.54	2.07	0.277
413	1.14	0.90	1.73	2.30	0.280
500	1.42	1.10	2.15	2.82	0.283
600	1.69	1.32	2.57	3.38	0.282
800	2.06	1.69	3.19	4.31	0.275
1000	2.33	1.89	3.59	4.82	0.276
1400	2.76	2.15	4.20	5.51	0.281
1800	3.27	2.39	4.87	6.16	0.288
2200	3.81	2.63	5.57	6.81	0.295
2600	4.32	2.88	6.23	7.49	0.300
2898	4.49	3.03	6.51	7.87	0.300
2898	6.2		6.2		0.5
3000	6.5		6.5		0.5
3500	8.1		8.1		0.5
4000	9.7		9.7		0.5
4500	11.1		11.1		0.5
4982	12.6		12.6		0.5

b) Stresses Above the Elastic Limit. (i) Crust. A significant number of experiments have been made to determine the modes of disintegration of various types of rock samples. The first experiments along these lines have been done early in the century to determine the modes of failure of such rocks as can be used as building materials. After BRIDGMAN[4] initiated the development of high pressure techniques, GRIGGS and others[5] investigated the deformation of many rock samples under extreme conditions. Special investigations have also been made to analyse

[1] BALAKRISHNA, S.: Current Sci. **26**, No. 12, 377 (1957).
[2] BAYUK, E. I.: Izv. Akad. Nauk SSSR., Ser. Geofiz. **1959**, No 6, 895 (1959).
[3] SILAEVA, O. I.: Izv. Akad. Nauk SSSR., Ser. Geofiz. **1959**, No 2, 221 (1959).
[4] BRIDGMAN, P. W.: The Physics of High Pressure; New Impression with Supplement. London: G. Bell 1949.
[5] GRIGGS, D. *et al.*: Bull. Geol. Soc. Amer. **62**, 853, 1385 (1951). — HANDIN, J.: Trans. A.S.M.E. **1953**, 315 (1953). — TURNER, F. J. *et al.*: Bull. Geol. Soc. Amer. **65**, 883 (1954). — HANDIN, J. and H. W. FAIRBAIRN: Bull. Geol. Soc. Amer. **66**, 1257 (1955). — ROBERTSON, E. C.: Bull. Geol. Soc. Amer. **66**, 1275 (1955). — MATSUSHIMA, S.: Zisin **8**, 173 (1956); **10**, 113 (1957); **12**, 45(1959). — HANDIN, J.: Quart. Colo. School Min. **52**, No. 3, 75 (1957). — PATERSON, M. S.: Bull. Geol. Soc. Amer. **69**, 465 (1958). — DJINGHEUZIAN, L. E., and M. A. TWIDALE: Quart. Colo. School Min. **54**, No. 3, 43 (1959). — MOGI, K.: Bull. Earthquake Res. Inst.

the destruction of rocks by explosions[1]. All these experiments must be taken as investigations belonging in to the present category of "short"-duration stresses.

From these investigations it appears that most crustal rocks fracture in a manner as postulated by MOHR's theory; i.e. one observes two preferred failure planes inclined at an angle of $\varphi < 45°$ toward the direction of maximum pressure and parallel to the intermediate principal stress. The value of the stress under which the samples break would be expected to be given by an envelope in MOHR's diagram. As this is rather difficult to tabulate, an idea of the magnitude of the stresses involved can be obtained by listing the maximum stress differences (i.e. the strength) that the material can support in a state of simple shear. This has been done in Table 17. The tensile strength is very low.

Table 17. *Strength of Materials*[2]

Material	Confining dressure (atm.)	Strength (cgs) (order of magnitude)
Marble	1	10^9
Marble	10000	10^{10}
Sandstone	1000	10^9
Granite	2000	10^{10}
Solenhofen limestone	1	10^9
Solenhofen limestone	10000	10^{10}
Quartz	1	10^{10}
Quartz	20000	10^{11}

A different mode of fracture is observed under pure tension. The tensile strength is much below the compressive strength of rocks, presumably owing to the presence of Griffith cracks. It is unlikely, however, that such a state exists in the Earth anywhere but on the very surface.

Modifications of the failure patterns discussed above may be due to pore water pressure in the rocks; this will be discussed in more detail in connection with the theory of faulting.

(ii) Mantle. Stress-changes of short duration above the elastic limit occur in the mantle at the focus of an earthquake during the actual shock. The fact that the phenomenon of an earthquake is at all possible,

37, 155 (1959). — ROBINSON, L. H.: Quart. Colo. School Min. **54**, No 3, 177 (1959). — WUERKER, R. G.: Min. Eng. **11**, No 10, 1022 (1959). — Quart. Colo. School Min. **54**, No 3, 3 (1959). — ZELENY, R., and E. L. PIRET: Quart. Colo. School Min. **54**, No. 3, 33 (1959). — GRIGGS, D., and J. HANDIN (ed.): Rock Deformation [Geol. Soc. Amer. Memoir 79 (1960)]. — HANDIN, J.: Trans. Amer. Geophys. Un. **41**, 162 (1960). — LUCKE, O.: Einige Ergebnisse der physikalischen Erforschung der Materie unter sehr hohem Druck. Berlin: Akademie Verlag 1960.

[1] RINEHART, J. S.: Quart. Colo. School. Min. **54**, No. 3, 61 (1959). — LANGEFORS, U.: Quart. Colo. School Min. **54**, No 3, 219 (1959). — JOHANSSON, H.: Quart. Colo. School Min. **54**, No 3, 297 (1959). — BOND, F. C., and B. B. WHITNEY: Quart. Colo. School Min. **54**, No 3, 77 (1959).

[2] After BENIOFF and GUTENBERG: In: The Internal Constitution of the Earth, ed. GUTENBERG New York: Dover Publications 1951.

indicates that the materials of the mantle has some resemblance to an "ordinary" solid. The earthquakes seem to correspond to sudden shear fractures (like an orange being sliced) wherein it is very characteristic that most of the displacements along the fracture surfaces are essentially parallel to the surface of the Earth; the "fractures" of course, may be the expression of a plastic instability[1], rather than of an ordinary fracture.

There are indications that modes of sudden displacement different from shear fracture may occur in the mantle. The seismic waves emanating from the focus of an earthquake do not always permit an interpretation in terms of "an orange being sliced". Thus, there are indications that, in certain cases, the distribution of waves is that as would correspond to a multipole of higher order (cf. Sec. 7.23). The physical significance of this in terms of an envisageable fracture pattern is not yet quite clear; it is probable that the observations must again be explained in terms of a plastic instability[1].

Finally, something may be said with regard to the imperfections of elasticity connected with the energy loss of seismic waves and of the Earth's free oscillations[2]. JEFFREYS'[3] stated that this energy loss should be due to scattering, since the damping increases with frequency. If it were due to anelasticity, it would have to increase with period. However, in spite of JEFFREYS'[3] remarks, newer calculations have yielded the result that the observed damping cannot completely be explained by scattering, and that a residual anelasticity is in fact involved[4–6].

3.63. Stresses of Intermediate Duration. The next task is to investigate the rheological behavior of the Earth under stresses of intermediate duration. As outlined above, we assume "intermediate duration" to mean from 4 hours to 15000 years, the "typical" duration being about 3 years. As might be expected, it is much more difficult to obtain an idea of the rheological behavior of the Earth in this time range than for "short" stresses. The information which we do have, thus, comes only from two sources: (a) the damping of the variation of latitude, and (b) from the release of stresses in earthquake sequences.

a) The Damping of the Variation of Latitude. An analysis of the observations on latitude variation yields that, after an annual term (probably due to the movement of air masses) has been taken out, a period

[1] OROWAN, E.: Geol. Soc. Amer. Memoir **79**, 323 (1960).
[2] BENIOFF, H. *et al.*: J. Geophys. Res. **66**, 605 (1961).
[3] JEFFREYS, H.: The Earth (1952 ed.), p. 107.
[4] MENZEL, H.: Publ. Bur. Centr. Seism. Int. A **19**, 125 (1954).
[5] FÖRTSCH, O.: Ann. Geofis., Roma **9**, 469 (1956).
[6] KNOPOFF, L., and G. J. F. MACDONALD: J. Geophys. Res. **65**, 2191 (1960). — Rev. Mod. Phys. **30**, 1178 (1958).

of approximately 420 days exists (Chandler wobble). On the assumption that this motion is due to random exciting impulses of unknown origin, it is possible to say that the damping causes the motion to decrease to a value of $1/e$ of the original in between 3[1] and 13[2] periods[3]. It had originally been thought that this damping would be caused by motions in the core of the Earth (JEFFREYS[1]). However, in a renewed discussion of the problem, BONDI and GOLD[4] claim that the motion of the core can*not* account for the observed damping. The damping would therefore have to be due to a Kelvin effect in the mantle of the Earth. Under this assumption, it is possible to obtain an estimate for the value of η/μ occurring in the constitutive equation of a Kelvin body. Indeed, assuming

$$T = 2\pi/\omega = 420 \text{ days} \qquad (3.63\text{–}1)$$

and the damping as of the order of 10 periods

$$1/\alpha = -4200 \text{ days} \qquad (3.63\text{–}2)$$

yields for the ratio of the Kelvin constants according to (3.43–9)

$$\eta_K/\mu_K \cong 2 \text{ days} = 1.72 \times 10^5 \text{ sec}. \qquad (3.63\text{–}3)$$

From the above considerations, the two "Kelvin constants" cannot be obtained separately. One may be tempted to use for μ_K the "rigidity" for the mantle as deduced from seismology and confirmed by a discussion of the Earth's bodily tides (JEFFREYS[1]), viz. (from Table 16)

$$\mu_K \cong 2 \times 10^{12} \text{ cgs} \qquad (3.63\text{–}4)$$

which would yield

$$\eta_K \cong 3 \times 10^{17} \text{ cgs} \qquad (3.63\text{–}5)$$

and then to say that the "rigidity" and the "viscosity" of the Earth's mantle has the "values" shown in (3.63–4/5). However, this leads to a fallacy. The names "viscosity" and "rigidity" have no properly defined meaning *per se*, unless it is specified to which rheological equation they refer. There might be some justification to use the elastic μ for the time range presently under consideration, particularly as the discussion of the Earth's tides gave the same result, but then the viscosity η_K obtained above in (3.63–5) is a "Kelvin constant" and has nothing whatever to do with the *creep* viscosity occurring in the rheological equation of a Maxwell body.

[1] JEFFREYS, H.: The Earth, 3rd ed. London: Cambridge Univ. Press 1952.

[2] WALKER, A. M., and A. YOUNG: Month. Not. Roy. Astron. Soc. **115**, 443 (1955).

[3] A recent investigation by RUDNICK [RUDNICK, P.: Trans. Amer. Geophys. Un. **37**, 137 (1956)] puts the period of the Chandler wobble as equal to 432 ± 2.6 days and the decay time equal to 9.5 years.

[4] BONDI, H., T. GOLD: Month. Noth. Roy. Astron. Soc. **115**, 41 (1955).

The above result that the Earth behaves as a Kelvin body with the constants (3.63–4/5) in the time range under consideration depends, as indicated, on the assumption that the motion in the core cannot account for the damping. This, however, is by no means established[1].

b) Earthquake sequences. After an earthquake occurs in some part of the world, if often happens that the first, "principal" shock is followed by a series of *aftershocks* of diminishing magnitude. BENIOFF[2] has studied a series of such aftershock sequences regarding the time dependence of their magnitudes. The "magnitude" as defined by GUTENBERG and RICHTER[3] permits an interpretation in terms of "strain release". The total energy of an (elastically) strained body is proportional to the square of the strain, hence the strain release in a sudden shock should be proportional to the square root of the energy release J. The energy release J (in ergs) was originally thought to be connected with the earthquake magnitude M by the equation[3]

$$\log J = 12.0 + 1.8M. \tag{3.63–6}$$

BENIOFF used the "old" equation of GUTENBERG and RICHTER[3] as given above for his studies; if a newer equation[4] is used, the essence of the results is not changed, as the interest lies solely with the relaxation times involved. If the magnitude of each shock of an aftershock sequence is known, it is therefore possible to plot a quantity

$$S = \sum J_i^{\frac{1}{2}} \tag{3.63–7}$$

which is proportional to the cumulative strain release. If this be done, one obtains graphs of the type shown in Fig. 66.

Such graphs suggest the following interpretation. The principal shock occurs when the strain accumulation has reached a critical value. This starts a break along an existing or a new fault. As soon as this has happened, the fault will be "greased" so that any strain accumulating can be released, apart for "bumps" that cause the motion to be somewhat discontinuous. One can thus picture an earthquake aftershock sequence to be the expression of the strain release of a body in which, after loading, the stress has been suddenly removed. (This interpretation

[1] In this instance it may for instance be noted that MELCHIOR [MELCHIOR, P. J.: Rend. Accad. Naz. Lincei; Cl. fis. mat. nat. Ser. VII, **19**, 137 (1955)] doubts entirely the validity of the usual analyses of the variation of latitude. He questions the applicability of the basic equations of motion. See also MELCHIOR, P. J.: Progr. Phys. Chem. Earth **2**, 212 (1957) and JEFFREYS, H.: Month. Not. Roy. Astron. Soc. **116**, 362 (1956).

[2] BENIOFF, H.: Bull. Seism. Soc. Amer. **41**, 31 (1951).

[3] GUTENBERG, B., and C. F. RICHTER: Bull. Seism. Soc. Amer. **32**, 163 (1942).

[4] See the discussion in Sec. 2.23.

of aftershock sequences is in accordance with the "strain rebound theory" of earthquake generation; cf. Sec. 7.25).

The resemblance of curves of the type shown in Fig. 66 with the strain-release curves of a Kelvin-body is rather striking and would be even more so if the aftershock sequence would not have been plotted on semilogarithmic paper. The maximum S reached in Fig. 66 corresponds to the unstrained position of the Kelvin body, the zero value of S

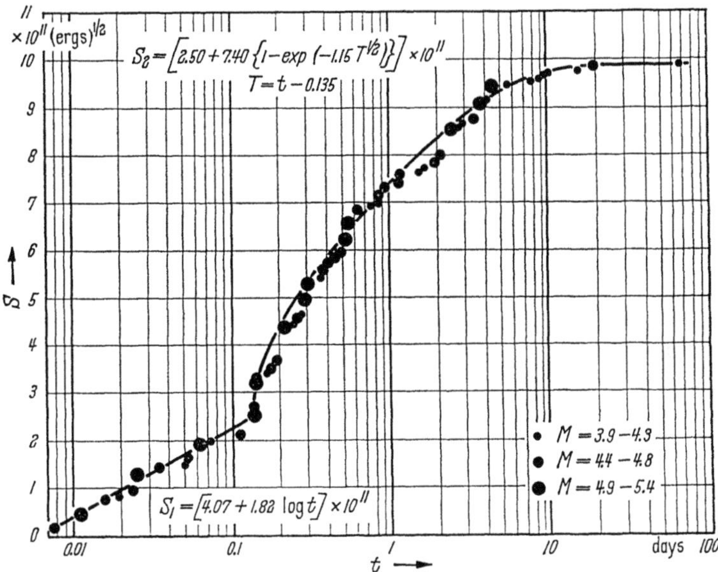

Fig. 66. Accumulated elastic strain rebound increments (times a constant) of the Long Beach aftershock sequence. After BENIOFF[1]

to the loaded position. Naturally, the curve of Fig. 66 cannot be expected to come to a complete levelling-off as, in addition to the strain release, it must be expected that a constant build-up of stress occurs at a slow rate which, presumably, never comes to a stop. Thus, after a few years, one would expect the same pattern to repeat itself.

The study of strain-release in earthquake aftershock sequences thus yields that the latter occurs in a fashion that can be explained in terms of elastic afterworking of a Kelvin body. Moreover, the relaxation time (i.e. the time in which the strain drops to $1/e$ times the initial value) turns out to be of the order of 2 days. This, however, is equal to the ratio η/μ of the Kelvin constants. It thus turns out that the interpretation of earthquake sequences in the above fashion leads to the same value for η/μ as that obtained from the damping of the variation of latitude. It would be strange indeed if this coincidence would be accidental. We

[1] BENIOFF, H.: Bull. Seism. Soc. Amer. **41**, 31 (1951).

can thus summarize the findings of this paragraph as follows: Under stresses of intermediate duration, indications are that the mantle of the Earth behaves like a Kelvin body with the Kelvin constants of the order of magnitude indicated above.

A final remark should be made regarding the possibility of fractures. The occurrence of earthquakes makes it evident that the Kelvin-type behavior subsists only up to a certain limit of stress. If this limit is exceeded, a fracture type process (an earthquake) occurs. What the limiting stress is and what the mode of fracture is (i.e. whether it is, e.g. governed by MOHR's criterion), no-one can say. Caution, however, should be used in applying to the Earth strength values from short-duration experiments with rocks or such like. In other words, one can *not* just take over the results from laboratory experiments to draw conclusions regarding the mode of fracture expectable in the Earth under intermediate-duration stresses. One direct piece of evidence is that most earthquakes show displacements essentially parallel to the surface, as was mentioned in Sec. 2.24. However, since the stress state producing the earthquakes is not known either, the field is wide open for speculations. MOHR's criterion is just one of many that *could* be valid.

Attempts have also been made at a direct measurements of the creep properties of geological materials[1-6]. Whilst certain results were obtained, it is quite reasonable to say that these are, in fact, explained by assuming that the materials were stressed beyond their elastic limit. Just as in the short-time range, it must be assumed that some kind of yielding or flow must take place also in the intermediate time range if the strength limit is exceeded.

Finally, a remark should be added explaining why we did not take into consideration any evidence resulting from tidal friction, which, at first glance, would fall into the time interval at present under consideration. JEFFREYS[7] has shown that the damping of the Chandler wobble and the tidal friction cannot be due to the same cause. If the time constant obtained from tidal friction is applied to the Chandler wobble, assuming the two effects as being due to the same cause, the wobble should be damped out in a few days instead of in some 12 years. Since

[1] See e.g. GRIGGS, D., J. HANDIN (ed.): Rock Deformation. Geol. Soc. Amer. Memoir No. 79 (1960).

[2] TSUTOMU, M.: J. Fac. Sci. Hokkaido Univ. 1958, Ser. VIII, **1**, No. 2, 103 (1958).

[3] IIDA, K. *et al.*: J. Earth Sci., Nagoya **8**, 1 (1960).

[4] HARDY, H. R.: Quart. Colo. School Min. **54**, No. 3, 135 (1959).

[5] BUCHHEIM, W.: Freiberger Forschungsh. C **45**, 68 (1958).

[6] WUERKER, R. G.: Quart. Colo. School Min. **54**, No. 3, 3 (1959).

[7] JEFFREYS, H.: The Earth, 3rd ed., p. 227.

JEFFREYS has also shown that friction in shallow seas can very well account for tidal friction, it is not necessary to try to resolve the discrepancy in time constants by assuming some rheological peculiarities.

3.64. Stresses of Long Duration. We finally come to the crucial problem of trying to elucidate the rheological behavior of the material in the Earth's crust and mantle under stresses of long duration. By "long" duration we mean, as stated above, from 15000 years upwards, with a "characteristic" duration of 100 million years. Needless to say, the behavior in such time intervals is only very vaguely known.

At the lower limit of this time interval is the rising of some ancient shields since the melting of the overlying ice caps which were present during the last ice age. It is generally assumed that the ice started to melt approximately 20000 years ago and that, e.g. Fennoscandia, has been (and still is) rising since that time owing to the principle of isostasy, buoyancy providing the driving force (cf. Sec. 8.52).

A discussion of the postglacial uplift of Fennoscandia with regard to its possible significance regarding the viscosity in the Earth's mantle and crust, has been given by GUTENBERG[1] and by VENING MEINESZ[2]. The results are summarized in an article by BENIOFF and GUTENBERG[3]. Accordingly, the uplift of Fennoscandia proceeds at a rate of the order of centimeters per century; "viscosities" calculated therefrom are of the order of

$$\eta_M = 10^{22}\,\text{cgs}. \tag{3.64-1}$$

(Note that the "Maxwell"-viscosity has no relation to the "Kelvin"-viscosity mentioned in the last section.)

A similar value for η_M was obtained by HASKELL[4] who gave a formal solution for the motion of a highly viscous fluid if a cylindrical body lighter than the fluid is rising in it, and comparing the latter to the Earth. The particular manner in which the above values for η_M were arrived at, will be presented in detail in Sec. 8.52. No value can be given for the other Maxwell constant (μ_M), since all the processes that can be investigated, are steady-state processes ($\dot{\tau}=0$). The term containing μ_M may therefore possibly be neglected; i.e. μ_M may possibly be set equal to ∞.

The above interpretation of the uplift of Fennoscandia has been criticized by LYUSTIKH[5] who claims that the figure of viscosity calculated in this manner has no physical meaning. This is because,

[1] GUTENBERG, B.: Bull. Geol. Soc. Amer. **52**, 721 (1941).

[2] VENING MEINESZ, F. A.: Proc. Kon. Akad. Wet. Amsterdam **40**, 654 (1937).

[3] BENIOFF, H., and B. GUTENBERG: In: Internal Constitution of the Earth, p. 382. New York: Dover Publ. Co. 1951.

[4] HASKELL, N. W.: Physics **6**, 265 (1935).

[5] LYUSTIKH, E. N.: Izv. Akad. Nauk SSSR., Ser. Geofiz. **1956**, 360.

according to LYUSTIKH, the uplift of Fennoscandia took place also before the last glaciation and therefore cannot be caused by isostatic forces.

However, if we accept that the uplift of Fennoscandia is an expression of creep during the past 20000 years, then it is more likely that there will be even more creep in longer time intervals. It is a general experience that creep becomes more pronounced the longer the stresses apply. Thus, we may assume that the Earth's mantle and crust show creep effects with a relaxation time of the order of 20000 years. However, this cannot possibly be the whole story. It is well known that mountain ranges can persist over several millions of years, in fact, the eventual disappearance of the mountains is not at all due to a creep effect, but to the gradual wearing down by the action of wind and water. This implies that the material, to a certain depth at least, also must have a non-vanishing yield stress. Introduction of a yield stress into the equation of a Maxwell body produces that of a Bingham material. Noting that in all processes that are at present amenable to discussion, no indication as to the prevailing Maxwell-rigidity μ_M can be obtained, it may thus be permissible to neglect the term containing the latter and therefore to use the Bingham equation as shown in (3.43–10).

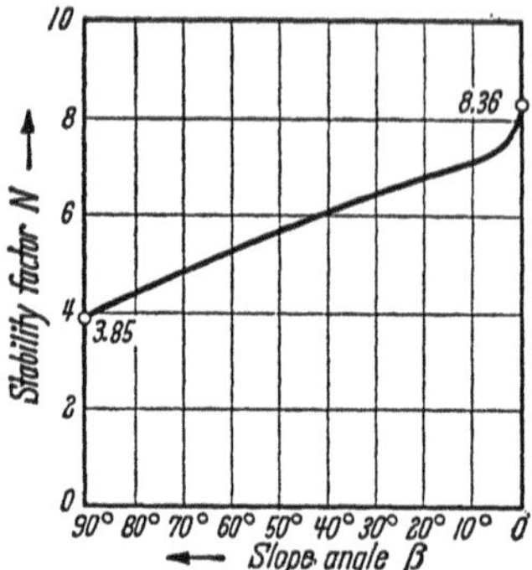

Fig. 67. Graph of stability factor *versus* slope angle. After TERZAGHI[1]

It remains to attempt to give an estimate for the yield stress ϑ_{ik} $(i \neq k)$ occurring in the Bingham equation. Since the *presence* of a yield stress is obviously ascertained from the existence of mountains, it may be expected that it is possible to extract some information regarding its *value* from the same source.

The stability of mounds of various materials has been discussed extensively by TERZAGHI[1] who postulated the following formula (Terzaghi equation):

$$H = \frac{\vartheta}{\varrho g} N. \tag{3.64–2}$$

Here, ϑ is the yield stress (to shear), H the critical height of a mound which can just support its own weight, g is as usual the gravity acceleration and N is a function of the angle β of the slope of the mound. This function is shown graphically in Fig. 67 (assuming vanishing internal friction).

[1] TERZAGHI, K.: Theoretical Soil Mechanics. London: Chapman & Hall 1943.

If we assume, as a crude approximation, that mountains also satisfy Eq. (3.64–2), then we obtain an estimate for the yield stress by putting $H=8$ km, $\varrho=3$, $g=980$ cm/sec and $N=6$ (corresponding to a slope angle of about 45°). This yields:

$$\vartheta = 4\times 10^{9}\,\text{cgs}. \tag{3.64–3}$$

This value is estimated for the surface of the Earth. It is an upper limit if it is assumed that mountains would be higher if they could support themselves. At depth, the yield stress may possibly be lower, in conformity with the trend of most materials to become weaker if the pressure and the temperature increase. It is very significant that the value arrived at in (3.64–3) is of the same order of magnitude as the apparent yield stress of rocks as found in short-duration experiments in the laboratory.

Thus, combining the phenomenology of the uplift with that of the persistence of mountains suggests that, under stresses of long duration, a rheological equation similar to that of a Bingham body should apply. However, there is a further puzzling factor. This is that the orogenetic phenomena occur in bands of rather narrow width. In these bands, stresses of intermediate time-duration are created so that the well-known tectonic effects of folding, faulting and earthquakes occur. The direction of the displacements in these bands seems to be in no simple relation to the boundaries of the bands, but is generally parallel to the surface of the Earth. For the explanation of this, one can think of the following possibilities:

(i) The bands occur because the crust of the Earth is built up of "stable masses", viz. the continents and ocean blocks, which are in "collision" due to contraction of the Earth's interior, continental drift or due to some other cause. It would naturally be the boundaries of the blocks that would be subject to the greatest deformations. A difficulty with this explanation is that the orogenetic bands are known to extend to great depths as evidenced by deep-focus earthquakes, whereas it is generally assumed that the "stable blocks" do not extend much below the Mohorovičić discontinuity[1]. It thus appears that the "collisions" could have no more than a triggering effect for the stress concentration, but could not be its actual cause.

(ii) The bands occur because the material is thixotropic: At the edges of the stable blocks a stress concentration is created in the upper regions, which causes motions in the intermediate time intervals. These, in turn, change the rheological equation so as to make the material more

[1] Note, however, that GUTENBERG's low velocity layer in the mantle may indicate that orogenesis reaches, infact, to a depth of about 140 km [cf. Sec. 2.14].

yielding. However, this "explanation" is actually nothing but an *ad hoc* hypothesis and therefore just a re-statement of the observations.

(iii) The bands occur because the rheological equation for long-term stresses allows this type of instability to occur. It is quite possible that the Bingham equation has integrals in which the deformation in narrow bands exceeds all bounds so that the solutions become unstable. The actual position of the bands could be triggered by the "stable masses" as indicated under (i). Unfortunately, nobody has investigated the Bingham equation regarding this possibility, but there is an analogy with BIJLAARD's bands of plastic instability (cf. Sec. 3.23). These bands of plastic instability occur in thin plates under tension; this has been experimentally demonstrated on thin steel plates. However, although the superficial appearance of the plastic bands bears some resemblance to features on the Earth, the reader should be warned not to carry this facile analogy too far. The mechanical conditions in thin steel plates under tension have almost certainly no relation to conditions in the Earth's crust and mantle. At best, it might be possible to make similar intuitive arguments in the two cases and, perhaps, to arrive at similar results. The underlying physical processes, however, will almost certainly be fundamentally different.

3.65. Summary. It thus appears that the behavior of the material in the Earth's crust and mantle is substantially different in the three time ranges which we have been considering.

a) In the short range (up to 4 hours) the material behaves as an elastic solid with a rigidity 2×10^{12} cgs and a modulus of elasticity E of about 5×10^{12} cgs. If the elastic limit is exceeded, the material undergoes a type of brittle fracture.

b) In the intermediate time range (4 hours to 15000 years) the material still behaves as an elastic solid with similar elastic constants, but with a significant after-effect. The relaxation time is about 2 days, the "viscosity" (Kelvin constant) η_K about 3×10^{17} cgs. It appears that there is a critical stress which, if exceeded, causes a fracture-type process. The limits of strength referring to this fracture-type process are not known. The process itself is probably very similar to that discussed under (a) as, once a fracture starts to move, this occurs in a time interval characteristic of (a).

c) In long-duration stresses, creep is the significant phenomenon; i.e. the material yields indefinitely at a constant rate if the stresses stay constant. The relaxation time appears to be of the order from 20000 years upward. The "viscosity" (Maxwell-constant) η_M is probably about 10^{22} cgs. From the relative persistence of mountain chains it may also be assumed that a low yield stress is present so that the rheological equation is essentially that of a Bingham body.

For the convenience of the reader, the results regarding the rheology of the Earth are summarized in Table 18.

Table 18. *Rheological Behavior of the Earth's Crust and Mantle*

Time interval	Applicable to	Behavior	Constants
Up to 4 hours ("short")	Seismic wave transmission	Elastic solid with strength limit	μ (rigidity) 2×10^{12} cgs. E (Young's modulus) 5×10^{12} cgs. Strength limit 10 cgs.
4 hours to 15000 years ("intermediate")	Chandler wobble Earthquake aftershocks Faulting Earth's tides	Kelvin body with strength limit	μ_K ("rigidity") 2×10^{12} cgs. η_K ("viscosity") 3×10^{17} cgs. Strength limit: present, value unknown
Longer than 15000 years ("long")	Large scale orogenesis Uplift of Fennoscandia Polar wandering	Bingham body	μ_M ("rigidity") term containing it may possibly be neglected η_M ("viscosity") 10^{22} cgs. ϑ (yield stress) 4×10^9 cgs. at surface

Any investigation of geodynamic problems should take the above facts into consideration. In general, the creation of an orogenetic system will take place during millions of years; it is therefore *not* permissible to apply "rigidities", "viscosities" etc. calculated from short- and intermediate-duration data to global considerations concerned with the distribution of orogenetic elements. Much unsatisfactory work has had its origin in this fundamental mistake.

On the other hand, the local tectonics is governed by the laws of the intermediate time-range: During the overall orogenetic process, the local stresses may be assumed to reach large values in periods substantially less than 15000 years. If a certain critical stress-difference is exceeded, a fracture-like process occurs which is felt as an earthquake and can possibly be recognized physiographically as a fault.

The laws of the short time range seem to have no application to geodynamics. They seem to be of importance solely in the propagation of elastic waves through the Earth.

IV. Effects of the Rotation of the Earth

4.1. The Figure of the Earth

4.11. The Ellipticity of the Earth. The Earth is a celestial body and is as such subject to gravitation. If all other forces were absent, the Earth would exhibit the form of a sphere and its surface would be perfectly smooth. We have already discussed some of the surface irregularities, but the principal departure from a spherical shape is due to the rotation of the Earth.

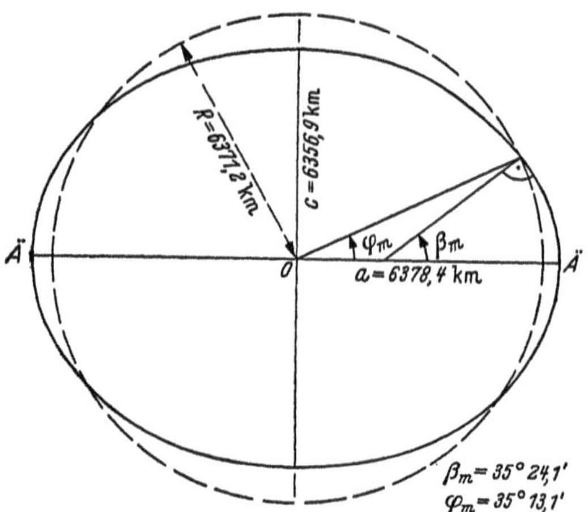

Fig. 68. Ellipsoid and sphere approximating the Earth; φ geocentric latitude, β geographic latitude; φ_m, β_m corresponding "average" latitudes. After JUNG[1]

The rotation of the Earth, owing to the centrifugal force it creates upon any body connected with the Earth, effects that the equilibrium figure (a "*geoid*", such as is represented by the surface of the oceans) is that of a spheroid of rotation rather than that of a sphere. It is customary to approximate that spheroid by an ellipsoid. The difference between an ellipsoid, properly chosen, and the true equilibrium spheroid is so small that it can usually be neglected. The data for the approximating ellipsoid are determined from geodesy[1]. The values obtained and accepted for further calculations are listed in Table 19 (see Fig. 68).

A further consequence of the ellipticity of the Earth is that the geographic latitude φ and the geocentric latitude φ' are not identical. The two are connected by the formula[2]

$$\tan(\varphi - \varphi') = \nu \sin 2\varphi' + O(\nu^2) \qquad (4.1\text{–}1)$$

where ν is the ellipticity (cf. Fig. 68).

[1] JUNG, K.: Handbuch der Physik, vol. **47**, p. 543. Berlin-Göttingen-Heidelberg: Springer 1956.

[2] JEFFREYS, H.: The Earth, 3rd ed., p. 130. London: Cambridge Univ. Press 1952.

Table 19. *The Earth**. (International ellipsoid)

Radius of Equator a	6378388 m
Ellipticity ν	1/297
Polar radius c	6356911.946 m
$\frac{1}{4}$ length of equator	10019148.441 m
$\frac{1}{4}$ length of meridian	10002288.299 m
Average radius	6371229.315 m
Radius of sphere with same surface	6371227.709 m
Radius of sphere with same volume	6371221.266 m
Surface	510100933.5 km²
Volume	1083319780000 km³

* After JUNG[1].

In addition to geodetic measurements, gravity measurements also give one a means to determine the equilibrium figure of the Earth, because the latter must represent an equipotential surface of the gravity field. In such measurements, it turned out that the Earth is slightly pear-shaped. It is interesting to note that gravity measurements at sea produce an exactly oppositely shaped "pear" than measurements from artificial satellites. The attendant problems have not yet been entirely resolved[2–4].

The various methods indicated above have been applied to the determination of the geoid. The results of these investigations have been summarized on several occasions in the literature[5–7]; for our purposes, an understanding of the general features of the shape of the Earth will suffice.

On several occasions, we mentioned the *rotation* of the Earth, implying it to be steady. However, there are in fact many deviations in the rotation of the Earth from a steady rate; much of the pertinent information hereon has been collected by SPENCER-JONES[8] and by MUNK and MACDONALD[9].

4.12. Theory of the Equilibrium Figure of the Earth. The fact that the equilibrium figure of the Earth is not a sphere permits one to draw certain conclusions regarding the density distribution in its interior. However, such considerations are beyond the scope of the present study.

[1] JUNG, K.: Handbuch der Physik, vol. **47**, p. 543. Berlin-Göttingen-Heidelberg: Springer 1956.

[2] O'KEEFE, J. A. *et al.*: Science **129**, 565 (1959).

[3] MUNK, W. H., and G. J. F. MACDONALD: J. Geophys. Res. **65**, 1169 (1960).

[4] MUSEN, P.: J. Geophys. Res. **60**, 403 (1961).

[5] HIRVONEN, R. A.: Adv. Geophys. **5**, 93 (1958).

[6] BOMFORD, G.: Geophys. J. Roy. Astr. Soc. **3**, 83 (1960).

[7] HEISKANEN, W. A.: J. Geophys. Res. **65**, 2827 (1960).

[8] SPENCER-JONES, H.: Handbuch der Physik, vol. **47**, p. 1. Berlin-Göttingen-Heidelberg: Springer 1956.

[9] MUNK, W. H., and G. J. F. MACDONALD: The Rotation of the Earth. London: Cambridge Univ. Press 1960.

The calculation of the equilibrium figure of the Earth, however, will be sketched below.

The general theory of the equilibrium figure of celestial bodies has been ably summarized by JARDETZKY[1, 2]. Since it is quite involved, we present here only a simplified version given by MILANKOVITCH[3].

Thus, if it be assumed that a liquid celestical body is rotating, then it is possible to determine its shape from the fundamental hydrodynamical equation

$$d\boldsymbol{v}/dt = \boldsymbol{f} - (1/\varrho)\,\mathrm{grad}\, p\,, \tag{4.12–1}$$

where $\boldsymbol{v}$ is the velocity-vector, $\boldsymbol{f}$ the specific mass-force due to gravity, p the pressure and t time. In case that the liquid is assumed as highly viscous (as must be the case with the Earth), it is to be expected that, through the interaction of neighboring particles, an equilibrium figure will develop which rotates "en bloc" like a solid body at a uniform angular velocity ω. It is then possible to describe the body by using a system of axes represented by the principal axes of inertia (X, Y, Z; moments of inertia A, B, C). For the dynamical calculation, the body may be considered as being at rest if the centrifugal force $\boldsymbol{F}$ per unit mass

$$\boldsymbol{F} = \omega^2 \boldsymbol{R} \tag{4.12–2}$$

is added to the corresponding gravitational force $\boldsymbol{f}$, $\boldsymbol{R}$ being a vector in a direction normal to the axis of rotation and of a magnitude equal to the distance of the particle under consideration from that axis. Eq. (4.12–1) can thus be written (for equilibrium)

$$\boldsymbol{f} + \boldsymbol{F} = \frac{1}{\varrho}\,\mathrm{grad}\, p\,. \tag{4.12–3}$$

The gravitational force can be expressed as the gradient of a potential U

$$\boldsymbol{f} = \mathrm{grad}\, U \tag{4.12–4}$$

where U can be found by a straigthforward integration over the body; assuming symmetry of rotation, i.e. $A = B$, which may be expected to be true for the Earth, yields (cf. MILANKOVITCH[3], p. 115):

$$U = \varkappa \frac{M}{r} + \frac{1}{2} \varkappa \frac{X^2 + Y^2 - 2Z^2}{r^5} (C - A) \tag{4.12–5}$$

---

[1] JARDETZKY, W. S.: Theories of Figures of Celestial Bodies. New York: Interscience 1958.

[2] See also: LEDERSTEGER, K.: Z. Vermessungsw. **84**, No. 3, 73 (1959).

[3] MILANKOVITCH, M.: Kanon der Erdbestrahlung und seine Anwendung auf das Eiszeitenproblem. Éd. spéc. Acad. Roy. Serbe Tome 133. Sec. Sci. Math. et Nat., Tome 33. 633 pp. Belgrade 1941.

where $\varkappa$ is the gravitational constant, M the total mass of the body, r the distance of the point under consideration from the center of the body and X, Y, Z the co-ordinates of that point in the system of principal axes of inertia (see above). Similarly, the centrifugal force can be expressed as the gradient of a potential U'

$$\boldsymbol{F} = \operatorname{grad} U' = \operatorname{grad} (\tfrac{1}{2} \omega^2 R^2) \tag{4.12–6}$$

(R being the magnitude of $\boldsymbol{R}$). Finally, setting

$$W = U + U' \tag{4.12–7}$$

permits (4.12–1) to be written as follows:

$$\operatorname{grad} p = \varrho \operatorname{grad} W. \tag{4.12–8}$$

According to earlier remarks, the equilibrium figure of the Earth must be a surface of constant W

$$W = W_0. \tag{4.12–9}$$

Introducing polar co-ordinates (ψ=longitude, φ=geocentric latitude) yields for W

$$W = \varkappa \frac{M}{r} + \frac{\varkappa}{2r^3} (C - A)(1 - 3\sin^2\varphi) + \frac{\omega^2 r^2}{2} \cos^2\varphi. \tag{4.12–10}$$

At the surface of the Earth, one can replace r by the equatorial radius a and one obtains (to the first order of approximation)

$$r = \varkappa \frac{M}{W_0} \left[1 + \frac{C - A}{2a^2 M} (1 - 3\sin^2\varphi) + \frac{\omega^2 a^3}{2\varkappa M} \cos^2\varphi\right]. \tag{4.12–11}$$

For $\varphi = 0$, one must obtain $r = a$, which yields

$$a = \frac{\varkappa M}{W_0} \left[1 + \frac{C - A}{2a^2 M} + \frac{\omega^2 a^3}{2\varkappa M}\right] \tag{4.12–12}$$

and for $\varphi = \pi/2$, $r = c$ if c denotes the polar radius. Thus

$$c = a\left[1 - \left(\frac{\omega^2 a^3}{2\varkappa M} + \frac{3}{2} \frac{C - A}{a^2 M}\right)\right]. \tag{4.12–13}$$

If we set

$$\nu = \frac{\omega^2 a^3}{2\varkappa M} + \frac{3}{2} \frac{C - A}{a^2 M} \tag{4.12–14}$$

we obtain

$$\nu = \frac{a - c}{a} \tag{4.12–15}$$

and

$$r = a(1 - \nu \sin^2\varphi) \tag{4.12–16}$$

which shows that the equilibrium figure of the Earth is (to the first order of approximation) indeed an ellipsoid with ellipticity ν as given by (4.12–14).

Finally, it may be remarked that the derivative of the potential function W at any one point gives the value of gravity at that point.

4.2. The Polfluchtkraft

4.21. Concept of the Polfluchtkraft. Another consequence of the rotation of the Earth has been the concept of a "polfluchtkraft". The concept has been developed by a combination of the theory of the Earth's rotation with the notion of isostasy.

If the Earth is assumed to be an essentially viscous, deformable body, then any masses (such as continents according to the theory of isostasy) whose density is less than that of the mantle, would float upon it. It has been postulated that such floating masses, owing to a force originating in the rotation of the Earth, would tend to drift towards the equator. This force has been termed pole-fleeing-force, or "polfluchtkraft".

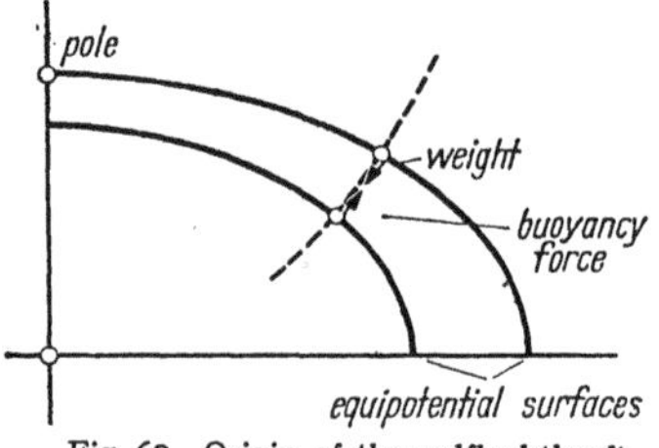

Fig. 69. Origin of the polfluchtkraft

The existence of a polfluchtkraft was first postulated by Eötvös[1] who noted that the direction-line of the vertical (i.e. the force-line of gravity) viewed in a meridional plane is curved in a rotating ellipsoidal Earth, the pole being located on its concave side. Furthermore, the center of gravity of the floating mass in which the weight is acting, must lie higher than the center of gravity of the displaced substratum (metacenter) in which the buoyancy-force is acting. The buoyancy-force as well as the weight are acting in the direction of the tangent to the corresponding force-line of the gravitational field; because of the latter's curvature (mentioned above), Eötvös reasoned that the two forces would not have the same direction and thus could not cancel each other, but would have a small resultant towards the equator. This is the polfluchtkraft. The situation is illustrated in Fig. 69.

In the wake of Eötvös' qualitative argument, there have been many attempts to calculate the polfluchtkraft analytically. Such attempts have been made notably by Epstein[2], Lambert[3], Ertel[4] and Milankovitch[5]. We shall give below (in Sec. 4.22) one of these calculations. The underlying model is essentially identical with that envisaged by Eötvös, and hence a polfluchtkraft is duly deduced.

---

[1] Eötvös, R. v.: Verh. 17. Allg. Konf. d. Internat. Erdmessung, I. Teil, 1913, p. 111. See also: Eötvös, R. v.: Gesammelte Arbeiten. Budapest: Academy 1953.

[2] Epstein, P. S.: Naturwissenschaften 9, No. 25, 499 (1921).

[3] Lambert, W. D.: Amer. J. Sci. 2, 129 (1921).

[4] Ertel, H.: Gerlands Beitr. Geophys. 43, 327 (1935).

[5] Milankovitch, M.: Kanon der Erdbestrahlung (l.c.).

However, rather severe criticisms against this type of deduction have been voiced in the meantime. We shall deal with them in Sec. 4.23. The criticisms are mainly directed against the models which were assumed in the various deductions of the polfluchtkraft and seem to be rather pertinent. The reality of a polfluchtkraft must therefore be severely questioned.

4.22. ERTEL'S Theory. We shall discuss now ERTEL's analytical theory of the polfluchtkraft. ERTEL[1] considered a light mass ("continent") floating upon a rotating liquid sphere. Denote the center of the buoyant force by A, the center of gravity of the light mass by S (see Fig. 70). Furthermore, let $\boldsymbol{A}$ and $\boldsymbol{g}$ denote the vectors of the buoyant force and of the gravitational force, respectively, and F_A and F_S the equipotential surfaces passing through A and S, respectively. The latter will diverge toward the equator. The line through A and S is assumed to be normal to F_A. Hence, the center of the buoyant force may be shifted into S according to a well-known theorem in mechanics. In S, the buoyant force $\boldsymbol{A}$ may be resolved into its components normal (A') and tangential (P_1) to P_S. The condition of floating for the light mass then yields:

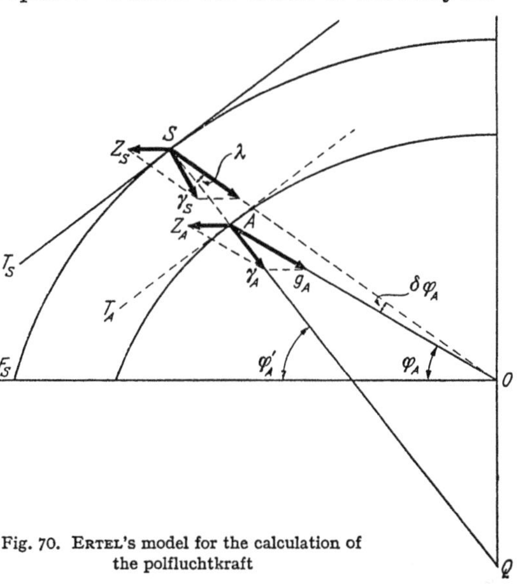

Fig. 70. ERTEL's model for the calculation of the polfluchtkraft

$$A' + g = 0 \tag{4.22-1}$$

whence it follows that P_1 is a force effecting a drift of the light mass towards the equator. This, however, presumes that S is moving in parallel to F_S. If the latter is not the case and S is assumed to move parallel to F_A, then it is possible to resolve $\boldsymbol{g}$ into components normal (g') and tangential (P_2) to F_A. The condition of floating yields in this case:

$$A + g' = 0, \tag{4.22-2}$$

[1] ERTEL, H.: Gerlands Beitr. Geophys. **32**, 38 (1931); **43**, 327 (1935). — SCHEIDEGGER, A. E.: Handbuch der Physik, vol. **47**, p. 277. Berlin-Göttingen-Heidelberg: Springer 1956.

and thus P_2 appears as the force effecting the drift towards the equator. In reality, S will move neither in parallel to F_A nor in parallel to F_S but somehow in between. However, the resulting tangential forces in the two limiting cases will differ only in higher order terms than are considered here; up to a significant order, the true drifting force may be identified with either P_1 or P_2. It is easier to calculate P_2 which may be done as follows.

Referring to Fig. 70, denote the geographical latitude of A by φ'_A, the geocentric latitude by φ_A. Furthermore, $\boldsymbol{g}_A$ and $\boldsymbol{Z}_A$ are the accelerations due to gravity and the centrifugal force in A which, combined, yield the total acceleration γ_A. If one sets $QA=a$, one has:

$$Z_A=\omega^2 a \cos\varphi'_A \tag{4.22–3}$$

where ω is the angular velocity of the Earth.

The total acceleration γ_A in A has no component parallel to the tangential plane T_A, since γ_A is normal to the equipotential surface F_A. This is expressed by the equation:

$$Z_A \sin\varphi'_A - g_A(\varphi'_A-\varphi_A)=0. \tag{4.22–4}$$

In S, one has the following value for the tangential (to T_A) component b_φ of γ:

$$b_\varphi=Z_S \sin\varphi'_A - g_S(\varphi'_A-(\varphi_A+\delta\varphi_A)); \tag{4.22–5}$$

for, when passing from A into S along the normal n of F_A, Z_A changes into $Z_S=Z_A+\delta Z_A$, g_A into $g_S=g_A+\delta g_A$, φ_A into $\varphi_S=\varphi_A+\delta\varphi_A$, whereas φ'_A remains unchanged. Thus, one can rewrite Eq. (4.22–5) as follows:

$$b_\varphi=\delta Z_A \sin\varphi'_A + g_A\,\delta\varphi_A - \delta g(\varphi'_A-\varphi_A). \tag{4.22–6}$$

If one sets

$$\operatorname{arc} QSO=\lambda,\qquad OA=r,\qquad AS=\delta n, \tag{4.22–7}$$

one has in view of $\lambda=\varphi'_A-(\varphi_A+\delta\varphi_A)$:

$$\delta n(\varphi'_A-\varphi_A-\delta\varphi_A)=r\,\delta\varphi_A; \tag{4.22–8}$$

or:

$$(r+\delta n)\,\delta\varphi_A=\delta n(\varphi'_A-\varphi_A). \tag{4.22–9}$$

As δn is much smaller than r, this is sufficiently approximated by

$$\delta\varphi_A=\delta n(\varphi'_A-\varphi_A)/r. \tag{4.22–10}$$

Substituting (4.22–10) into (4.22–6), one has

$$b_\varphi=\delta Z_A \sin\varphi'_A + (g_A/r-\partial g_A/\partial n)\,\delta n\,(Z_A/g_A)\sin\varphi'_A. \tag{4.22–11}$$

The expression $(g_A/r-\partial g_A/\partial n)$ can be rewritten by means of the gravitational potential Φ. Firstly, one can set

$$g_A=-\delta\Phi/\delta n, \tag{4.22–12}$$

and secondly, use the fact that

$$\operatorname{lap} \Phi = -4\pi\varkappa\varrho \qquad (4.22\text{–}13)$$

where $\varkappa$ is the gravitational constant and ϱ the density. In polar coordinates, and with sufficient accuracy for the present calculation, one can rewrite this as follows:

$$\frac{\partial^2 \Phi}{\partial n^2} + \frac{\partial \varphi}{\partial n}\frac{2}{r} = -4\pi\varkappa\varrho, \qquad (4.22\text{–}14)$$

which yields for g_A:

$$-\partial g_A/\partial n = 2g_A/r - 4\pi\varkappa\varrho. \qquad (4.22\text{–}15)$$

Therefore:

$$g_A/r - \partial g_A/\partial n = 3g_A/r - 4\pi\varkappa\varrho; \qquad (4.22\text{–}16)$$

and hence:

$$b_\varphi = \delta Z_A \sin\varphi'_A + \left[\frac{3g_A}{r} - 4\pi\varkappa\varrho\right]\delta n \frac{Z_A}{g_A}\sin\varphi'_A. \qquad (4.22\text{–}17)$$

From (4.22–3) one has

$$\delta Z_A = \omega^2 \cos\varphi'_A\, \delta n \qquad (4.22\text{–}18)$$

and also:

$$Z_A = \omega^2 r \cos\varphi_A. \qquad (4.22\text{–}19)$$

If all this be substituted into (4.22–17), one obtains:

$$b_\varphi = \frac{\omega^2}{2}\,\delta n \sin 2\varphi'_A + \frac{3}{2}\left[1 - 4\frac{\pi\varkappa\varrho r}{3g_A}\right]\omega^2\,\delta n\, 2\cos\varphi_A \sin\varphi'_A. \qquad (4.22\text{–}20)$$

The difference between φ_A and φ'_A may be neglected, since it is small. Thus

$$2\cos\varphi_A \sin\varphi'_A = \sin 2\varphi'_A, \qquad (4.22\text{–}21)$$

whence:

$$b_\varphi = 2\omega^2 \sin 2\varphi'_A\, \delta n\left[1 - \frac{\pi\varkappa\varrho r}{g_A}\right]. \qquad (4.22\text{–}22)$$

If a mean density $\bar{\varrho}_m$ is defined by:

$$g_A = \tfrac{4}{3}\pi\varkappa\bar{\varrho}_m r, \qquad (4.22\text{–}23)$$

then it is seen that this cannot be very much different from the mean density ϱ_m of the whole sphere which, therefore, may be used *en lieu* of $\bar{\varrho}_m$. Hence:

$$b_\varphi = 2\omega^2\,\delta n \sin(2\varphi'_A)\left[1 - \frac{3}{4}\frac{\varrho}{\varrho_m}\right]. \qquad (4.22\text{–}24)$$

Finally, if m is the mass of the floating part, one has for the drifting force:

$$K_\varphi = 2m\omega^2\,\delta n \sin(2\varphi'_A)\left[1 - \frac{3}{4}\frac{\varrho}{\varrho_m}\right]. \qquad (4.22\text{–}25)$$

This is ERTEL's expression for the polfluchtkraft.

4.23. Criticisms. As mentioned in Sec. 4.21, the reality of the polfluchtkraft is not at all as established as might appear from analytical deductions. PREY[1] has voiced a series of pertinent criticisms which can be summarized by stating that one should consider an extended floating mass, since a mass with small horizontal dimensions, i.e. a pencil-shaped body as considered in Sec. 4.22, could never attain equilibrium at all, but would simply tip over. Only a spherical body could attain equilibrium in this fashion. If an extended body is considered, it can attain equilibrium by tilting a little. The geometry assumed in Fig. 70 is therefore, according to PREY, entirely inadequate.

PREY substantiated his criticisms by explicitly calculating the equilibrium position of an extended "continent" floating upon a denser substratum. Because the geometry of such an arrangement is rather complicated, the calculations are correspondingly lengthy and involved. Starting from a position in which the floating mass is bounded above and below by equipotential surfaces of the gravity field, he showed that equilibrium can be attained by moving it by a very small amount *towards* the pole and tilting it. The necessary displacements are extremely small indeed.

Thus, if the continent is rigid, there exists an equilibrium position in which no forces are acting. In this equilibrium position, internal shear stresses will be present because the boundaries of the continent above and below do no longer coincide with the equipotential surfaces of the gravity field. If the continent is not assumed as rigid, there is therefore the possibility that it might yield under those stresses to adapt itself to a condition where its boundaries would again coincide with the equipotential surfaces. Hence it could be driven again into a new equilibrium position by moving slightly further towards the pole and tilting. This process could conceivably repeat itself and result in a net drift of the (deformable) continent *towards* the pole.

The relation of PREY's calculations with respect to those of EÖTVÖS (and ERTEL) can therefore be stated as follows. Disconnected floating "pencils" (the model used by EÖTVÖS) are subject to a polfluchtkraft (as long as they stay upright) and *drift towards the equator*, but if a continent is large and strong enough so that tilting can take place, there is a net displacement towards the pole. If, in addition, the continents are able to yield slowly to shear stresses, an actual *drift towards the pole* will be the net result. Thus, if continents are weak enough to yield slowly, but not so weak as to yield instantly, there will be a poleward motion. The whole matter hinges on the question whether one can assume instantaneous isostatic equilibrium everywhere on the Earth (EÖTVÖS) or whether this is not permissible (PREY). It would appear that PREY's model is closer to reality than that of EÖTVÖS and his followers.

[1] PREY, A.: Gerlands Beitr. Geophys. **48**, 349 (1936).

4.3. The Question of Stability of the Earth's Axis of Rotation

4.31. The Problem. As outlined in the first two Chapters of this book, many geological and geophysical observations point towards the likelihood of a change in the direction of the axis of rotation during geological time.

The postulate of such a change poses certain difficulties from a dynamical viewpoint: The Earth is a body rotating around the principal axis of the moment-of-inertia-tensor which corresponds to the largest moment of inertia; such an axis is a stable axis of rotation,—at least for a rigid body. In order to justify the heuristic inference of polar wandering dynamically, one must therefore seek to refute the apparent *a priori* impossibility of its occurrence.

Attempts to do this have been based upon various considerations. We shall deal with these below.

4.32. Effects of Circulations on a Rigid Earth. Let us first consider[1] the effects of displacements of matter on the surface of the Earth, in particular of such displacements which have the form of circulations. Such circulations are produced and maintained entirely by forces within the Earth and neither affect its total moment of inertia nor the position of its center of gravity. Circulations can be characterized by their angular momentum vector relative to the rotating Earth. The total angular momentum of the Earth, then, can be separated into the angular momentum without the circulations plus the relative angular momenta of the circulations.

During the motion of the Earth along its orbit, the total angular momentum of the Earth remains a constant vector in space. Thus, during the occurrence of a particular circulation, the instantaneous axis of rotation displaces itself. The point on the Earth's surface where the latter is pierced by the total angular momentum vector is not fixed. This point is not the pole of instantaneous rotation, but would be the pole if the circulation would come to a stop. GOGUEL calls this the "permanent pole" (designated *PP* in the accompanying Fig. 71) of the Earth. Since the angular momentum of the circulation is fixed with regard to the Earth (at least as long as the circulation remains stationary), the displacement of the pole from the permanent pole is constant. The angle ε of this displacement (cf. Fig. 71) can be calculated for various cases from the knowledge of the total moment of inertia of the Earth, equal to 6.77×10^{44} cgs units and from the angular momenta of the possible circulations. GOGUEL considered winds and ocean currents.

With regard to winds, let us represent a cyclonic movement by assuming that, within a radius of 2000 km, a wind of 48 km/h affects

[1] GOGUEL, J.: Ann. Géophys. 6, 139 (1950).

the atmosphere to a height of 5000 m. Since the density of air can be, on the average, assumed as approximately equal to 10^{-3} g/cm^3, the angular momentum in question turns out to be equal to 12.5×10^{30} cgs units. The position of the cyclones and anticyclones is strongly influenced by the distribution of continents and oceans. GOGUEL assumes that the resulting angular momentum of all the wind movement is, on the average, constant and equal to one-half of the angular momentum of one single cyclone. In this fashion, he obtains an angle ε equal to about 10^{-10} radians.

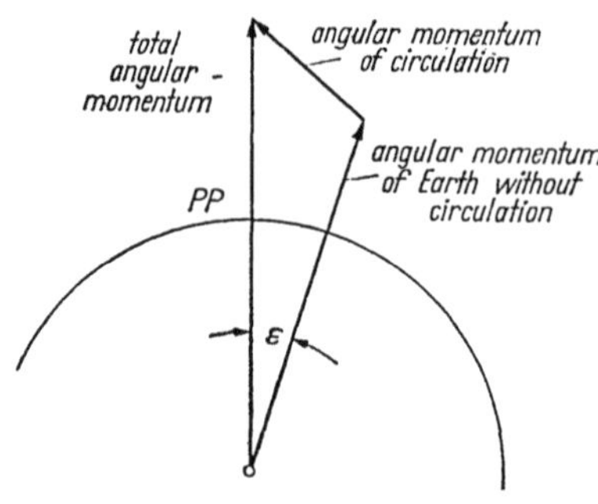

Fig. 71. GOGUEL's[1] decomposition of the Earth's angular momentum

Now GOGUEL assumes that an angle ε between the instantaneous and permanent poles would induce a polar wandering of $2\pi R\varepsilon$ per day at right angles to the connecting line between the poles, R being the radius of the Earth. This wandering would be an example of the nutation of a symmetric top as seen in the body-fixed system. It would correspond to about 1.46 m/year, a very large value indeed. However, if the Earth is assumed to be an ellipsoid instead of a sphere, the equatorial bulge brings polar wandering to a halt.

A similar conclusion can be reached if one considers circulations in the sea. Thus, following GOGUEL, let us represent an ocean current by assuming a circular trajectory of 5000 km diameter, 500 m depth and 100 km width, with a velocity of 2 knots. GOGUEL calculates the angular momentum of such a current as equal to 10^{31} cgs units, which is of the same order of magnitude as that obtained above for a single cyclone. One can estimate, then, the resulting angular momentum from the known ocean currents; GOGUEL obtains 3×10^{30} cgs units, in the direction of the meridian of 7° E. The angle ε of the of the pole turns out to be about 1/4 of that calculated for the cyclones and the polar wandering induced in this manner in a perfectly spherical Earth should therefore be about 1/4 of that calculated above. The ellipticity of the Earth, however, will again make polar wandering impossible.

The above deductions have been made for *a solid* Earth. If the Earth is able to *yield*, then matters are entirely different. It has been pointed out by INGLIS[2] that in this case circulations have a very similar effect as any other asymmetry in the crust of the Earth. The influence of such asymmetries upon the position of the axis of rotation in a *yielding* Earth

[1] GOGUEL, J.: Ann. Géophys. **6**, 139 (1950).
[2] INGLIS, D. R.: Rev. Mod. Phys. **29**, 9 (1957).

will be discussed in the Sections just following. However, in spite of this possible effect of circulations, their quantitative influence will in any case be much smaller than that due to geographical asymmetries;— simply because of the orders of magnitude involved.

4.33. Polar Wandering in a Yielding Earth. In order to obtain any possibility of polar wandering at all, one has to consider a model of the Earth where it is assumed that the latter is able to *yield*. In such an Earth, "polar wandering" means of course a shift of the Earth with respect to the axis of rotation; the actual motion of the latter in space may thereby remain relatively small.

The first attempt to obtain polar wandering assuming a yielding Earth seems to have been an essay by DARWIN[1]. The result was that polar wandering is possible for a "fluid" Earth;—but it has been pointed out by LAMBERT[2] that there was an algebraic error in DARWIN's calculations. JEFFREYS[3] rectified the error in DARWIN's calculation and obtained as a consequence of DARWIN's physical assumptions that polar wandering was impossible.

However, entirely non-analytic reasoning by GOLD[4] would make it plausible that significant polar wandering in an Earth which is capable of yielding, should be expected. MUNK[5], inspecting the physical basis of DARWIN's calculations, noted that one of the assumptions was probably incorrect. This can be demonstrated as follows.

The problem of polar wandering on a plastic Earth involves three types of poles: The pole F of rotation, the pole F' of the moment of inertia tensor (pole of figure) corresponding to the *instantaneous* shape of the Earth at the moment under consideration, and the geoidal pole F'' corresponding to the pole of the ellipsoid approximating the Earth's instantaneous shape.

DARWIN now made the following assumptions which are implicit in his equations:

(a) the velocity of wandering of the rotation pole F is proportional to its separation from the pole of figure F',

(b) the geoidal pole F'' moves at a rate proportional to the separation of the rotation-pole F from the pole of figure F'.

According to MUNK[5], the assumption (b) does not seem warranted. It seems much more likely that the geoidal pole F'' moves at a rate proportional to its separation from the rotation pole F, for it is the

---

[1] DARWIN, G. H.: Phil. Trans. Roy. Soc. **167**, Pt. 1, 271 (1877).

[2] LAMBERT, W. D.: Bull. U.S. Nat. Res. Counc. No. 78, Chap. 16 (1931).

[3] JEFFREYS, H.: The Earth, 3rd ed., p. 343. London: Cambridge University Press. 1952.

[4] GOLD, T.: Nature, Lond. **175**, 526 (1955).

[5] MUNK, W.H.: Nature, Lond. **177**, 551 (1956).

magnitude of this separation which determines the stresses in the Earth and hence the rate of flow altering the shape. It thus appears that assumption (b) should be replaced by the following

(b′) the geoidal pole F'' moves at a rate proportional to its separation from the pole of rotation.

It thus turns out, quite intuitively[1], that polar wandering should be quite a rapid process as soon as a small asymmetry has arisen somewhere in the Earth's crust.

The mathematical expression for a special case of the above reasoning has been achieved by MILANKOVITCH[2] long before the intuitive arguments of GOLD and MUNK became available. MILANKOVITCH considered the special case where the Earth is so quickly adjusting its shape that the rotation pole and the geoidal pole coincide.

Following MILANKOVITCH, we approximate the equilibrium figure of the Earth by an ellipsoid of rotation whose meridian is given by the equation

$$r = a(1 - \nu \sin^2 \varphi) \tag{4.33–1}$$

where r is the length of the radius vector, a the equatorial radius of the ellipsoid, ν the eccentricity and φ the geocentric latitude. The point corresponding to $\varphi = 90°$ is the geoidal pole F''. If we denote the principal moments of inertia of the ellipsoid by A, B, C, then we have because of symmetry properties

$$B = A. \tag{4.33–2}$$

According to general theorems of mechanics, the moment of inertia T with reference to any arbitrary axis ζ through the center of the ellipsoid is given by

$$T = A \cos^2\alpha = B \cos^2\beta + C \cos^2\gamma \tag{4.33–3}$$

where $\cos\alpha$, $\cos\beta$, $\cos\gamma$ are the direction cosines of the axis ζ. In the present case, the last equation reduces to

$$T = A + (C - A) \cos^2\gamma. \tag{4.33–4}$$

This defines a scalar field $T = T(\alpha, \beta, \gamma)$ describing the dependence of the moment of inertia of the "equilibrium Earth" as a function of the direction of the axis with regard to which it is taken.

---

[1] GOLD, T.: Nature, Lond. **175**, 526 (1955).

[2] MILANKOVITCH, M.: Glas. Acad. R. Serbe **152**, 39 (1932). — Handbuch der Geophysik, Bd. 1, Abschn. 7, Kap. 25, S. 438. 1933. — Publ. Math. Univ. Belgrade **1**, 129 (1932). — Glas. Acad. R. Serbe **154**, 1 (1933). — MILANKOVITCH, M.: Kanon der Erdbestrahlung und seine Anwendung auf das Eiszeitproblem; Éd. spéc. Acad. R. Serbe Tome 33, Belgrade. 1941. 633 pp.

However, it must be assumed that the Earth is not in an equilibrium condition. The moment of inertia referring to the axis ζ is therefore not T, but, say, J:

$$J = T + \Omega \tag{4.33-5}$$

where now Ω is that part of the moment of inertia which is due to the deviation of the Earth's surface from an equilibrium figure. It must thus be expected that Ω as a function of α, β, γ is not symmetrical with regard to the axis of rotation.

Owing to the asymmetry of Ω, the pole F' of J (i.e. the pole of figure; this corresponds to an extremal value of J) does not coincide with the geoidal pole F'' of the equilibrium figure, but must be somewhere near it. The co-ordinates ξ, η of F' with respect to the geoidal pole F'' can be found from the equation expressing that there is an extreme value for J for those co-ordinates:

$$\frac{\partial J}{\partial \xi} = 0; \quad \frac{\partial J}{\partial \eta} = 0, \tag{4.33-6}$$

or

$$\frac{\partial T}{\partial \xi} + \frac{\partial \Omega}{\partial \xi} = 0; \quad \frac{\partial T}{\partial \eta} + \frac{\partial \Omega}{\partial \eta} = 0. \tag{4.33-7}$$

It is convenient to use as co-ordinates ξ, η orthogonal co-ordinates in the plane tangent to a unit sphere engendered by the variable axis ζ at the point where it is penetrated by the axis through F''; the origin of these co-ordinates being at that point of penetration.

Since the pole of figure F' is very near the geoidal pole F'', it is possible to neglect powers higher than the first of ξ and η. We can thus express the last equation as follows:

$$\left.\begin{aligned} \frac{\partial T(0,0)}{\partial \xi} + \xi \frac{\partial^2 T(0,0)}{\partial \xi^2} + \frac{\partial \Omega(0,0)}{\partial \xi} + \xi \frac{\partial^2 \Omega(0,0)}{\partial \xi^2} &= 0 \\ \frac{\partial T(0,0)}{\partial \eta} + \eta \frac{\partial^2 T(0,0)}{\partial \eta^2} + \frac{\partial \Omega(0,0)}{\partial \eta} + \eta \frac{\partial^2 \Omega(0,0)}{\partial \lambda^2} &= 0. \end{aligned}\right\} \tag{4.33-8}$$

Furthermore, we have in virtue of (4.33-4):

$$\frac{\partial T}{\partial \gamma} = -(C-A)\sin 2\gamma, \tag{4.33-9a}$$

$$\frac{\partial^2 T}{\partial \gamma^2} = -2(C-A)\cos 2\gamma, \tag{4.33-9b}$$

and hence, since $d\xi = d\gamma$, one obtains finally

$$\xi = \frac{1}{2(C-A)} \frac{\partial \Omega(0,0)}{\partial \xi}; \quad \eta = \frac{1}{2(C-A)} \frac{\partial \Omega(0,0)}{\partial \eta}. \tag{4.33-10}$$

The vector $\boldsymbol{a}$ of displacement of the pole of figure F' with regard to the geoidal pole F'', in a plane tangent at the penetration point of the axis through F to the sphere engendered by unit vectors along the variable axis ζ, is therefore given by

$$\boldsymbol{a} = \frac{1}{2(C-A)} \operatorname{grad} \Omega. \tag{4.33-11}$$

As outlined earlier, we now assume that the Earth can yield so fast that the pole of rotation F and the geoidal pole F'' always coincide. This is a special case of assumption (b′) above implying that the constant of proportionality implied by that assumption is very large. Finally, assumption (a) yields for the velocity $\boldsymbol{v}$ by which the pole of rotation F (which is now identical to F'') moves:

$$\boldsymbol{v} = c' \boldsymbol{a} \tag{4.33-12}$$

where c' denotes a proportionality constant. It follows that the equation governing polar wandering is

$$\boldsymbol{v} = c \operatorname{grad} \Omega, \tag{4.33-13}$$

where c is again a certain constant. The last equation represents what has been known in the German literature for a long time under the name of "MILANKOVITCH's theorem".

The above form of the dynamic condition of polar wandering is obviously well suited to the model of the Earth where one assumes an essentially fluid substratum that can assume an equilibrium position instantly, with all the deviations concentrated in a thin crust. Then, the field Ω is independent of time and the possible polar paths are those corresponding to the last equation. The changing of the elevation of points of the crust owing to the adjustment of shape to the instantaneous equilibrium figure of the Earth is therefore automatically taken into account. The polar paths can be calculated and are as shown in Fig. 72. Here it has been assumed that the axes X, Y, Z correspond to $\Omega_1, \Omega_2, \Omega_3$, respectively, the latter being the eigenvalues of Ω with $\Omega_1 < \Omega_2 < \Omega_3$.

The next problem is to calculate the monent of inertia for any position of the axis of rotation. Supposing the variation of density with depth to be uniform for all continents (designated by $\varrho'(r)$) and for all oceans ($\varrho(r)$), the excess Q of continental over oceanic inertia equals (ϑ = colatitude, φ = longitude with regard to the axis under consideration)

$$Q = \iiint (\varrho' - \varrho)\, r^4 (\sin^2\vartheta \sin\varphi + \cos^2\vartheta) \sin\vartheta \, dr\, d\vartheta\, d\varphi = I q \tag{4.33-14}$$

with

$$I = \int (\varrho' - \varrho)\, r^4\, dr \tag{4.33-15}$$

and

$$q = \iint (\sin^2\vartheta \sin\varphi + \cos^2\vartheta) \sin\vartheta\, d\vartheta\, d\varphi, \tag{4.33-16}$$

where the integrals over the angles are evaluated over the continental area only. MUNK[1] evaluated the values for q for various axes assuming

[1] MUNK, W. H.: Geophysica 6, No. 3, 335 (1959).

standard crustal sections; the resulting "q-topography" is shown in Fig. 73. The possible polar paths are the orthogonals to the q niveau lines.

MILANKOVITCH did not attempt to give actual values for the integrals, but only took a "different" areal mass density for continents as compared with oceans. He then estimated the path of the pole for the present

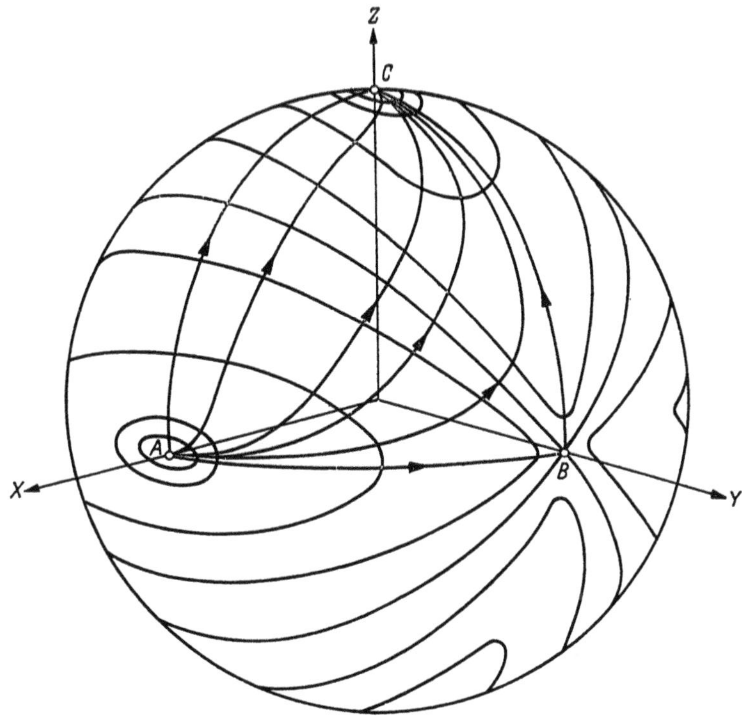

Fig. 72. Possible paths of polar wandering after MILANKOVITCH[1]

distribution of continents and obtained the picture shown in Fig. 74. A similar path would be obtained from MUNK'S q-topography, except that the direction of the motion would be reversed. It had already been pointed out by GUTENBERG[2] that, for an isostatically adjusted Earth, the center of gravity of the lighter masses would lie above that of the heavier displaced material, so that the time-arrow implied in Fig. 74

[1] MILANKOVITCH, M.: Kanon der Erdbestrahlung und seine Anwendung auf das Eiszeitproblem. Éd. spéc. Acad. Roy. Serbe Tome 33, Belgrade 1941.

[2] GUTENBERG, B.: In: The Internal Constitution of the Earth, p. 203. New York: Dover 1952.

should point into the opposite direction. This would put the present pole into a position which is about as far removed from its stable position as it could be. This seems hardly satisfactory. However, it seems that the assumption of the sign as accepted by MILANKOVITCH would not be as absurd as it appears at first glance. In the isostatic model of the Earth, the inertia of the crust depends on the second order term resulting from the slightly larger radial distance of continents and

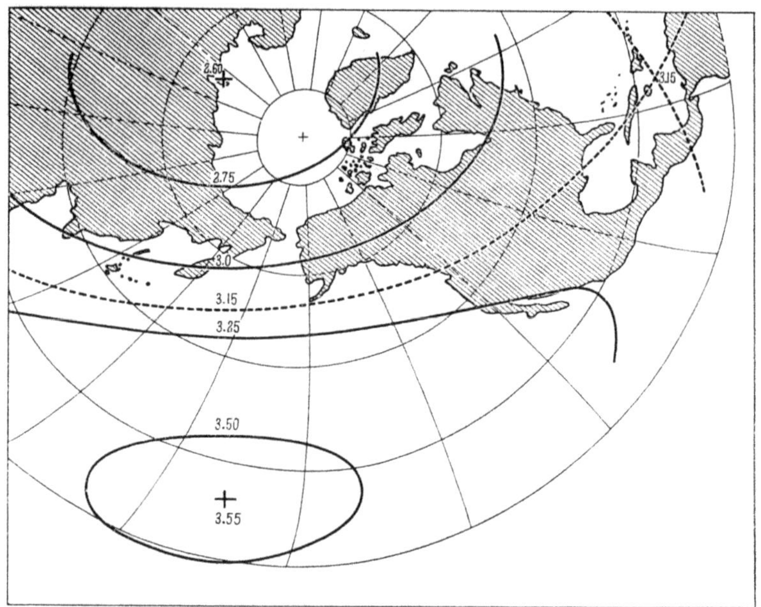

Fig. 73. MUNK's[1] q-topography

mountains as compared with oceans from the center of the Earth. If it is supposed that the isostatic balance does not hold precisely, but that there is erosion of continental matter and sedimentation on the ocean floor which is not compensated, then this represents a first order effect which might reverse the sign of the polar wandering so as to be in conformity with MILANKOVITCH's assumption.

The length of the time units in Fig. 74 depends on the constant in the fundamental equation, and hence on the "viscosity" of the Earth.

The curve in Fig. 74 has been calculated by assuming the present-day distribution of mass upon the Earth. However, it is well known that large tectonic movements have been taking place during the interval

[1] MUNK, W. H.: Geophysica 6, No. 3, 335 (1959).

since the Earth's creation. It is therefore to be expected that the pole may wander indefinitely as new sets of possible paths come into being with each redistribution of mass.

It may be noted that a displacement of the pole relative to the continents can also be regarded as a shift of the continents in their position, i.e. as "continental drift", rather than as "polar wandering".

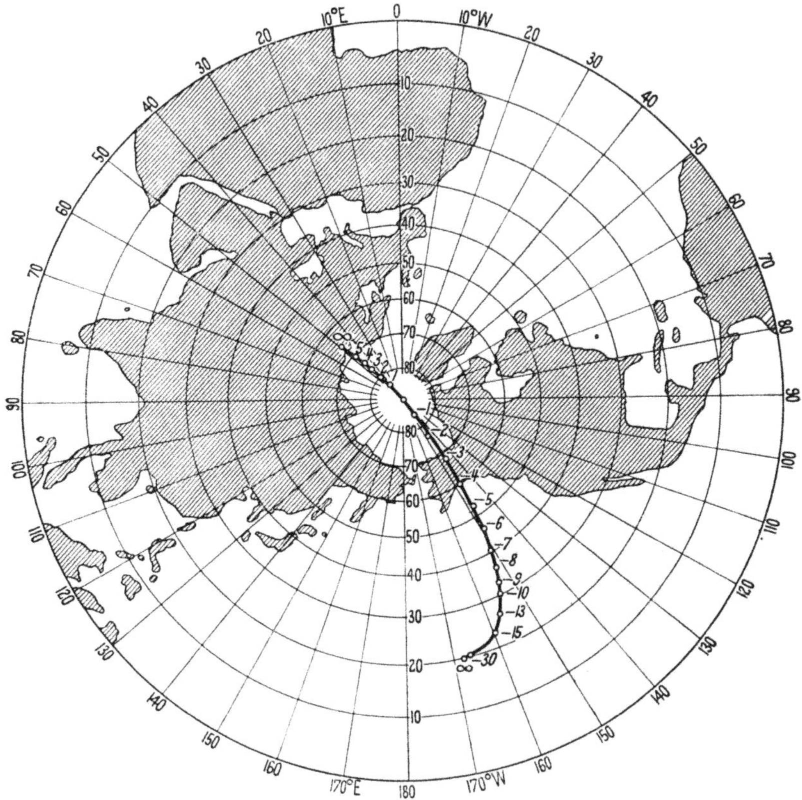

Fig. 74. Path of polar wandering for the present mass-distribution of the Earth as calculated by MILANKOVITCH[1]

Assuming that the pole tends to move away from the continental masses is the same as assuming that a "polfluchtkraft" is operative. Since the sign of the latter is uncertain, it need not surprise the reader that the same is true for the direction of "polar wandering".

[1] MILANKOVITCH, M.: Kanon der Erdbestrahlung und seine Anwendung auf das Eiszeitproblem, Éd. spéc. Acad. Roy. Serbe Tome 33, Belgrade 1941.

4.4. Other Effects of the Earth's Rotation

4.41. Tidal Forces. In conclusion of the Chapter on the effects of the Earth's rotation, we may consider two further possibilities in which forces might originate that could conceivably have a bearing upon orogenesis.

The first is the force due to tidal effects caused by the Sun and by the Moon. It has been investigated by many people (for a recent summary see MELCHIOR's[1] monograph). For our purposes, we quote here a calculation of JEFFREYS[2] who came up with the conclusion that at the utmost tidal forces might result in a drag at the bottom of the crust of the order of 40 dynes/cm^2. This force reverses its sign with every tide and can therefore have an orogenetic significance only in such a way that, on the average, the tendency persists to create a bulge at the equator. The long-time effect of the tides is therefore similar to that of the polfluchtkraft, but the magnitude of this tidal force is very small indeed.

A somewhat greater effect is caused by the tidal forces in an indirect manner, viz. by their moving the waters in the oceans around. It has been shown by JOBERT[3] that the bending of the Earth's crust due to the different amounts of water overlying the crust below the oceans at various times is not negligible. However, the bending of the crust caused in this manner is, of course, again periodic so that no long-term effects (other than possibly due to fatigue fractures) may be expected.

In addition, there is a westward stress due to the secular effect of tidal friction, but its magnitude is only of the order of 10^{-4} dynes/cm^2. If this stress is assumed to act upon a whole continent, it will cause a compressive stress therein greater than the westward stress roughly in proportion to the ratio of the area of the continent to its cross section[4]. The latter may be approximately equal to 100:1. This would make the stress equal to about 10^{-2} dynes/cm^2. This is much less than any stress that could have an orogenetic significance. In fact, it has been calculated by JEFFREYS[2] that, even if there were no Bingham-yield stress present in the Earth, to make America move westward to its present distance from the Old World, would take 10^{17} years. To produce the same effect in, say, 3×10^7 years, one would require a tidal friction so high that it would stop the Earth's rotation within a year. In spite of these

[1] MELCHIOR, D. J.: Les marées terrestres. Monographie No. 4, Observatoire Royal de Belgique 1954. See also the Proceedings of the various Internat. Symposiums on Earth tides, held under the auspices of the U.G.G.I.

[2] JEFFREYS, H.: The Earth, 3rd ed. London: Cambridge Univ. Press 1952.

[3] JOBERT, G.: C. R. Acad. Sci., Paris **244**, No. 2, 227 (1957).

[4] GRIGGS, D.: Amer. J. Sci. **237**, 611 (1939).

formidable objections, NADAI[1] has held tidal friction responsible for the (hypothetical) drift postulated in the continental drift theory (cf. Secs. 1.31 and 6.3).

In spite of the apparent insignificance of tidal forces with regard to orogenesis, these forces are not negligible. They have presumably caused the Earth's rotation to slow down appreciably since the latter's creation[2]. Also, estimates of the Earth's rigidity have been obtained from them, valid for the appropriate time interval (cf. Sec. 3.63).

4.42. Coriolis Force. The second of the two effects of the rotation of the Earth mentioned in Sec. 4.41 is the Coriolis force. The fact that the Earth's axis of rotation is inclined towards the plane of the ecliptic has the effect that the axis precesses. The forces causing this precession have a different action on parts of the Earth of different density, hence a component results that might conceivably make the continents move.

However, JEFFREYS[3] analyzed this force, too, and found that it cannot cause stresses greater than 60 dynes/cm². Moreover, it is also mainly alternating in direction and therefore cannot be of any more significance than the tidal forces which have been shown to be negligible as far as geodynamics is concerned.

---

[1] NADAI, A.: Trans. Amer. Geophys. Un. **33**, 247 (1952).
[2] PARIISKY, N. N.: Proc. Third Int. Symp. Earth Tides (U.G.G.I.), p. 19 (1959).
[3] JEFFREYS, H.: The Earth, 2nd ed., p. 304.

V. Continents and Oceans

5.1. Primeval History of the Earth

5.11. The Problem of Continents and Oceans. The present distribution of continents and oceans poses many puzzling problems. Continents are undoubtedly the most permanent features of the Earth's surface and therefore their existence probably dates back to the Earth's early history. This very fact makes an explanation of their presence rather difficult: nothing certain is known about the origin of the Earth, and therefore any attempt at establishing a theory of continents is tied up with speculations.

It may be noted that most theories of the existence of continents have a counterpart in corresponding theories of the existence of mountains, i.e. in theories of orogenesis. Since usually any change in the distribution of continents will cause orogenesis, this is not too unexpected. Nevertheless, it seems worth while to separate the two topics into separate Chapters for the purposes of the present study. We shall treat here those aspects of geodynamics which have more specifically a bearing upon continental effects, and treat in the next Chapter those aspects that deal more specifically with orogenesis. However, there will of necessity be many cross-references.

5.12. The Origin of the Earth. Questions regarding the origin of the Earth have a bearing upon the problem of continents and oceans and it is therefore necessary to give here a brief review thereof.

The Earth is one of many planets and its origin is intimately tied up with that of the solar system. Unfortunately, the problem of how the solar system was created has still not yet been satisfactorily solved although attempts have not been lacking. In fact, the variety of ideas which has been advanced to explain the origin of the solar system is almost as great as the variety of ideas advanced to explain orogenesis. It would thus easily be possible to write a separate monograph on that subject.

In the present context we are only interested in the possible connection between the theories of the origin of the Earth and the latter's subsequent development. For this purpose it is quite sufficient to confine ourselves to an account of the broad implications of the theories without going into much detail. A useful account along these lines has been given for instance by SMART[1] to which the reader is referred for the details.

Accordingly, the theories of the origin of the solar system can be split into two categories: Uniformitarian theories and cataclysmic theories. We shall briefly discuss these two categories.

a) Uniformitarian Theories. The prototype of a uniformitarian theory is the nebular condensation hypothesis of LAPLACE. In it is assumed that the whole solar system was a gaseous nebula at its beginning. Such a nebula would have a rather dense core with a very thin atmosphere reaching to beyond the present boundaries of the solar system. Owing to gravitational attraction, the nebula would slowly contract which, in turn, would cause any initial rotation to become more rapid. Eventually the rotation would become so fast that at the outer boundary of the nebula a gaseous ring would be thrown off. The latter, in time, would condense to form the first planet. This process would repeat itself until all the planets were formed. The original core of the nebula would form the Sun.

Although the above theory of LAPLACE has some apparent success, there are in fact many severe difficulties. The most serious one is that the distribution of angular momentum in the solar system is at complete variance with any distribution that would be consistent with the theory: In the solar system, most of the angular momentum is found in the distant planets; LAPLACE's theory yields the reverse.

Essentially the same ideas as those of LAPLACE's have also been defended by KANT. The latter author, however, did not assume that the nebula was subject to a primeval rotation, but tried to deduce that such a rotation would automatically develop. The difficulties in the theory are thereby not lessened.

A modern revival of the uniformitarian theory has been proposed by von WEIZSÄCKER[2, 3] who assumed that the Sun was formed from one of the interstellar dust-clouds which are fairly common in the Milky Way. The planetary system was thereby formed as part of the process by condensation of the dust particles into the required number of larger masses. The dust cloud is assumed to be in a state of turbulent motion and endowed with a definite angular momentum. At a late stage of the process it would have a disc-like shape. WEIZSÄCKER has shown that certain internal states of motion in the disc are more stable than others;—in fact that a pattern of vortices may develop which would be capable to persist in quasistationary motion for a considerable length of time. At the boundaries of the vortices the dust would collect and

[1] SMART, W. M.: The Origin of the Earth, 239 pp. London: Cambridge University Press 1951.

[2] WEIZSÄCKER, C. F. v.: Z. Astrophys. **22**, 319 (1944).

[3] CHANDRASEKHAR, S.: Rev. Mod. Phys. **18**, 94 (1946).

form the nuclei for the future planets. At this stage of the evolution the quasistationary pattern of vortices may disappear as the planets could grow by themselves by further accretion of matter. WEIZSÄCKER estimates that the time required for a planet to grow to its final size was about 10^8 years whereas the Sun might have been "finished" in 10^7 years. The difference of a factor 10 might just have been sufficient to allow for most of the lighter elements to escape from the planets and thus to account for the different composition of the Sun (consisting mostly of hydrogen) from that of the rest of the solar system.

Finally, KUIPER[1] modified the previous discussions by assuming gravitational instability within a disc-shaped solar nebula as a source of gaseous spheres (protoplanets) which eventually would contract to form planets.

b) Cataclysmic Theories. The basis of cataclysmic theories of the origin of the solar system is the assumption of some catastrophe. A good example of such a theory is the hypothesis, due to HOYLE[2], that the Sun was part of a binary system and that the Sun's companion blew up as a supernova. According to HOYLE, a slight eccentricity of the explosion in the Sun's companion would produce the correct distribution of angular momentum in the solar system. Furthermore, the nuclear chemistry of a supernova explosion (being completely unknown) could account for the different composition of Sun and planets.

Other cataclysmic theories include the tidal theory of JEFFREYS[3] in which it is envisaged that the Sun was disrupted by a tidal resonance effect with a passing star. The fragments would ultimately form the planets. Difficulties arise in this theory from two sides. First, if it is assumed that the actual fragments of the Sun formed the planets, there is the difference in chemical constitution of the Sun and of the planets to be accounted for. This is only possible by assuming a tremendous thinning out of the fragments just after the catastrophe and a subsequent slow recondensation which would provide time for the light elements to escape. Second, it is again not easy to account for the distribution of angular momentum in the solar system. To avoid this difficulty it has been assumed that the tidal effect occurred with a hypothetical binary companion of the Sun instead of with the Sun itself. Similar problems occur if the catastrophe is assumed to be a head-on collision rather than a tidal resonance effect.

In conclusion, it may be remarked that all theories of the origin of the solar system have two principal difficulties to cope with: First there

[1] KUIPER, G. P.: Chapter 8 in Astrophysics, ed. by Hynek. New York: McGraw-Hill Publ. Co. 1951.

[2] HOYLE, F.: Proc. Camb. Phil. Soc. **40**, 265 (1944).

[3] JEFFREYS, H.: The Earth, 2nd ed. London: Cambridge University Press 1929.

is the distribution of angular momentum in the solar system which is chiefly (98%) concentrated in the planets and not in the Sun, Jupiter making the biggest single contribution. Second, there is the difference in composition between the Sun and the planets. The former consists chiefly of hydrogen, the latter of heavier elements. The uniformitarian theories differ from the cataclysmic ones mainly by the likelihood of the occurrence of the postulated process. In all uniformitarian theories, the acquiring of a planetary system is part of the normal evolution of any star, whereas in cataclysmic theories this would be an extremely rare and unique occurrence. Agreement as to the correct theory has obviously not yet been achieved.

5.13. The Earth's Early Thermal History. The various theories of the origin of the Earth discussed in Sec. 5.12 of this book are of importance in geodynamics because of their implications regarding the Earth's thermal history. The layered structure of our globe suggests that the Earth might have gone through a molten state at one stage of its life; for in a molten body the differentiation into a dense core and progressively less dense upper strata is most naturally accomplished. Most discussions of the Earth's thermal history, therefore, start with a hot Earth which is gradually cooling down. This has also been done in the discussion of the Earth's thermal history in Sec. 2.63.

It is therefore of some interest to investigate whether there are any cosmological indications that the Earth has gone through a molten stage, in the light of the various theories of the origin of the solar system. In the cataclysmic theories, where it is assumed that the planets were formed from the fragments of a star, it would presumably be natural to assume that these fragments were hot. In this instance, it should not be overlooked, however, that the fragments could possibly have been spread out after the explosion into a gas cloud of very low density; the gas cloud, in turn, would condense later to form the planets. Any dispersed gaseous matter in the stellar space would of necessity be cooled down very rapidly to form "ice crystals" which creates the problem of explaining the subsequent condensation in such a fashion so as to arrive at a hot Earth. In this instance, the problem is the same as that encountered in the uniformitarian theories of creation of the solar system. Here, no cataclysm is assumed in the first place and a way must be found to condense a cold cloud of interstellar matter in such a fashion that it becomes hot;—at least if it is desired to start the development of the Earth from a hot sphere.

There are two ways in which the Earth could have heated up during or immediately after its formation, the latter being assumed to have taken place from a could of dispersed material: first by a conversion of mechanical energy into heat during the contraction of the cloud, and second by the effect of radioactivity of the constituent elements in the Earth.

The process of condensation or contraction of a cool gas cloud to a planet has been discussed by HOYLE[1]. In his investigations, HOYLE came to the conclusion that such cold condensation would convert sufficient mechanical energy into heat to melt a planet of the size and composition of the Earth. However, a scrutiny of his argument shows that it depends very sensitively on the speed of condensation, a quantity which is not very well known. Accreted material strikes a condensation surface of radius s and mass M with a kinetic energy E due to gravitation of

$$E = \varkappa \frac{M}{s} dm \tag{5.13–1}$$

where $\varkappa$ is the gravitational constant (as usual) and dm the mass of the striking body. If the condensation proceeds at such a speed that the mass of the "planet" is doubled in time t, then the addition of energy per unit time is

$$\dot{E} = \varkappa \frac{M}{s} \frac{M}{t} \tag{5.13–2}$$

which is per unit surface

$$\frac{\dot{E}}{4\pi s^2} = \frac{\varkappa M}{s\, 4\pi s^2} \frac{4}{3} \pi \varrho \tau s^3 \frac{1}{t} = \varkappa \frac{M \varrho}{3t} \tag{5.13–3}$$

where ϱ is the density of the "planet". The surface temperature of the planet will presumably become so high that an equal amount of energy per unit time is lost by radiation into space as is acquired from the bombardment of the accreting particles. The energy per unit time and surface at temperature T (°K) is given by the Stefan-Boltzmann law

$$\frac{\dot{E}}{\text{surface}} = \sigma T^4 \tag{5.13–4}$$

with $\sigma = 5.67 \times 10^{-5}$ cgs units. Equating the right hand sides of Eq. (5.13–3) and (5.13–4) leads to

$$\sigma T^4 = \varkappa \frac{M \varrho}{3t} \tag{5.13–5}$$

and hence to

$$T = \left(\frac{\varkappa M \varrho}{3 t \sigma}\right)^{\frac{1}{4}}. \tag{5.13–6}$$

For a planet like the Earth ($\varrho = 5.6$, $M = 5.98 \times 10^{27}$, all cgs units) this leads to

$$T = 1.9 \times 10^6 / t^{\frac{1}{4}}\ \text{°K} \tag{5.13–7}$$

---

[1] HOYLE, F.: Month. Not. Roy. Astron. Soc. **106**, 406 (1946).

which shows that the temperature T is sensitive to t. HOYLE chose $t=10^5$ years ($=3.15\times10^{12}$ sec) and obtained

$$T \cong 1400\ °\mathrm{K}. \tag{5.13-8}$$

This, he argued, would be sufficient to melt the Earth. However, it seems very doubtful whether the accretion process could have proceeded at such a fast pace. Judging by the Weizsäcker theory, a characteristic agglomeration time t of 10^8 years would be much more reasonable. This would lower the temperature by a factor 5.6 to yield

$$T \cong 250°\mathrm{K}. \tag{5.13-9}$$

This is certainly not sufficient to melt the Earth as it is below the freezing temperature of water.

It seems therefore that, except in the case of an extremely rapid condensation, there is not enough heat from mechanical energy available to melt the Earth if it was formed from a cold cloud. We shall therefore investigate the second possibility of heat production, viz. that due to the presence of radioactive matter in rocks. A study of this possibility has been undertaken by URRY[1]. If it is taken into account that radioactive matter decays exponentially, then it is reasonable to expect that the density of heat-generating matter within the Earth must have been much greater during the primeval days than it is at present. URRY'S result is that the radioactivity would be sufficient to melt the Earth. The same conclusion was arrived at by BIRCH[2] using newer values for the radioactive decay constants: an initially cold Earth containing as little as 0.1 % potassium would eventually melt,—at least partially. The time required to reach the molten state might not exceed 10^8 years.

From these investigations it would appear as likely that even had the Earth been formed by cold accretion, it would have melted early in its life. However, it is possible that the above authors overestimated the effect of radioactivity. In his study of the thermal history of the Earth mentioned in Sec. 2.63. JACOBS[3] showed that the final solution of the heat condictivity equation is the superposition of the cooling of a non-radioactive Earth from its "initial" temperature (assumed by JACOBS as high) plus the heating-up of an originally cold Earth due to radioactivity. If one inspects JACOBS' solution for the latter case, it becomes evident that no very high temperatures are reached (except at the surface; see Fig. 75). For his calculations, JACOBS assumed the present layered distribution of elements inside the Earth, but he made allowance for the time variation of radioactivity. The end result of the

[1] URRY, W. D.: Trans. Amer. Geophys. Un. **30**, 171 (1949).
[2] BIRCH, F.: J. Geophys. Res. **56**, 107 (1951).
[3] JACOBS, J. A.: Pub. Bur. Centr. Séismol. Int. A **19**, 155 (1956).

heating-up does not yield the commonly accepted present temperature distribution inside the Earth, but the calculations do show that there is indeed the possibility that the Earth as a whole is slowly heating up rather than cooling down, and that it has never been molten to any considerable extent. Unreasonable as this possibility might appear (particularly with regard to the explanation of the origin of the layered structure) it should however be noted that it cannot entirely be ruled out.

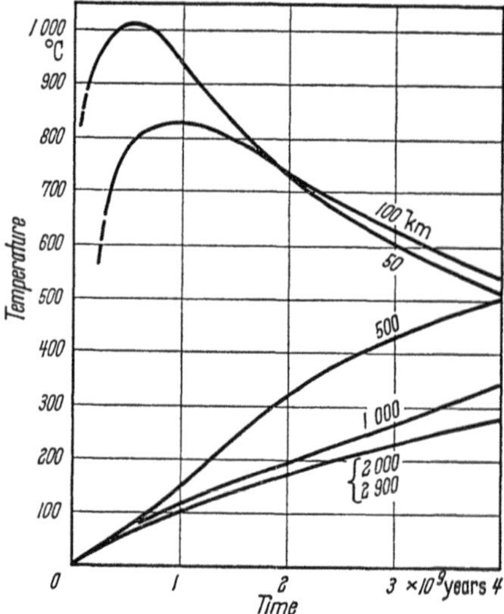

Fig. 75. Temperature distribution at various depths in a radioactive Earth, starting from zero temperature. After JACOBS[1]

In addition to the investigations mentioned above, many more attempts at elucidating the early thermal history of the Earth have been published in the literature[2-7]. In spite of many contentions to the contrary, it must be admitted, however, that it is, todate, not known whether the Earth started as a hot or as a cold body, and whether it is now or has been up to now, heating up or cooling down.

5.14. The Birth of the Moon. Another event that has been connected with the Earth's early history is the birth of the Moon. In conformity with the various theories of the origin of the solar system, a corresponding number of theories of the origin of the Moon has been proposed. Of these the uniformitarian theories have obviously little bearing upon the Earth's physiography, but the cataclysmic ones do: for, the cataclysm would involve the Earth as parent body. Thus, the removal of the Moon from the Earth has been thought to have occurred in a similar fashion as the hypothetical removal of solar matter from the Sun to form the planets. One of the difficulties existing in the theory of the

---

[1] JACOBS, J. A.: Pub. Bur Centr. Séismol. Int. A **19**, 155 (1956).

[2] BIRCH, F. P.: Geophysics **19**, 645 (1954).

[3] FESENKOV, V. G.: Astron. Zhur. **34**, No. 1, 105 (1957).

[4] LYUBIMOVA, E. A.: Geiphys. J. Roy. Astron. Soc. **1**, 115 (1958). — 21. Int. Geol. Congr. (Norden), Dokl. Sov. Geol. **2**, Sec. 2, 14 (1960).

[5] MACDONALD, G. J. F.: J. Geophys. Res. **64**, 1967 (1959).

[6] SAFRONOV, V. S.: Izv. Akad. Nauk SSSR., Ser. Geofiz. **1959**, No. 1, 139 (1959).

[7] RINGWOOD, A. E.: Geochim. et Cosmochim. Acta **20**, 241 (1960).

solar system is absent in the case of the Moon as the composition of the latter is very similar to that of the Earth's crust. In order to cause the hypothetical separation of the Moon from the Earth, one could perhaps invoke a resonance effect in the oscillations of the Earth with the tidal forces exerted by the Sun[1], or else an internal "explosion"[2].

Such a cataclysmic birth of the Moon would have far-reaching consequences upon the physiographic appearance of the Earth. These consequences have been discussed for the first time by FISHER[3]; they have been re-stated more recently e.g. by BOWIE[4] and by ESCHER[5]. Accordingly, it is assumed that the Earth already had a solid crust when the Moon was torn off. Consequently, a tremendous wound would be left which would be represented now by the Pacific Ocean. The "suction" caused by the wound would help to break up the remaining part of the crust and would cause the pieces to move toward the Pacific. This, in turn, might explain why all the continents can be fitted together as discussed in Sec. 1.31.

A different view from that outlined above has been taken by HAARMANN[6] who supposed that the Moon was torn off from the Earth not where there is now the Pacific Ocean, but in Central Asia. The wound would have rapidly filled-in with sediments and risen to the present high elevation owing to the action of isostasy.

An investigation into the physical possibility of a separation of the Moon from the Earth must be directed at the hypothetical resonance effect. JEFFREYS[7] does this and shows that the envisaged process is really quite untenable as there are three major difficulties. First, during the oscillations there would be a considerable discontinuity of velocity at the core boundary which would lead to energy-loss owing to friction. This, in turn, would restrict the maximum possible amplitude of the oscillations to much less than that required to cause disruption. Second, the linear theory of harmonic oscillations is really not applicable for a calculation of the resonance effect because of the fact that the amplitudes involved are large. As soon as nonlinear terms are taken into account, however, resonance does no longer occur in the customary fashion and it is very doubtful whether a critical amplitude could be attained at all. Third, any body ejected from the Earth should of necessity return to it which is, in case of the Moon, contrary to experience.

[1] DARWIN, G. H.: Phil. Trans. Roy. Soc., Part ii, p. 532 (1879).

[2] QUIRING, H. L.: Gerlands Beitr. Geophys. **62**, 81 (1952). — Z. dtsch. geol. Ges. **105**, 203 (1953). — Neues Jb. Geol. Paläont. Min. **3**, 140 (1961).

[3] FISHER, O.: Nature, Lond. **25**, 243 (1882).

[4] BOWIE, W.: Sci. Monthly **41**, 444 (1935).

[5] ESCHER, B. G.: Bull. Geol. Soc. Amer. **60**, 352 (1949).

[6] HAARMANN, E.: Die Oszillationstheorie. Stuttgart: Ferdinand Enke 1930

[7] JEFFREYS, H.: The Earth, 3rd ed. London: Cambridge Univ. Press; 1962 p. 234

It seems therefore that there are almost insurmountable difficulties in any theory claiming that the Moon was formed from the Earth and therefore also in any theory attributing the present distribution of continents to such a cause.

5.2. Evolution and Growth of Primeval Continents

5.21. The Hypothesis of Laurasia and Gondwanaland. One of the attempts at explaining the present distribution of continents and oceans is by assuming the break-up and drift of one or two primeval continents. If *one* primeval continent is assumed, it is usually called *Pangea*, if *two* primeval continents are assumed, they are usually referred to as *Laurasia* and *Gondwanaland*. If one assumes two primeval continents, they are supposed to have originally been formed at the (then) poles of the Earth, Laurasia in the North, Gondwanaland in the South. After their formation, these primeval continents are supposed to have broken up and possibly to have grown and the pieces to have drifted into the present position of the land masses.

Most of the evidence put forth in support of Laurasia and Gondwanaland is of a physiographic nature; i.e. it is geological, botanical, climatological etc. This physiographic evidence has already been listed in Sec. 1.31. We shall concern ourselves below with the mechanical aspects of the problem.

5.22. The Notion of Continental Drift. The search for a mechanism that might have broken up Laurasia and Gondwanaland (or, for that matter, a *single* primeval continent) and that might have caused the pieces to move into the present position of the continents, led to the continental drift hypothesis of WEGENER[1]. Accordingly, the continents are light masses floating upon a denser substratum in isostasy. The forces causing the drifting are not very well defined, one usually assumes that the polfluchtkraft (cf. Sec. 4.2) would be able to do this. The continental drift hypothesis is intimately tied up with problems of mountain building since the same hypothetical forces that could cause the shifts of the continents also would effect a crumpling-up at their margins and therewith cause mountain building. Because of the great importance of continental drift in the analysis of orogenesis, its more detailed discussion and the difficulties connected with it (particularly with regard to the forces which are supposed to cause the drifting) will be relegated to Sec. 6.3.

[1] WEGENER, A.: Die Entstehung der Kontinente und Ozeane. 3. Aufl. Braunschweig: F. Vieweg & Sohn 1922.

5.23. Continental Spreading. A modification of the continental drift theory has been proposed by GUTENBERG[1] who assumed that there is essentially only *one* continuous continental block which is spreading apart. Instead of having the continents *break* apart they are therefore supposed to have *flowed* apart (hence the German name "Fliesstheorie" for the continental spreading theory). The forces pulling the original continent apart so as to make it spread would be the same as those active in the drift theories. The basic difference in the theory of drift and that of spreading is therefore concerned with the rheology of the material in question. In the drift theory one has rigid masses undergoing brittle fracture, whereas in the spreading theory one has plastic or ductile deformation in the continents.

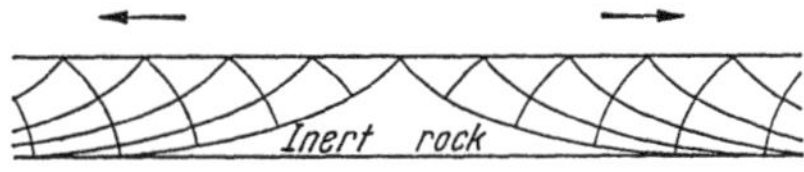

Fig. 76. Slip line field in a slab under gravity (after EVISON[2])

As with the continental drift theory, it is not clear what the forces are that are supposed to cause the spreading. An attempt at finding such forces has been made by EVISON[2] who investigated the effect of gravity on a yielding mass. In this fashion, he did not arrive exactly at a process for moving continents around, but he established a mechanism for the growth of continents. The slip line field in a slab stressed beyond its elastic limit is shown in Fig. 76; it is clear that such a type of slip could be advocated for an explanation of the growth of continents.

5.24. Volcanic Growth of Continents. A theory of the evolution of continents similar to that of continental spreading has recently been proposed by BUCHER[3] and WILSON[4]. Accordingly, the undisturbed state of the Earth was one where there was no crust, the surface of the Earth being marked by what is now the Mohorovičić discontinuity. WILSON assumes that at the present time there is on top of the Mohorovičić discontinuity a layer of basalt 5 km thick. Superimposed upon this there are the continental blocks, 30 km thick and consisting of andesitic material. WILSON assumes that this state has been reached by a process of continental growth. He estimates the present volume of the basaltic layer as equal to $2.5 \times 10^9 \text{km}^3$ and that of the andesitic layer as equal to $3.7 \times 10^9 \text{km}^3$. These volumes, according to WILSON, were formed by outpourings of lava during volcanic activity. In order to substantiate this possibility,

[1] GUTENBERG, B.: Gerlands Beitr. Geophys. **16**, 239 (1927); **18**, 281 (1927). — Bull. Geol. Soc. Amer. **47**, 1587 (1936).

[2] EVISON, F. F.: Geophys. J. Roy. Astr. Soc. **3**, 155 (1960).

[3] BUCHER, W. H.: Trans. Amer. Geophys. Un. **31**, 495 (1950).

[4] WILSON, J. T.: Nature, Lond. **179**, 228 (1957).

he uses an estimate of SAPPER'S[1] according to which the volume of lava poured out in recent times is 0.8 km^3/year. At that rate, the whole crust could have been formed in approximately 8×10^9 years. Since the age of the Earth is known to be not greater than about 4.5×10^9 years, of which probably only 3×10^9 years were available for geological processes to take place, it is necessary to assume that the volcanic activity was higher during the Earth's early history than it is at present. In view of the uncertainty regarding the thermal history of the Earth (cf. 2.63), this is perhaps not an entirely unreasonable assumption. However, SAPPER'S figure is highly at variance with VERHOOGEN'S discussion of volcanism as presented in Sec. 8.43, which is based upon geological recognition of lava-outpourings. According to VERHOOGEN, the total lava flow since the beginning of the Cambrian should not have exceeded 30×10^6 km^3 which yields an average outpouring of 0.05 km^3 per year (taking the beginning of the Cambrian as roughly 600×10^6 years ago). This raises the time required to form the crust by volcanic activity by a factor of 16.

Regarding the difference in chemical composition of the ocean bottoms and the continental areas, WILSON accounts for this by concluding that "basalts are formed by partial melting of the shallow layers (tens of km deep) and that the andesites are a differentiate from deep layers (hundreds of km deep)". At the edges of continents, the fractures are assumed as deep, hence andesitic materials are added to the growing continents; in mid-ocean the fractures are shallow, hence the mid-ocean ridges are basaltic. Processes of magma-differentiation of the postulated type have been considered in connection with the undation theory of orogenesis (cf. Sec. 6.62); they are, as far as can be judged at present, still somewhat speculative.

Thus, according to the volcanic growth theory, continents are expanding by the addition of andesitic material at their margins produced by volcanic activity originating in very deep layers. The nuclei of continents presumably were formed by the accidental creation of very deep fissures; once a continent has been started, the creation of further deep fissures is supposedly a self-supporting process. The basic acting force is assumed to be that of volcanism. The extrusion of lava from below the original surface of the Earth (i.e. the present Mohorovičić discontinuity) would of necessity cause the latter to contract and thereby to provide an orogenetic force equal to that assumed in any contraction theory. The orogenetic patterns incidental to the theory of volcanic growth of continents should therefore be equal to those postulated in the thermal contraction theory.

[1] SAPPER, K.: Vulkankunde. Stuttgart 1927; S. 269.

5.25. Meteorite Impact Hypothesis. A completely different hypothesis of the origin of oceans has recently been proposed by HARRISON[1] and by GILVARRY[2]. These authors assumed that one or more explosions, due to the impact of meteorites or even asteroids, took place. Naturally, it is extremely difficult to estimate the size of such a projectile as the mechanism of explosions of such a gigantic scale is only imperfectly understood. HARRISON estimates that for meteorites with radii of 100 km and 500 km penetrating some hundreds of km into the Earth, the crater radii would be 3000 and 7000 km, respectively. He assumes that a large crater rim was thrown up which, immediately after its formation, started to sink down because of the action of isostasy. Similarly, the crater bottom started to rise for the same reason to the level of the present abyssal plains.

Needless to say, the above hypotheses are highly speculative.

5.3. Primeval Convection

5.31. The Formation of Continents by Convection. The concept of Laurasia and Gondwanaland or Pangea as primeval continents outlined in Sec. 5.2, immediately poses the question as to how the latter may have originated. One of the possibilities that has been advocated to this end is the hypothesis of the formation of continents by convection[3].

In this hypothesis it is assumed that in the early days of the Earth's history, the latter was well-nigh liquid. The heavier material sank to the center to form the core, and what is now the mantle proceeded to cool (heat being lost into the universe) by thermal convection. One school of scientists assumes that these convection currents in the mantle of the Earth are still operative to the present day, thereby providing a force for orogenesis (cf. Sec. 6.4). An alternative opinion, however, is that such convection currents were possible only in the primeval days of the Earth. Whatever the solution to this question may be, it seems agreed that thermal convection would be a possible means of creating primeval continents.

The creation of a continent by convection can be envisaged to have occurred in one of two ways. First, if one has a rising current, then the material just above it will be brought to a higher elevation than the surrounding material owing to the effects of mechanical dragging. It thus would form a "continent". However, the "continent" would be able to subsist only for so long as the corresponding convection current is operative. In order to adopt this view, it must therefore be assumed

[1] HARRISON, E. R.: Nature, Lond. **188**, 1064 (1960).
[2] GILVARRY, J. J.: Nature, Lond. **190**, 1048 (1961).
[3] HILLS, G. F. S.: The Formation of Continents by Convection. London: E. Arnold & Co. 1947.

that rising convection currents are operative everywhere underneath continents up to the present day. A second way by which continents may be formed is by assuming the latter as much lighter than the liquid. The continents would thus correspond to "scum" (in isostatic equilibrium) on a liquid which accumulates over the *descending* branch of a convection current. After the convection stopped, the "scum" (i.e. the continents) would simply remain in its prior position or possibly get dragged around and broken up due to incidental causes (see Fig. 77).

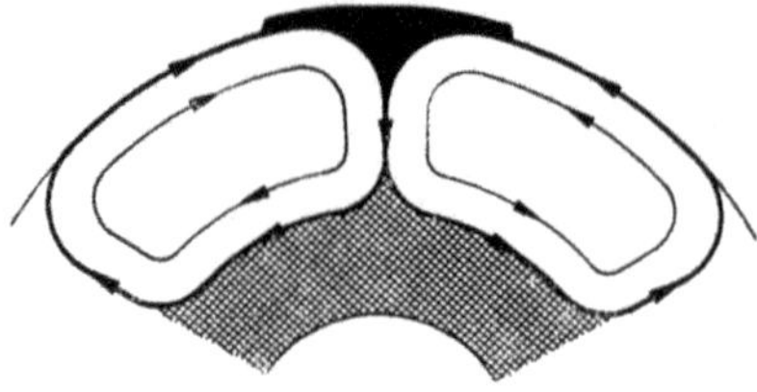

The hypothesis of the existence of Laurasia and Gondwanaland as primeval continents implies that the primeval convection currents would have had such a geometrical arrangement that two continents formed at the (then) poles of the Earth. If the scum-theory is adopted, this means that the currents must have been descending at the poles, otherwise they would have to be rising at the poles.

However, the primeval existence of Laurasia and Gondwanaland is not at all certain. A different system of primeval continents is arrived at from the observation that four old shields, at present, have a position roughly at the corners of a tetrahedron. If it is not conceded that continents may have moved around much during the history of the Earth, an explanation for the position of these continents may be sought in the assumption of an octahedral system of convection currents. This has been proposed by VENING MEINESZ[1]. Fig. 78 shows the system of currents. The creation of this system could be made plausible by the remark that a regular pattern is most likely to occur. The octahedron is the only regular surface in which an even number of sides touch in one corner, and this is a necessary condition in a convection current system.

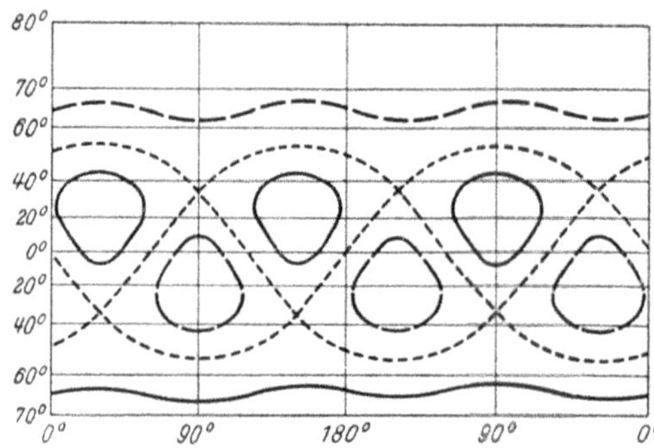

Fig. 78. Octahedral arrangement of convection currents (inward currents dotted, outward currents solid). After VENING MEINESZ[1]

[1] VENING MEINESZ, F. A.: Versl. K. Akad. Wet. **53**, No. 4 (1944).

It is thus seen that there is at least the *possibility* of explaining the existence of continents in terms of convection currents. We shall investigate below the physical aspects of such currents in somewhat greater detail.

5.32. Physical Aspects of Convection Currents. The investigations of JEFFREYS, LOW and others discussed in Sec. 3.33 show indeed that convection currents are possible in viscous media, provided certain fundamental requirements are fulfilled. Let us therefore investigate the

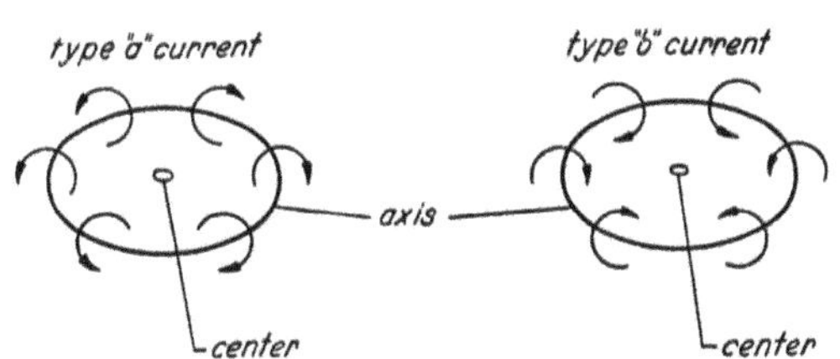

Fig. 79. Possible types of convection currents

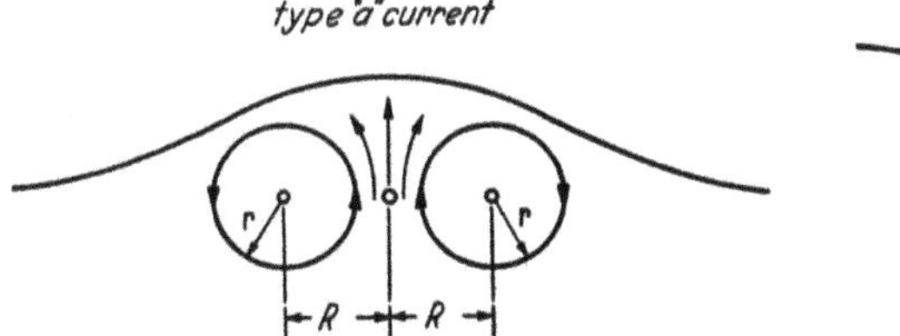

Fig. 80. Material accumulated on top of an outward, type *a* convenction current

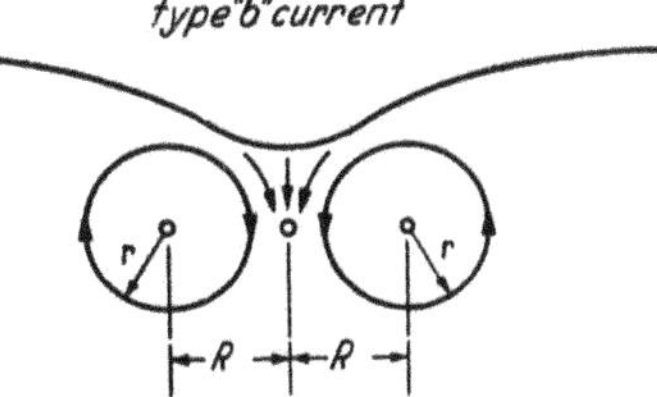

Fig. 81. Hole created by an inward, type *b*, convection current

implications of the hypothesis of convection currents somewhat further from a physical standpoint and see what they might do and what they might not do[1].

Since heat must be conveyed from the bottom to the top, the material must flow in that direction. Furthermore, owing to general hydrodynamic requirements, the axis of a convection current must be closed. Therefore, the patterns that are possible (type ***a*** und type ***b***) are easily shown to be of the general type of those shown in Fig. 79 where the axial rings lie approximately parallel with the surface of the Earth.

The effect on the surface of a type ***a*** current would be that material would accumulate over the current (see Fig. 80). In type ***b***, a hole would result (see Fig. 81). If the radius R of the convection current becomes large, then ring-shaped structures would result. If the accumulation of material were to be attributed to the presence of scum, then the effects of the currents would be exactly reversed from those shown in Fig. 80 and 81.

[1] SCHEIDEGGER, A. E.: Trans. Amer. Geophys. Un. **33**, 585 (1952).

We can calculate the rotating speed (t = time of revolution) for such a convection current which would be necessary to give a certain heat flow. For we note that the heat H transported by such a revolving torus is

$$H = 2\pi^2 r^2 R\, n\, \Delta T/t. \tag{5.32–1}$$

The volume of the torus is $2\pi^2 r^2 R$, the temperature difference between the two levels is ΔT, and the specific heat per unit volume is n. Thus the heat flow per unit area (denoted by h) is

$$h = \frac{H}{\pi(R+r)^2} = 2\pi \frac{r^2 R}{(R+r)^2}\, n\, \frac{\Delta T}{t}. \tag{5.32–2}$$

Hence

$$t = 2\pi r^2 \frac{R}{(R+r)^2}\, n\, \frac{\Delta t}{h}. \tag{5.32–3}$$

We must now make reasonable assumptions for the quantities occurring in the last equation. Thus we assume that r is of the order of 100 km, R of 1000 km, n of the order of 1, ΔT of the order of 1000 degrees and h of the order of 12×10^{-7} cal cm^{-2} sec^{-1} (i.e. the present-day value). Then we obtain

$$t \cong 1.5 \times 10^8 \text{ years} \tag{5.32–4}$$

and we see that the time of revolution will be of the order of several times 10^8 years. The speed of the material where it moves fastest is

$$2\pi r/t = 4 \text{ mm/year}. \tag{5.32–5}$$

The speed of the material is thus of the order of a few millimeters per year.

The stresses that occur due to viscous drag can also be estimated. Denoting the shear stress by τ, the coëfficient of viscosity by η, one has in approximation[1]

$$\tau = \eta\, r\, d\omega/dr. \tag{5.32–6}$$

ω being the angular velocity of the torus. We have[1] $\omega = A/r^2$. A is a constant of integration which has to be determined so that for $r = R$ we have $\omega = 2\pi/t$. Hence we find

$$d\omega/dr = 4\pi/(tR) \tag{5.32–7}$$

and

$$\tau = 4\pi\,\eta/t. \tag{5.32–8}$$

This equation connects the stresses with the coëfficient of viscosity. If the material is to behave as though it was rigid in certain instances,

[1] See Lamb, H.: Hydrodynamics, 6th ed. New York: Dover Publications 1945. p. 587.

then the limiting shear stress τ_0 (beyond which rupture will occur) must be rather low (see Sec. 3.65); but the viscosity rather high (at least 10^{22} poise; cf. Sec. 3.65). Our last equation indicates that these requirements are indeed compatible. Choosing

$$\tau_0 \cong 10^9 \text{ dynes/cm}^2 \tag{5.32–9}$$

yields

$$\eta < \tau_0 \times t/(4\pi) \sim 4 \times 10^{23} \text{ poise} \tag{5.32–10}$$

with the value found above for t.

Thus, from a standpoint of heat flow values, there is no difficulty in assuming slow convection currents, moving at a rate of about four millimeters per year.

5.33. Analytical Theory. The above discussion does not yet provide any thermo-mechanical explanation for the existence of convection currents. A much-quoted analytical attempt to elucidate the mechanics of these convection currents within the Earth has been made by PEKERIS[1]. For his calculation, PEKERIS chose a model of the Earth in which there is a temperature variation with depth as well as a zonal temperature variation over the surface of the Earth. The zonal temperature variation is a much more effective cause of convection currents than that presented by a temperature variation with depth. The mean temperature variation (due to exposure to solar radiation) from the equator to the poles is about 60° C and penetrates to great depth since it is independent of time. Another possible source of zonal temperature variations is that the bottom of the oceans is uniformly at a temperature of about 2° C and, in addition, that there is only a small crustal layer over the latter.

PEKERIS analyzed two particular models. He assumed an axis of rotational symmetry (not necessarily coincident with the axis of rotation of the Earth) and calculated the flow patterns as a function of colatitude Θ (with respect to that axis) and radial distance from the center. The assumed zonal temperature variation in the first model is from 100° to −100° if Θ varies from 0 to 180°, and in the second model from 100° to −50° to 100° if Θ varies from 0° through 90° to 180°. PEKERIS found rising convection currents underneath the temperature maxima (i.e. at $\Theta=0°$ in his first, and at $\Theta=0°$ and 180° in his second model) and descending currents underneath the temperature minima (i.e. $\Theta=180°$ in the first and $\Theta=90°$ in the second model). The result for the second model is shown in Fig. 82.

In detail, we give here PEKERIS' treatment of the second model. In polar co-ordinates (r, Θ) the equations of steady motion (neglecting

[1] PEKERIS, C. L.: Month. Not. Roy. Astron. Soc., Geophys. Suppl. **3**, 343 (1935).

quadratic terms in the velocities) are[1]

$$\frac{\partial p}{\partial r} = \frac{\eta}{3}\frac{\partial s}{\partial r} + \eta\left[\operatorname{lap} u - \frac{2u}{r^2} - \frac{2}{r^2}\frac{\partial v}{\partial \Theta} - \frac{2v\cot\Theta}{r^2}\right] - g\varrho, \qquad (5.33\text{–}1)$$

$$\frac{1}{r}\frac{\partial p}{\partial \Theta} = \frac{\eta}{3r}\frac{\partial s}{\partial \Theta} + \eta\left[\operatorname{lap} v + \frac{2}{r^2}\frac{\partial u}{\partial \Theta} - \frac{v}{r^2\sin^2\Theta}\right] \qquad (5.33\text{–}2)$$

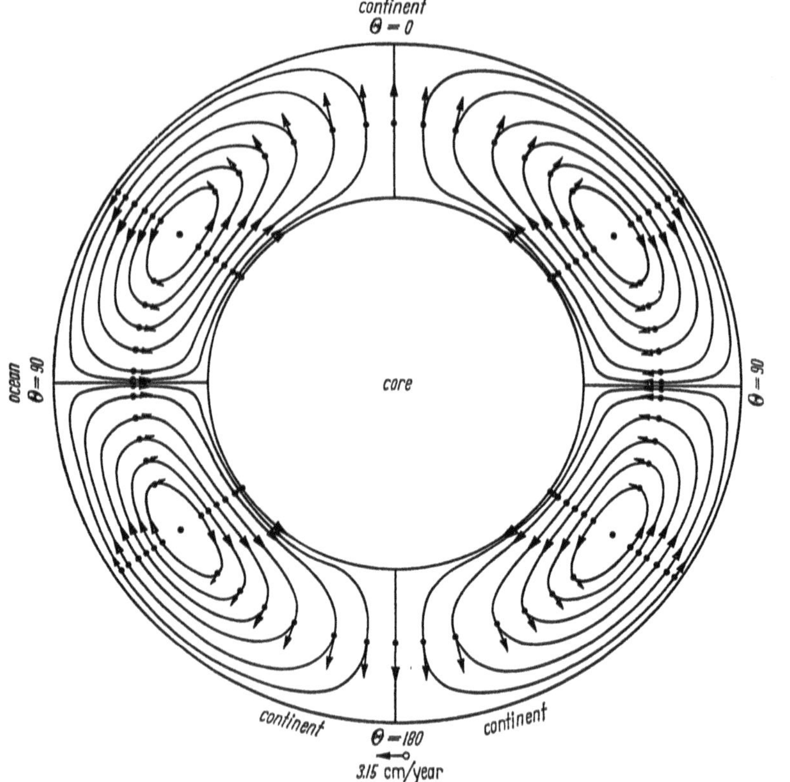

Fig. 82. Convection streamlines and flow velocities in one of PEKERIS'[2] models

where u, v are the radial and zonal velocities, respectively, s is the divergence of the velocity, p is the pressure, η the viscosity, ϱ the density and g the gravity acceleration. The density is assumed to vary as follows

$$\varrho = \varrho_0(1 - \alpha T) \qquad (5.33\text{–}3)$$

where α is the coëfficient of thermal expansion, T the temperature and ϱ_0 is a constant. The continuity condition yields

$$s = -\frac{1}{\varrho}\left(u\frac{\partial\varrho}{\partial r} + \frac{v}{r}\frac{\partial\varrho}{\partial\Theta}\right) = \alpha\left[u\frac{\partial T}{\partial r} + \frac{v}{r}\frac{\partial T}{\partial\Theta}\right]. \qquad (5.33\text{–}4)$$

[1] Cf. e.g. BASSET, A. B.: A Treatise of Hydrodynamics. Cambridge 1888.

[2] PEKERIS, C. L.: Month. Not. Roy. Astron. Soc., Geophys. Suppl. 3, 343 (1935).

Furthermore, the temperature obeys the equation

$$\varrho\, c\left[\frac{\partial T}{\partial t}+u\frac{\partial T}{\partial r}+\frac{v}{r}\frac{\partial T}{\partial \Theta}\right]=k\,\mathrm{lap}\,T+P \qquad (5.33\text{–}5)$$

where P denotes the rate of generation of heat owing to radioactivity.

Pekeris proceeds now in such a fashion that all quantities are developed into a series in α, where terms of an order higher than the first are eventually to be neglected. Thus

$$u=\alpha u_1+\alpha^2 u_2+\cdots, \qquad (5.33\text{–}6a)$$

$$v=\alpha v_1+\alpha^2 v_2+\cdots, \qquad (5.33\text{–}6b)$$

$$s=\alpha s_1+\alpha^2 s_2+\cdots, \qquad (5.33\text{–}7)$$

$$p=-\int^{r} g\,\varrho_0(1-\alpha\,\tau_0)\,dr+\alpha p_1+\alpha^2 p_2+\cdots. \qquad (5.33\text{–}8)$$

Furthermore, the solution is to be expressed in a series of spherical harmonics:

$$\left.\begin{aligned}
u_1&=\sum_1^\infty \Phi_n(r)\,P_n(\cos\Theta),\\
v_1&=\sum_1^\infty \Psi_n(r)\,\frac{\partial}{\partial\Theta}P_n(\cos\Theta),\\
s_1&=\sum_1^\infty \sigma_n(r)\,P_n(\cos\Theta),\\
p_1&=\sum_1^\infty \pi_n(r)\,P_n(\cos\Theta).
\end{aligned}\right\} \qquad (5.33\text{–}9)$$

In virtue of the identity

$$\frac{1}{\sin\Theta}\frac{\partial}{\partial\Theta}\left(\sin\Theta\frac{\partial P_n}{\partial\Theta}\right)=-n(n+1)P_n \qquad (5.33\text{–}10)$$

the continuity condition yields

$$\sigma_n=\frac{1}{r^2}\frac{d}{dr}(r^2\Phi_n)-n(n+1)\frac{\Psi_n}{r} \qquad (5.33\text{–}11)$$

and

$$\Phi_n=\frac{2}{r}\Phi_n-\frac{1}{r}n(n+1)\Psi_n=0. \qquad (5.33\text{–}12)$$

Substituting this into the second of the equations of motion (5.33–2) and using the identity

$$\mathrm{lap}\,v_1-\frac{v_1}{r^2\sin^2\Theta}=\sum_1^\infty\frac{\partial P_n}{\partial\Theta}\left[\ddot{\Psi}_n+\frac{2}{r}\dot{\Psi}_n-\frac{1}{r^2}n(n+1)\Psi_n\right], \qquad (5.33\text{–}13)$$

this yields

$$\pi_n=\eta\left[\frac{1}{3}\dot{\Phi}_n+\frac{8}{3r}\Phi_n+r\ddot{\Psi}_n+2\dot{\Psi}_n-\frac{4}{3r}n(n+1)\Psi_n\right]. \qquad (5.33\text{–}14)$$

Using the assumption that $\eta=$ const., and setting $\nu=\eta/\varrho_0$ one finally ends up with the following differential equation:

$$\left.\begin{aligned}\frac{1}{n(n+1)}\left(r^2\overset{\cdots\cdot}{\Phi}_n+8r\dddot{\Phi}_n+12\ddot{\Phi}_n\right)-2\ddot{\Phi}_n-\frac{4}{r}\Phi_n\\ -\frac{2}{r^2}\Phi_n+\frac{1}{r^2}n(n+1)\Phi_n=\frac{g}{\nu}\tau_n.\end{aligned}\right\}\quad(5.33\text{–}15)$$

This, together with (5.33–12) enables one to determine the Φ and Ψ, i.e. the velocities, and hence to draw the stream lines. The solution depends on the initially assumed temperature perturbation. In accordance with the model under discussion, it is assumed that $\tau_1=0$, $\tau_2\neq 0$; thus one has from (5.33–15):

$$r^2\overset{\cdots\cdot}{\Phi}_2+8r\dddot{\Phi}_2-\frac{24}{r}\dot{\Phi}_2+\frac{24}{r^2}\Phi_2=\frac{6g}{\nu}\tau_2. \quad (5.33\text{–}16)$$

In particular, assuming

$$\tau_2=133.3\frac{b^3}{r^3}\frac{r^5-a^5}{b^5-a^5} \quad (5.33\text{–}17)$$

where b is the outer and a the inner radius of the convective shell, PEKERIS obtained

$$\Phi_2=\frac{c_{-4}}{r^4}+\frac{c_{-2}}{r^2}+c_1r+c_3r^3+E\left[\frac{r^4}{24}-\frac{a^5}{4r}\right], \quad (5.33\text{–}18)$$

$$\Psi_2=\frac{1}{6}r\dot{\Phi}_2+\frac{1}{3}\Phi_2=-\frac{c_{-4}}{3r^4}+\frac{c_1r}{2}+\frac{5c_3r^3}{6}+E\left(\frac{r^4}{24}-\frac{a^5}{24r}\right) \quad (5.33\text{–}19)$$

with

$$E=133.3\frac{b^3g}{\nu}\frac{1}{b^5-a^5} \quad (5.33\text{–}20)$$

and

$$\left.\begin{aligned}c_{-4}&=93.588E,\\ c_{-2}&=168.409E,\\ c_1&=9.89664E,\\ c_3&=-0.449201E.\end{aligned}\right\}\quad(5.33\text{–}21)$$

The stream lines corresponding to this are those that have been shown in Fig. 82.

Similar calculations as those just mentioned have also been made by CHANDRASEKHAR[1], by UREY[2] and by LATYNINA[3] with corresponding results.

According to PEKERIS, a rising current creates a continent, a descending one an ocean. The system of currents, thus, would be self-

[1] CHANDRASEKHAR, S.: Phil. Mag. **43**, 1317 (1952).
[2] UREY, H. C.: Phil. Mag. **44**, 227 (1953).
[3] LATYNINA, L. A.: Izv. Akad. Nauk SSSR., Ser. Geofiz. **1958**, 1085 (1958).

perpetuating if continents would automatically always stay hotter than oceans. This is, however, an unsettled question[1].

One must therefore seek a mechanism that might keep the continents warmer than the oceans. It has been claimed that this might have been achieved by solar radiation. It is true that in moderate zones, the surface temperature of land is higher than that of the sea, but in the Arctic the reverse is true. If differences in solar irradiation would have caused convection currents, it must surely be assumed that the primary zonal differences that started the whole process going, must have been caused by the fact that less heat was received by the (then) polar regions than by the equatorial regions. This is the exact opposite to the picture envisaged by PEKERIS in his second model. One would therefore expect a ring-shaped continent to have formed around the equator; this continent then would also be at a higher temperature than the surrounding ocean which would, as outlined above, help to perpetuate the original system of convection currents. Such a result, however, would not at all agree with the phenomenologically suggested pattern of the arrangement of original continents at the poles, and it seems, therefore, that little is gained by making any elaborate calculations.

The situation might be saved by taking recourse to the scum-theory of continent formation according to which continents are ultimately formed over descending branches of convection currents. This is contrary to PEKERIS' picture and one would therefore have to exchange "continents" and "oceans" in Fig. 82. Under these conditions, the system of currents would be self-perpetuating if the "scum" (the future continents) would always stay *cooler* than the exposed naked substratum. Since this condition must have been maintained in primeval days long before the emergence of continents (due to isostasy) and long before the condensation of primeval steam into water, its existence is again pure speculation.

5.34. Possibility of Present-Day Convection. We have shown above (Sec. 5.32) that the average heat flow values observed to-day can be reconciled with the existence of present-day convection currents. The actual thermodynamics of such currents, if they exist at the present time, is, however, not yet quite clear. In this instance, all that was said in Sec. 5.33 for primeval currents, also applies to modern currents, except that one has less latitude for adjusting the required numerical parameters (see also Sec. 6.4).

Nevertheless, because of the possibility of the existence of such currents, one might want to speculate regarding their present-day effects. Most of these effects will bear upon the theories of orogenesis, and a

[1] See also e.g. the critisism by LYUSTIKH, E. N.: Izv. Akad. Nauk SSSR. 1957, 604 (1957).

detailed discussion will therefore be relegated to Chap. VI. However, there might be some continental effects that should be properly dealt with here.

Thus, an important investigation by RIKITAKE and HORAI[1] deals with a possible explanation of the heat flow anomalies observed over mid-ocean ridges (see Sec. 2.61) in terms of present-day continental convection currents. The result of this investigation is that the larger the diameter of a current, the greater should be the heat flow through the crust above the ascending branch. Thus, the Pacific being a large feature presumably associated with a large convection current, its heat flow anomalies should be high.

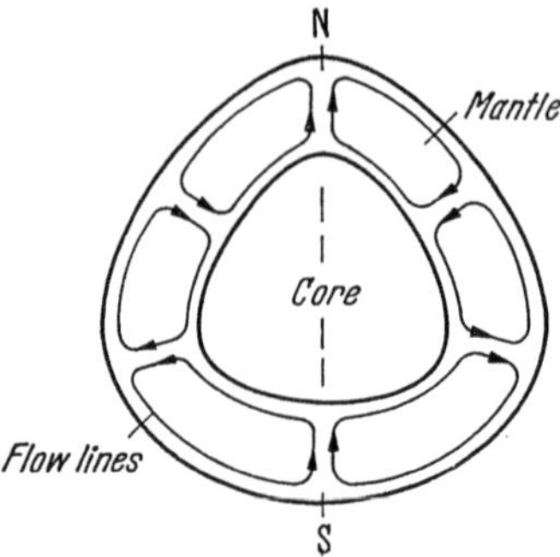

Fig. 83. A global system of convection currents (the NS axis is an axis of symmetry) that could cause the Earth to be pear-shaped. After LICHT[2]

Another possible effect of present day convection currents, if they exist, has been pointed out by LICHT[2]. We have noted in Sec. 4.11 that there are indications that the Earth might be slightly pear-shaped. Such a pear-shape could easily be associated with a global system of convection currents as shown in Fig. 83. LICHT calculated the perturbation of the gravitational field due to such a system (note that the density of the material in the ascending branches of the currents is less than in the descending branches) and came up with some values that are not inconsistent with those obtained from satellite observations.

5.4. Tetrahedral Shrinkage

5.41. Principles. Another theory to account for the morphological facts about continents and oceans which, at the same time, aims at an explanation of the tetrahedral arrangement of the former, is a theory assuming a particular type of shrinkage. It is based upon the assumption that a tendency exists for a contracting sphere to shrink tetrahedrally, simply because the tetrahedron has minimum volume for a given surface of all *regular* bodies. The case for this theory has recently been re-stated by WOOLNOUGH[3].

Thus, if it be assumed that the Earth had at one time cooled enough so that the outermost layer had become a solid skin incapable of changing its area, then a tetrahedral shape might be considered as the logical outcome of such a process. The corners of the tetrahedron would

[1] RIKITAKE, T., and K. HORAI: Bull. Earthquake Res. Inst. **38**, 403 (1960).
[2] LICHT, A. L.: J. Geophys. Res. **65**, 349 (1960).
[3] WOOLNOUGH, W. G.: Bull. Amer. Ass. Petrol. Geol. **30**, 1981 (1946). The present discussion is after the author's Handbuch-article.

correspond to the continents, the faces to the ocean basins. A proper arrangement concerning the size of the tetrahedron would also explain the ratio 1:2 occupied by continents and by oceans.

An idea very similar to that presented above has been suggested long ago by DAVISON[1]. Accordingly, the Earth is contracting in its upper layers only (due to cooling) which are therefore in a state of internal tension. Owing to the pressure of the continents, DAVISON assumes that the amount of stretching under them must have been very much less than under the great oceanic areas. This would tend to make the ocean basins subside even further and present a physical cause for their permanence. Any orogenetic effects would be most pronounced at the junction of the oceans with the continents, thereby leading to the idea of continental growth.

5.42. Criticism. The chief criticism of the tetrahedral shrinkage theory is[2] that the topmost "skin" of the Earth simply does not have such properties which would prevent it from changing its area under the action of tangential forces. It is thus quite inconceivable that it would retain its area upon a shrinking interior; at the very least it would either thicken in spots or else become folded over in the manner of nappes. The evidence of folding seems to show that adjustment of an outer shell to a collapsing interior would take place continually or in a rapid sequence of diastrophisms rather than in a slow settling to the form of a tetrahedron. Furthermore, the theories of deformation of such an outer shell seem to indicate that buckling would be the mechanism determining the adjustment of a *rigid* shell to a collapsing interior. It has been shown[3,4] that the deformation of a buckling sphere is symmetrical about a diameter and that the deviations of the shape are given by a series of spherical harmonics along parallels of latitude associated with the diameter of symmetry. This obviates the postulate of tetrahedral shrinkage.

5.5. Formation of Continents by Expansion

5.51. General Principles. Several theories have been proposed in which it has been assumed that the Earth is subject to *expansion*. The idea that there is some expansion of the Earth is not new[5]; it has

[1] DAVISON, C.: Phil. Trans. Roy. Soc. Lond. A **178**, 240 (1888).

[2] SCHEIDEGGER, A. E.: Handbuch der Physik, vol. **47**, p. 283. Berlin-Göttingen-Heidelberg: Springer 1956.

[3] ZOELLY, R.: Über ein Knickungsproblem an der Kugelschale. Diss. E. T. H. Zürich, 1915.

[4] LEUTERT, W.: Die erste und zweite Randwertaufgabe der linearen Elastizitätstheorie für die Kugelschale. Diss. E. T. H. Zürich, 1948.

[5] HALM, J. K. E.: J. Astron. Soc. S. Afr. **4**, 1 (1935). — HILGENBERG, O. C.: Vom wachsenden Erdball. Berlin, 1933.

currently been revived because of the recognition that many tensional features, such as the mid-oceanic rifts, are present on the Earth's surface[1-6]. As far as these latter features are concerned, the expansion hypothesis is a hypothesis of orogenesis and, in this context, will be treated in the next chapter.

Expansion, however, has also been advocated as a cause of the origin of continents[7, 8]. Thus, it has been assumed that the Earth was much smaller at the beginning that it is now, having a diameter of about one-half of its present one. Somehow, a "crust" was formed on it, which

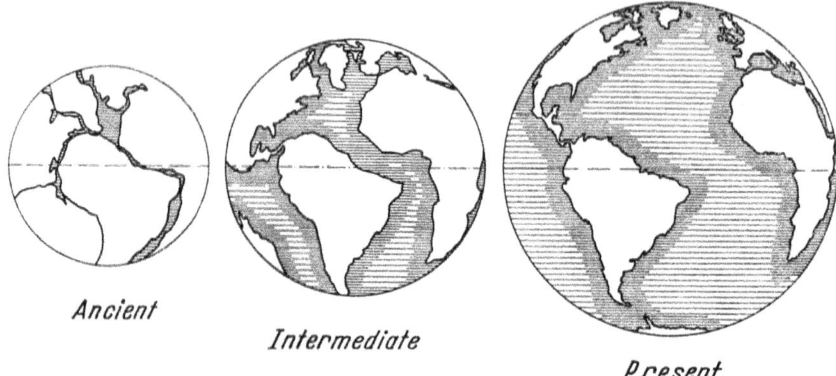

Fig. 84. How the Earth's oceans could have developed by global expansion (after HILGENBERG[7])

was everywhere some 30 km thick. Then, as the diameter grew, the original crust brocke up and its remnants are the present continents (see Fig. 84). Expansion was supposed to have started by ocean "cracks" like the Mid-Atlantic Rift. An increase by a factor two in diameter represents a surface increase by a factor four, which produces about the right order of relative area occupied by present-day continents. However, such a radius increase also produces a volume increase, and a corresponding density decrease, by a factor eight. Since the present-day average density of the Earth is about $5^1/_2$ g/cm^3, the average density before the start of the expansion must have been about 44 g/cm^3. It is very difficult to explain this. We shall investigate some of the possible causes below.

---

[1] HEEZEN, B. C.: Preprints Int. Oceanogr. Cong. 26 (1959).
[2] EGYED, L.: Geofis. Pura Appl. **33**, 42 (1956).
[3] CAREY, S. W.: J. Alberta Soc. Petrol. Geol. **10**, 95 (1962).
[4] CAREY, S. W., and J. A. O'KEEFE: Science **130**, 978 (1959).
[5] GROEBER, P.: Bol. Inf. Petrolif. No. **311**, 101; No. **312**, 181 (1959).
[6] WILSON, J. T.: Nature, Lond. **185**, 880 (1960).
[7] HILGENBERG, O. C.: Vom wachsenden Erdball. Berlin 1933.
[8] EGYED, L.: Geofis. Pura Appl. **45**, 115 (1960).

5.52. Thermal Theories. The most obvious effect that could cause expansion is a thermal one. One could imagine that, if the Earth started out as a hot liquid sphere, it might form a solid layer on the top like ice on water. This layer might in a way serve as an insulator so that immediately after its formation the interior would heat up again since the heat created by radioactivity could now no longer escape. Although the above idea seems appealing at first glance, it is in fact quite untenable because (unlike ice and water) most rocks are more dense in the solid state than in the molten state. If it is to stay on top, the solid crust cannot therefore simply be the solid phase of the substratum but must be composed of a different substance. If this be assumed, however, then the thermal history for various Earth-models can be calculated, as has been reported in Sec. 2.63. Accordingly, it is possible that the whole Earth was remelted early in its history but it does not seem reasonable that it would have melted later on. Looking at the values given earlier in Table 15, one must admit, however, that the possibility exists that there is a net gain in heat and hence a slight expansion of the Earth as a whole during its history. It does not seem possible, though, that this could effect more than the formation of e.g. deep ocean trenches,—although any exact calculations do admittedly not exist.

A similar idea has been proposed by MATSCHINSKI[1] who assumed that during solidification of the crust (consisting of lighter material than the substratum) the latter would decrease its volume from V in the liquid state to $(1-\gamma)\,V$ in the solid state. This would create great tensions and hence produce the same effects as an expansion underneath. MATSCHINSKI calculated these tensions by considering the instant where the crust from R_1 upwards (R_1 measured from the center of the Earth) to its surface at distance R from the center is solidified, the material below being liquid. If the cavity of radius R_1 were empty, the crust would contract in accordance with the value of γ. However, since the cavity is filled with an almost incompressible liquid, it can sustain a very large pressure p_1 without contracting much. The contraction u_1 (at the radius R_1, say) is given by

$$u_1 = \frac{p_1 R_1}{3K} \tag{5.52-1}$$

if the bulk-modulus is denoted by K. Then, in a spherically symmetrical Earth, the equations of elasticity require[2]

$$\frac{\partial}{\partial r}(r^2 \tau_{rr}) = 2r\,\tau_{\lambda\lambda} \tag{5.52-2}$$

and

$$\tau_{rr} = 2\mu\left(\frac{\partial u}{\partial r} + \frac{1}{2}\Theta\right), \tag{5.52-3a}$$

$$\tau_{\lambda\lambda} = \tau_{\vartheta\vartheta} = 2\mu\left(\frac{u}{r} + \frac{1}{2}\Theta\right) \tag{5.52-3b}$$

[1] MATSCHINSKI, M.: Ann. Geofis., Roma 7, 1 (1954).

[2] Setting POISSON's ratio $m = 0.25$ which is a good value for rocks.

where, as usual, ϑ, λ denote spherical co-ordinates (co-latitude and longitude, respectively), Θ is given by

$$\Theta = \frac{1}{r^2} \frac{\partial}{\partial r} (r^2 u) \tag{5.52-4}$$

and u denotes the radial displacement. Hence one has

$$r^2 u'' + 2ru' = 2u \tag{5.52-5}$$

and

$$u = C_1 r + C_2/r^2. \tag{5.52-6}$$

The constants of integration C_1 and C_2 have to be determined from the boundary conditions. The displacement at the bottom of the crust must coïncide with the displacement at the top of the liquid; hence we have

$$\tfrac{1}{3} R_1 \gamma + u_1 \big|_{r=R_1} = u \big|_{r=R_1}. \tag{5.52-7}$$

The pressures must also be equal at that interface, thus:

$$p_1 \big|_{r=R_1} = \tau_{rr} \big|_{r=R_1}. \tag{5.52-8}$$

Hence the boundary conditions can be formulated as follows

$$\frac{1}{3} R_1 \gamma + \frac{R_1}{3K} \tau_{rr} \bigg|_{r=R_1} = u \bigg|_{r=R_1}, \tag{5.52-9}$$

$$\tau_{rr} \big|_{r=R_1} = 0. \tag{5.52-10}$$

MATSCHINSKI introduces the abbreviation

$$\varepsilon = \frac{\mu}{3K} \tag{5.52-11}$$

and hence one obtains from Eq. (5.51-10)

$$5C_1 - \frac{4C_2}{R^3} = 0; \quad \frac{4}{5} C_2 = R^3 C_1, \tag{5.52-12}$$

In virtue of Eq. (5.51-9), this yields

$$C_1 = \frac{4 R_1^3 \gamma}{3\{4R_1^3 + 5R^3 + 20\varepsilon(R^3 - R_1^3)\}}, \tag{5.52-13}$$

$$C_2 = \frac{5 R^3 R_1^3 \gamma}{3\{4R_1^3 + 5R^3 + 20\varepsilon(R^3 - R_1^3)\}}. \tag{5.52-14}$$

Finally, MATSCHINSKI obtained for the stresses

$$\tau_{rr} = \frac{\mu\gamma}{3} \frac{20 R_1^3}{4R_1^3 + 5R^3 + 20\varepsilon(R^3 - R_1^3)} \left\{1 - \frac{R^3}{r^3}\right\}, \tag{5.52-15}$$

$$\tau_{\lambda\lambda} = \tau_{\vartheta\vartheta} = \frac{\mu\gamma}{3} \frac{10 R_1^3}{4R_1^3 + 5R^3 + 20\varepsilon(R^3 - R_1^3)} \left\{2 + \frac{R^3}{r^3}\right\}. \tag{5.52-16}$$

This shows that τ_{rr} is a compressive stress, whereas $\tau_{\lambda\lambda}$ and $\tau_{\vartheta\vartheta}$ are tensions;—which is what was to be proven.

Although it is now ascertained that the solidification of the crust puts the latter into a state of tension, MATSCHINSKI's model is, in fact, somewhat artificial: it has to be implied that the whole crust solidifies at once, otherwise the solidification would take place from the bottom up (i.e. the solidified pieces would sink to the bottom as they are assumed as denser than the melt) and the model as envisaged would be impossible. This does not seem to be quite reasonable; if it has occurred at all, it must have occurred very early in the Earth's history. Under these circumstances, as is the case with all expansion theories, it could be thought that the tensions created would cause fractures which might be comparable to ocean trenches, but never a disruption of the crust into the present continents. In order to create the latter, the volume contraction upon solidification would have to be on an excessive scale indeed.

5.53. Chemical Theories. Calculations have also been made to investigate whether a chemical change or a phase change in the interior of the Earth could be held responsible for a large-scale expansion of the globe. Such calculations are simply based on a comparison of the energy required to cause the expansion with the energy available in chemical bonds. The energy required for expansion is simply the difference in gravitational potential energy in the small and in the big Earth.

The gravitational energy difference between a small Earth and a big one depends somewhat on the density distribution in the interior. BECK[1] found that for any reasonable density distribution, an expansion of the radius of about 100 km is possible, but for expansions of 1000 km or more this is not so. Similarly, COOK and EARDLEY[2] estimated that a uniform expansion of the radius of the Earth by 20% would require an amount of energy equal to that required to dissociate almost all the chemical bonds of the molecules constituting the Earth.

It appears therefore, that no chemical or similar sources could supply enough energy to cause an expansion of the Earth that would be required to create continents by this mechanism.

5.54. Cosmological Speculations. Expansion of the Earth has been attributed not only to thermal causes, but also to a slow change of the value of the gravitational "constant" postulated in some cosmological speculations. JORDAN[3], in a discussion of projective relativity theory, came up with the conclusion that the quantity $\varkappa$ in NEWTON's law of gravitational attraction

$$F=\varkappa\frac{m_1 m_2}{r^2} \tag{5.54-1}$$

[1] BECK, A. E.: J. Geophys. Res. **66**, 1485 (1961).

[2] COOK, M. A., and A. J. EARDLEY: J. Geophys. Res. **66**, 3907 (1961).

[3] JORDAN, P.: Schwerkraft und Weltall. Braunschweig 1952. — Naturwiss. **48**, 417 (1961).

(where F is the force; m_1 and m_2 are the two masses involved; r is the distance between them) should not be a fundamental constant as commonly assumed, but in fact should be a variable which has been slowly decreasing since the beginning of the universe. Similar postulates have also been made by DICKE[1, 2].

Accepting the above conclusion of JORDAN, JOKSCH[3] tried to account for the peculiar statistical composition of the hypsometric curve of the Earth explained in Sec. 1.32. Accordingly, soon after the Earth was created, it consolidated into various layers. In virtue of the decrease of $\varkappa$, the Earth then expanded, the uppermost layer being the first to be torn up owing to the tensions created by the expansion. The tearing-up would lead to a logarithmico-normal distribution of heights within that layer. The process, then, was repeated with a second and with a third layer. Each time a layer was torn, a logarithmico-normal distribution of heights was the result. By assuming three layers, JOKSCH accounted for the tripartite composition of the hypsometric curve.

BECK[4] has also estimated the energy available from a decrease of the gravitational constant and again came to the conclusion that not more than an increase of 100 km in the Earth's radius can be accounted for in this fashion. Thus, if expansion on the postulated scale occurred at all, a complety unknown energy source must be found.

5.6. Evaluation of Theories of Continents and Oceans

Looking over the various theories of the formation of continents and oceans, it is apparent that one has a series of contradicting opinions. There is, however, only one theory that is at all widely accepted: that of the formation of continents by convection. This does not necessarily contradict the possibility of subsequent continental growth or continental drift: it is very well possible that, during the Earth's early history, convection was a significant phenomenon, but that it died down later, leaving wide room for other effects to occur.

The other theories of the origin of continents seem to be somewhat more artificial. However, they certainly cannot be ruled out entirely. Much depends on the actual state of the Earth in its early history, i.e. on whether it had a cold or a hot beginning. As long as the answers to such fundamental questions are as uncertain as they are at present, a definite explanation of the origin of continents and oceans cannot be hoped for.

---

[1] DICKE, R. H.: Rev. Mod. Phys. **29**, 355 (1957).
[2] DICKE, R. H.: J. Wash. Acad. Sci. **48**, 213 (1958).
[3] JOKSCH, H. C.: Z. Geophys. **21**, 109 (1955).
[4] BECK, A. E.: J. Geophys. Res. **66**, 1485 (1961).

VI. Orogenesis

6.1. Fundamentals

6.11. General Remarks. The central aim of the science of geodynamics is to elucidate the mechanism of mountain building, called *orogenesis*. The object of a "theory of orogenesis" is to explain the physiographic and geophysical features of the Earth summarized in Chap. I and II of this book.

There are many theories of orogenesis in existence. In the present Sec. 6.1, which is preliminary to a discussion of these various theories, we shall analyze some of the fundamental concepts that are again and again referred to in the various geotectonic hypotheses.

6.12. The Volumes Involved in Orogenesis. Let us consider first the volumes involved in orogenesis. The *continental* orogenetic activity is at any one time concentrated in narrow belts that form a world-wide pattern which nearly follow two great circles (cf. 1.43). Thus, let us assume that in a single orogenetic cycle two-thirds of two great circles about the Earth are folded into mountains 2 km high and 300 km wide. The length L of a complete orogenetic system is thus

$$L=5.3\times10^4\ \text{km}. \qquad (6.12\text{–}1)$$

The volume V of an orogenetic system is thus

$$V=32\times10^6\ \text{km}^3. \qquad (6.12\text{–}2)$$

Second, we turn to a discussion of the possible origin of the various structural elements of the *ocean bottom*. The most prominent features, as we have mentioned earlier (Sec. 1.53), are the mid-ocean ridges. The volume V of the presently known system of ridges can be estimated as follows.

We take the height H of the ridges as being 3 km (above the abyssal plains):

$$H=3\ \text{km}. \qquad (6.12\text{–}3)$$

The width, d, has been measured as being on the average some 1600 km. If the cross-section A be assumed to be triangular, one obtains:

$$A=\tfrac{1}{2}\times3\times1600\ \text{km}^2=2400\ \text{km}^2. \qquad (6.12\text{–}4)$$

The length L of the system is approximately equal to the circumference of the Earth; hence

$$L = 40000 \text{ km}. \tag{6.12–5}$$

This yields for the volume V

$$V = 96 \times 10^6 \text{ km}^3. \tag{6.12–6}$$

If this be compared with the standard value of V for a continental orogenetic system (cf. 6.12–2), one observes that the oceanic ridge system is roughly 3 times larger than the former.

6.13. The Hypothesis of Crustal Shortening. One of the assumptions which is often at the basis of a theory of orogenesis, is that mountain building is due to crustal shortening. There is no doubt that at least an apparent crustal shortening occurred in *some* places. We have stated (Sec. 1.42) that for most continental mountain ranges, geological estimates of shortening are of the order of

$$s_A = 50 \text{ km}. \tag{6.13–1}$$

We denote this value of shortening by the subscript "A" to indicate that this is the geologically "apparent" shortening. The Alps are an exception; the observed values of crustal shortening are up to 320 km. It is difficult in any theory of orogenesis to arrive at such large values.

For marine mountain ranges (mid-ocean ridges), no values for crustal shortening have ever been postulated. In fact, it is much more likely that mid-ocean ridges are not connected with crustal shortening at all, although this is not yet entirely certain.

The crustal shortening cannot be entirely independent from the volumes in orogenesis. In fact, there must be a connection with the apparent shortening s_A across the orogenetic system. This connection is a most basic relationship in geodynamics.

By the term "apparent" it is already implied that there also should be a "true" shortening s_T. The apparent shortening is obtained by assuming that in a normal cross-section of a mountain range the length of a stratum (which is a curved line) is equal to the length of that section before it was folded, i.e. when it was flat on the ground, and comparing it with the width of the mountain range. The difference is the "apparent shortening" s_A. It is, however, not a foregone conclusion that the strata did not undergo an extension of their length during folding. The "true" shortening may therefore have been less than the "apparent" shortening. Let us assume that the extension of length was by the "extension"-factor γ, then we have

$$s_A = \gamma s_T. \tag{6.13–2}$$

Furthermore, during an orogenetic diastrophism, the surface only of the Earth is affected. Let us denote the (hypothetical) depth to which the shortening is felt by h. Then, if the total length of the orogenetic system is again denoted by L, the volume that appears as mountains is given by

$$s_T L h = V. \qquad (6.13\text{-}3)$$

Replacing the hypothetical true shortening by the measurable apparent shortening, and putting all the hypothetical quantities on one side of the equation, we obtain:

$$\frac{h}{\gamma} = \frac{V}{L s_A}. \qquad (6.13\text{-}4)$$

This is a basic relationship which every theory of orogenesis must fulfill. Such theories yield values for the hypothetical constants; the fact that these are not independent, has usually been overlooked.

An interesting outcome is observed if the numerical values obtained earlier are inserted into the basic relationship (6.13-4). One then obtains:

$$h/\gamma \cong 12\ \text{km} \qquad (6.13\text{-}5)$$

which is of the order of the thickness of the crust (as defined by the Mohorovičić discontinuity); in fact it is only a little less than the weighted mean thickness of an oceanic (5 km) and continental (35 km) (of frequency 2:1) crust (which would yield about 15 km). Thus, if it is assumed that γ is of the order of 1 (no significant extension of the strata), one can explain the geologically observed shortening and the volume of mountains by postulating that the apparent shortening approximately equals the true shortening and that the depth to which orogenesis is felt is determined by the Mohorovičić discontinuity. This leaves one with the difficulty of finding forces that can produce the required large shifts.

On the other hand, if γ is assumed to be significantly larger than 1 (of the order of up to 10), then it is easy to find possible forces to produce the required small shortening, but the depth to which orogenesis is felt becomes much larger and the explanation of large extension factors γ itself becomes problematic.

It is thus seen that the value of the extension factor γ is very characteristic for any theory of orogenesis and, in fact, enables one to make a classification of the latter. However, in the following survey we shall follow the historical classification rather than that indicated by various values of γ.

The above argument assumes that there is no density reduction in the material affected by orogenesis. If there is such a density reduction, possibly due to rock-metamorphism, say by the *metamorphosis factor* ζ, the basic equation reads[1]

$$\frac{h\zeta}{\gamma} = \frac{V}{L s_A}. \tag{6.13-6}$$

A further interesting remark can be made with regard to the *maximum speed* with which crustal shortening can take place. If crustal shortening is assumed to be due to the sliding of the crustal parts in question over the substratum, the work necessary to produce the motion is expended against the frictional resistance occurring at the sliding surface. The resistance W to the edgewise motion (with velocity v) of a circular disc of radius c in a viscous liquid (of viscosity η) has been calculated by LAMB[2]; it is given by the following expression

$$W = 6\pi \eta R v \tag{6.13-7}$$

with

$$R = \frac{16c}{9\pi} = 0.566c. \tag{6.13-8}$$

A *floating* disc experiences only half of this resistance, hence

$$W = 3\pi \eta R v = \frac{16}{3} c \eta v. \tag{6.13-9}$$

If the crustal parts are *sliding* over the substratum, a force as given by the last equation must act on these parts. This introduces stresses τ in the latter whose order of magnitude is

$$\tau = W/(2cH) \tag{6.13-10}$$

where H is the thickness of the crustal part in question. The stresses τ obviously cannot exceed the yield stress ϑ of the surface material:

$$\tau \leqq \vartheta \tag{6.13-11}$$

which, in turn, imposes a limit on the speed v with which the crustal shortening can proceed. The movements considered here belong into the "long" time range in the sense of Sec. 3.6; using the corresponding values for η, ϑ etc., one obtains (with $H = 40$ km corresponding to the depth of the Mohorovičić discontinuity in mountainous areas) for the maximum speed at which crustal parts can slide over the substratum

$$v = \frac{3W}{16c\eta} = \frac{3\vartheta\, 2cH}{16c\eta} = 6\times 10^{-7}\ \text{cm/sec} = 18\ \text{cm/year}. \tag{6.13-12}$$

---

[1] It may be noted, however, that rock metamorphism is generally connected with an *increase* in density; hence we have, in general, $\zeta < 1$. This is generally ignored in theories of orogenesis where, if metamorphism is considered at all, it is always assumed that $\zeta > 1$.

[2] LAMB, H.: Hydrodynamics, p. 605. New York: Dover Publ. Co. 1945.

Thus, in order to create crustal shortening of the order of 40 km (Rocky Mountains, cf. Sec. 1.42), at least about 200000 years would be required; in order to produce the shortening of 320 km quoted for the Alps, at least about 1.8 million years are necessary. These values constitute the absolute minima of the time necessary to produce the mountain ranges in question. It rules out any speculations that mountain building might have occurred by instantaneous catastrophes. It should be noted, however, that the above argument does *not* hold if it is assumed that the substratum is moving in unison with the crust (cf. Sec. 6.4 on the convection current hypothesis of orogenesis). In that case, speeds faster than those calculated above might be possible.

The above remarks refer to continental orogenesis. As stated earlier, marine "orogenesis" is probably not due to crustal shortening; however, notwithstanding this, it has been assumed on occasion in the literature that the marine ridges *are* due to crustal shortening and it is therefore necessary to repeat some of the above calculations for the marine "orogenetic cycles".

Thus, if it is assumed that the system of ridges was formed by crustal shortening, like continental mountains, then one can calculate the amount of such crustal shortening s_T that would have been involved. One has (cf. Sec. 6.12)

$$s_T = \frac{V}{L h} = \frac{A}{h} \tag{6.13–13}$$

where h is the depth to which the shortening is being felt. This depth is assumed to occur at various levels in the various theories of orogenesis. However, one might take as a likely value $h = 5$ km, which corresponds to the depth of the Mohorovičić discontinuity beneath oceans. One then obtains for the crustal shortening required [from (6.12–3/6)]:

$$s_T = \frac{A}{h} = \frac{2400}{5}\,\text{km} = 480\,\text{km}. \tag{6.13–14}$$

This is not too much different from what has been quoted for the Alps. In continental mountain building, however, it is usually assumed that the strata were not only folded, but also underwent a lateral extension by the extension factor γ. What is observed, then, is the apparent shortening, s_A, instead of the true shortening, s_T,

$$s_A = \gamma s_T \tag{6.13–15}$$

where γ can never be smaller than 1.

Thus, if the system of mid-ocean ridges is treated as an analogue to continental mountain systems, created by crustal shortening like the latter, the much larger volume of rock involved will effect a change in the basic relationships so as to cause difficulties.

A further modification of the above calculations is obtained if it is assumed that the "crustal shortening" is not only felt to the base of the Mohorovičić discontinuity, but to a depth of 140 km which corresponds to GUTENBERG'S low velocity layer in the mantle (cf. Sec. 2.14).

Let us therefore investigate the above relationships for the case that we assume for h the depth of the low-velocity layer[1]. Thus we set first of all $h=140$ km in Eq. (6.13–5), which applies to "continental" mountains. This yields

$$\gamma=\frac{140}{12}=11.7. \qquad (6.13\text{–}16)$$

Furthermore, calculating now s_T by means of Eq. (6.13–2), with $s_A=$ 50 km, yields

$$s_T=\frac{50}{11.7}=4.3 \text{ km}. \qquad (6.13\text{–}17)$$

The large value of γ which was found above is in itself rather interesting. It indicates that the apparent crustal shortening as observed by geologists is much larger than the true shortening so that a relatively minor thrusting in the rocks above the low velocity layer will produce a large mountain system.

Let us now investigate the corresponding results for marine orogenesis. In this case, no "measured" value for the apparent crustal shortening s_A has ever been postulated. Thus from the relationships (6.13–4) and (6.13–2) one can arrive at only a hypothetical value for the true shortening s_T which would be required if the ridges were produced by shortening. One obtains from Eqs. (6.13–4) and (6.13–2)

$$s_T=\frac{V}{Lh}. \qquad (6.13\text{–}18)$$

This yields with the appropriate values for V and L [cf. Eqs. (6.12–5) and (6.12–6), setting $h=140$ km:

$$s_T=\frac{96\times 10^6}{140\times 4\times 10^4}\text{ km}=17.1 \text{ km}. \qquad (6.13\text{–}19)$$

It will be recalled that the corresponding value was 480 km if the base of the crust (at 5 km depth in the oceans) was taken as depth to which orogenesis is felt (6.13–14). The new value obtained in Eq. (6.13–19) sounds almost reasonable in comparison with the former one.

6.14. Possible Magmatic Origin of Oceanic Features[2]. As noted, some physiographic aspects of the mid-ocean ridges suggest that the latter, in fact, are not folded like continental mountains, but are magmatic

[1] SCHEIDEGGER, A. E.: J. Geol. **68**, 177 (1960).

[2] This Section after SCHEIDEGGER, A. E.: J. Alberta Soc. Petrol. Geol. **6**, 266 (1958).

oozes which came forth through tensional cracks in the Earth's crust. Thus, a second way of looking at oceanic orogenesis is by assuming that the relief features were built up by an accumulation of volcanic material from below the Morohovičić discontinuity.

It is fairly certain that the archipelagic aprons in the Pacific are of volcanic origin. To estimate the total volume of volcanics that must have been produced to create the ridges as well as the aprons, one obtains:

$$\left.\begin{aligned} V_r\,(\text{of ridges}) &= 96\times 10^6\ \text{km}^3 \\ V_a\,(\text{of Pacific aprons}) &= 4\times 10^6\ \text{km}^3 \\ V_t\,(\text{total}) &= 10^8\ \text{km}^3. \end{aligned}\right\} \qquad (6.14\text{–}1)$$

The volume of aprons in the Atlantic has not been estimated, but in view of the above figures, it can be regarded as being small compared with the volume of the ridges. It thus appears that the volume contained in the aprons is entirely insignificant, compared with the volume in the ridges. If a mechanism can be found for extruding enough volcanic material to create the ridges, then there is obviously no difficulty in explaining the existence of islands and aprons on the same basis (as they may be regarded simply as sporadic occurrences of the same type as ridges).

The amount of volcanic material that is being produced on the Earth's surface has been estimated on several occasions. SAPPER[1] estimates that the volume of lava being poured out in recent years is 0.8 km³/year. This is highly at variance with an estimate of VERHOOGEN'S[2] based upon the geological evidence of extrusions, which is approximately 0.06 km³/year. Since the latter estimate is based on lava-outpourings of the geological past in continental areas only, one should probably assume that, on the average, twice this amount of lava is being poured out in oceanic areas, since the latter are of twice the magnitude of continental areas. This brings VERHOOGEN'S estimate for the oceanic areas to 0.12 km³/year, and for the whole Earth to 0.18 km³/year, which is still much less than SAPPER'S.

Therefore, if the (present) system of mid-ocean ridges is due to volcanic activity, with all the oceanic lava produced being contained in them (and in the aprons), it should have taken 550 million years (using VERHOOGEN'S estimate) to build them up, but by SAPPER'S estimate (which is roughly $4\frac{1}{2}$ times bigger) only 125 million years would be required to produce the same effect.

The above type of reasoning assumes that all the volcanic material somehow should end up in the mid-ocean ridges, which is certainly an oversimplification of the picture. However, it is probably correct to

---

[1] SAPPER, K.: Vulkankunde. Stuttgart 1927.

[2] VERHOOGEN, J.: Amer. J. Sci. **244**, 745 (1946).

say that all the *oceanic* lava will somehow end up in mid-ocean ridges and archipelagic aprons. One should perhaps add to this some 5000 to 10000 sea mounts whose volume, however, is small. Using the lesser of the estimates, (i.e. VERHOOGEN's) which, by considering proportionality as outlined above, yields 0.12 km^3/year of lava for the oceans, it would appear that about 830 million years were required to build up the oceanic features by volcanic activity. Even with the $4\frac{1}{8}$ times faster rate of SAPPER's, one would still require some 200 million years. Taking the sea mounts into account would increase both estimates.

The above rates are relatively slow and seem to indicate that, if the mid-ocean ridge systems are due to volcanic activity, they must have been built up much less rapidly than, for instance, continental mountain systems. Using the shorter of the two possible estimates (i.e. 200 million years), there is no difficulty in postulating that many different mid-ocean ridge systems have been formed during the Earth's life which since have been removed by erosional processes (possibly turbidity currents);—although these ridges could not have been as numerous as continental mountain systems. Using the lesser of the estimates, this is not possible, and it would then appear that the present-day example is the one and only mid-ocean ridge system that the Earth ever possessed. However, should it turn out that mid-ocean ridge are simply bulges in the oceanic crust, without crustal thickening, then the above argument would have to be modified.

6.15. Geosynclines. As noted in the chapter on physiography, orogenetic processes have often been associated with the notion of a *geosyncline*. The term seems to have been introduced by DANA[1] when he was investigating the Apallachians and indicates a great thickening of the sedimentary layers in a through in the Earth's crust which is destined to become a mountain chain. Many types of geosynclines have been discerned, particularly by STILLE[2], since the creation of the term. However, it should be noted that the one-time existence of a trough in the Earth's crust in places where there are now mountains, is quite hypothetical although it is beyond question that, in orogenetic belts, tremendous thicknesses of sediments are present. As possible present-day geosynclines, the Adriatic Sea[3], the Timor Through[4] and the Gulf Coast Region[5] have been quoted. Much of the *pro* and *con* for the very concept of "geosyncline" has recently been reviewed by KNOPF[6].

[1] DANA, J. D.: Amer. J. Sci. (3) 5, 423 (1876).
[2] STILLE, H.: Einführung in den Bau Amerikas. Berlin: Bornträger 1940.
[3] KOSSMAT, F.: Paläogeographie und Tektonik. Berlin: Bornträger 1936.
[4] KUENEN, P. H.: Sci. Res. Snellius Exp. 5, 54 (1935).
[5] BUCHER, W. H.: Trans. Amer. Geophys. Un. **32**, 514 (1951).
[6] KNOPF, A.: Amer. J. Sci. **258** A, 126 (1960).

Nevertheless, if it be assumed that geosynclines have existed, one immediately must ask himself what their significance is with regard to the physics of orogenesis.

First of all, as had already been indicated in Chap. I, it is impossible that geosynclines were *caused* by the weight of sediments deposited in the respective areas;—at least if the notion of isostasy is even remotely valid. This point has been particularly emphasized by HOLMES[1]. Thus, let us assume that the deposition of sediments (density $\varrho_s=2.4$) took place in a water (density $\varrho_w=1$) depth of $h_w=30$ meters, and proceeded until the water depth was completely filled in. If the maximum thickness of sediments deposited in this manner be h meters, then the ultimate amount of the depression of the crust into the mantle ($\varrho_m=3.4$) is $(h-h_w)$. Isostasy then requires

$$h\varrho_s=h_w\varrho_w+\varrho_m(h-h_w) \tag{6.15-1}$$

or

$$h=h_w\frac{\varrho_m-\varrho_w}{\varrho_m-\varrho_s} \tag{6.15-2}$$

which yields with the above values for the constants

$$h=30\times 2.4\text{ m}=72\text{ m}. \tag{6.15-3}$$

It is seen that the ultimate thickness of sediments that can be depressed in this fashion is but a tiny fraction of the sediment thicknesses surmised for the hypothetical geosynclines. Thus, the weight of the sediments cannot possibly be the *cause* of the formation of a trough.

The relation between geosynclines and isostasy has more recently been investigated by HSU[2]. HOLMES had assumed a constant thickness of the crust under a geosyncline, but it may be assumed that the thickness of the crust h_c (density ϱ_c) may vary. The isostatic relationship then requires, assuming that the deposition takes place in very shallow water ($h_w\sim 0$):

$$h_c^{\text{normal}}\varrho_c=h_s\varrho_s+h_c^{\text{geosyncline}}\varrho_c+h_m\varrho_m \tag{6.15-4}$$

where h_m denotes the deviation of the mantle-crust interface from the "normal" position; h_m is positive if the deflection is up, negative if it is down (for a mass deficiency above, h_m is positive). Assuming the "normal" crust to be 33 km thick, HSU obtained values for the sediment thicknesses in relation to crustal thickness, some of which are shown in Table 20.

It thus appears that geosynclines might simply be areas where the crust, for some reason or other (such as a tension crack) happens to be

[1] HOLMES, A.: Principles of Physical Geology. London: Nelson 1944, see p. 380 therein.

[2] HSU, K. J.: Amer. J. Sci. **256**, 305 (1958).

thinner than normal. It is then easily possible to accumulate the required sediment thicknesses. However, let it be noted once more that the thinning of the crust is not *caused* by the sediment deposition, but that the sediments were deposited there *because* the crust was thin.

Since it has been found that many rocks in mountain chains are highly metamorphosed, investigations have been directed towards estimating whether a temperature effect might be connected with the formation of a geosyncline. The most notable study along these lines was made by GROSSLING[1]. GROSSLING assumed that a geosyncline represents a subsidence of the whole crust and therefore based his calculations on surmising a standard continental crust underneath. Under these assumptions, he found a significant temperature increase in a geosynclinal area. However, in the light of HSU's investigations, it is doubtful whether a geosyncline can simply be regarded as crustal subsidence. It is not clear how GROSSLING's calculations would have to be modified to allow for a thinner crust.

Table 20. HSU's *calculation of isostasy in geosynclines*

$h_c^{geosyncl.}$ (km)	h_s (km) (for $\varrho_s = 2.4$)
5	13.8
10	11.4
15	8.9
20	6.4
25	3.95
30	1.48
$33 = h_c^{normal}$	0

6.2. The Contraction Hypothesis

6.21. Principles. The contraction hypothesis is one of the earliest attempts to explain the origin of geodynamic forces. It goes back at least to the time of DESCARTES. In its modern form, it is usually presented as stated by JEFFREYS[2]. Accordingly, it is assumed that the Earth began as a hot, celestial body. Early in its history, it differentiated into an iron core and an essentially silicate mantle. The mantle solidified outwards from its base at the liquid iron core and has since been cooling by conduction without convection currents. From the center of the Earth to within about 700 km of the surface there has not been time since the earth solidified for any appreciable cooling or change in volume to have taken place. Within the region from about 700 to 70 km, cooling by conduction is taking place and hence this layer is contracting and being stetched about an unchanging interior. Hence it is in a state of internal tension.

Near the surface the rocks have already largely cooled so that they are in thermal equilibrium with the heat provided by solar radiation. They are therefore not changing very much in temperature and the

[1] GROSSLING, B. F.: Bull. Geol. Soc. Amer. **70**, 1253 (1959).

[2] JEFFREYS, H.: The Earth. London: Cambridge University Press 1929.

cooling and contraction of the layer or shell beneath them puts the outermost shell into a state of internal compression above a level of no strain at 70 km depth.

Thus, the contraction hypothesis divides the Earth upon grounds of thermal and mechanical behavior into three shells: the non-contracting part of the Earth below a depth of about 700 km, the contracting part of the mantle above 700 km and below the level of no strain at about 70 km, and the "exterior" which is crumpling up due to the contraction below. These shells are not dependent upon the Earth's composition and hence should not be confused with such terms as core, mantle and crust. As the Earth cools, the boundaries between the shells move deeper into the Earth. The stress state assumed in the contraction hypothesis is shown in Fig. 85.

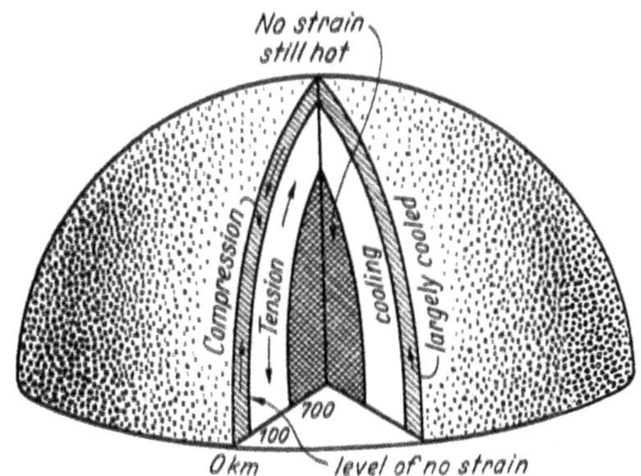

Fig. 85. Stresses in the Earth according to the contraction hypothesis

The level of no strain has been taken at 70 km depth, which is JEFFREYS' value. However, there is nothing magic about this number and it may be preferable to put it, say, at 140 km depth which would correspond to GUTENBERG's low velocity layer (cf. Sec. 2.14).

At first glance, the contraction hypothesis explains many physiographic facts about the Earth. The cross-section of an island arc can be envisaged as the outcome of a deep faulting process in the contracting layer of the Earth (see Fig. 86). This would be in conformity with the claim that the foci of deep-focus earthquakes lie approximately upon a surface dipping into the Earth beneath recent island arcs (cf. Sec. 2.22). The geological implications of this assumption, in conjunction with the assumed stress state, have been followed up by WILSON[1] and are shown in Fig. 87, which is self-explanatory. Similar results have also been arrived at by ASLANYAN[2].

The discussion so far deals only with the cross-sections of island arcs. For a consistent theory, the arcuate surface structure of the

[1] WILSON, J. TUZO: Proc. Geol. Ass. Can. 3, 141 (1950).

[2] ASLANYAN, A. T.: Исследование по теории тектонической деформации земли. Erevan: Iz-vo Akad. Nauk Armyan. SSR 1955.

orogenetic belts must also be obtained. Possible bases for such an explanation would be furnished[1, 2] either by assuming plastic yielding or yielding by creep, or else by assuming sliding fracture.

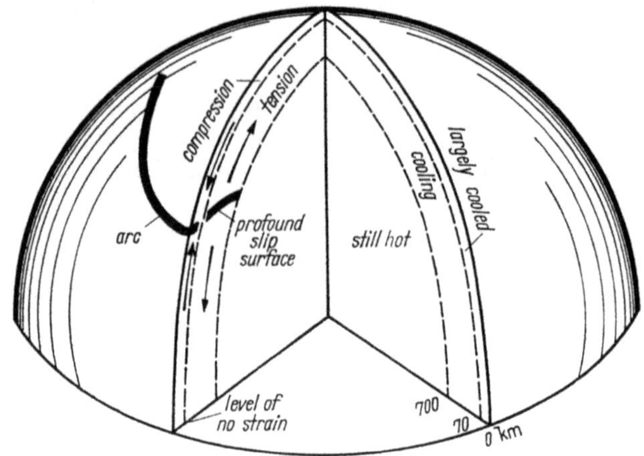

Fig. 86. Formation of an island arc in the contraction hypothesis. After WILSON[3]

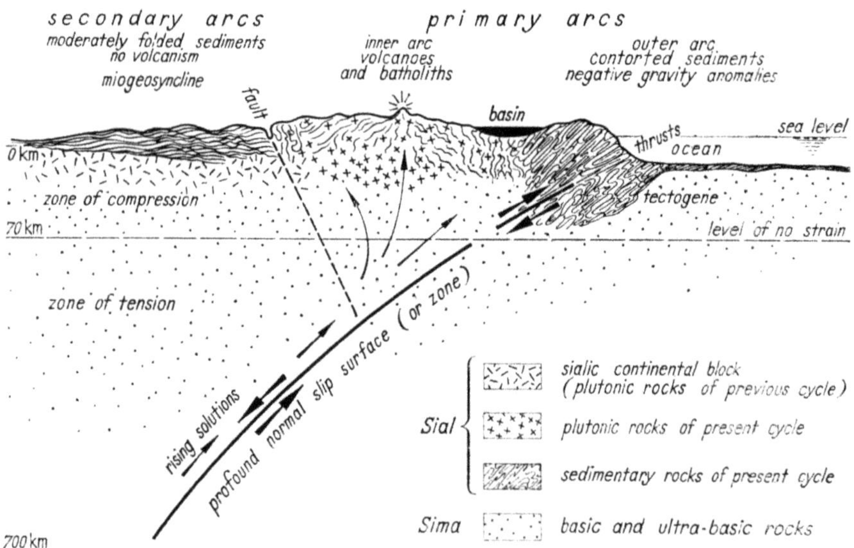

Fig. 87. Diagrammatic cross section of a double mountain range, as envisaged in WILSON's[3] development of the contraction theory (great exaggeration of vertical scale in upper part of the picture)

[1] SCHEIDEGGER, A. E., J. T. WILSON: Proc. Geol. Ass. Can. **3**, 167 (1950).

[2] The following is after the writer's discussion in Bull. Geol. Soc. Amer. **64**, 127 (1953).

[3] WILSON, J. TUZO: Proc. Geol. Ass. Can. **3**, 141 (1950).

According to the first assumption, an explanation of the shape of failure on the Earth would be that around a "weak point" (of symmetry), spiral-shaped slip-lines form (Fig. 57). Underneath each of those lines would be a surface dipping down into the Earth (Fig. 40). If the margins of continents are assumed to be weak zones owing to the tremendous amount of deposition taking place there (in conformity with the notion of geosynclines), corresponding to the stamp in Fig. 57, then island arcs and perhaps marginal ranges might correspond to spiral slip lines springing from the margin of continents. The fact that the material along the slip lines undergoes a different kind of deformation from that in other regions might account for earthquakes and volcanoes near island arcs.

From the theoretical standpoint, there are some objections to this theory. Slip lines form a double family of curves crossing each other, not just single lines, and there is no evidence on the Earth of two sets of arcs crossing each other approximately at right angles. Furthermore, each family of slip lines theoretically forms a great number of curves that cover a whole region quite densely. On the Earth, there are only single spirals at comparatively wide intervals. There is no evidence that several arcs start from one region of weakness. It might be possible to adjust the facts better to the theory by geological explanations, such as the assumption of a certain anisotropy in the Earth. However, it will require further geological and geophysical explanations and evidence to justify the assumption of a plastic slip phenomenon as explanation for spiral-shaped island arcs and mountain ranges. The same arguments apply to any explanation that arcs are due to yielding by creep.

RUUD[1] also assumes plastic failure as the cause of orogenesis, but postulates that the arcuate structures upon our planet correspond to the lines of weakness instead of the slip lines. Assuming circular symmetry, the primary orogenetic structures to be expected are ring-shaped craters like those on the Moon. In this instance, a major difficulty seems to be that the chains of island arcs on the Earth bear but little resemblance to the craters on the Moon. Also, the craters on the Moon are now generally believed to be of meteoritic origin.

Instead of assuming plastic yielding, let us now investigate fracture as a possible explanation for island arcs. For the sake of simplicity, we shall take MOHR's criterion as relevant (cf. Sec. 3.53). If circular symmetry is again assumed around a point, the greatest pressure must be assumed as vertical (or nearly so) because of the weight of the overlying material, and because below the level of no strain, the other principal stresses are tensions. MOHR's theory then predicts that the failing surfaces will be inclined at an angle of less than 45° to the radial

[1] RUUD, I.: Gerlands Beitr. Geophys. **52**, 123 (1938).

direction of the sphere and parallel to the intermediate principal stress. This can be either the tangential direction to a circle around the point of symmetry or the meridional direction. The first of these cases might explain circular island arcs and mountain ranges, for it leads to a conical fracture with a dip of about 45°. On the other hand, if island arcs are considered to be spiral-shaped, the theory would have to be modified. One might assume, for instance, that the stress state is symmetrical with respect to a point in a limited region only and that farther away there is an undisturbed stress state. Thus, starting in a "weak region", a sliding crack would develop in a circular manner before flattening out in an irregular strike. This is the pattern which suits, in fact, most of the island arcs and mountain ranges investigated numerically in Sec. 1.42.

Based upon the above considerations (sliding fracture), WILSON[1, 2] has extended the contraction hypothesis to present a theory of the growth of terrestrial features. Accordingly, continents, once nucleated by the processes discussed in Chap. V, keep growing throughout the ages. The coast lines represent weak zones around which circular fractures occur as discussed above. The cross section of each fracture is as shown in Fig. 87; the fractures are identified with island arcs. The island arcs become gradually transformed into mountains and the process then starts anew elsewhere. During the Earth's history, the level of no strain is assumed to move deeper into the Earth which may be the reason why very early mountain ranges had a different physiographic pattern as compared to recent ones (cf. Sec. 2.52). At any one time, a whole orogenetic system may be active; however, it constantly expands and changes until an entirely new physiographic picture may be formed. Thus, about 9 to 10 "distinct" systems may have been created since the solidification of the Earth's crust. The energy for the various processes, corresponding to the contraction hypothesis, would be provided by the cooling and contraction below the level of no strain.

A very recent modification of the contraction hypothesis has been suggested by BUCHER[3] and WILSON[4] according to which not cooling, but the extrusion of volcanic material from below the Mohorovičić discontinuity would be the cause of the shrinkage. The details of this hypothesis have been discussed in Sec. 5.24. The relevant mechanics with regard to orogenesis is the same as if *thermal* contraction were assumed to take place. The same may be said regarding a suggestion of SONDER's[5]

[1] WILSON, J. T.: Pap. & Proc. Roy. Soc. Tasmania p. 85 (1950).

[2] WILSON, J. T.: In: The Earth as a Planet, ed. KUIPER, p. 138. Chicago: University of Chicago Press.

[3] BUCHER, W. H.: In Geotektonisches Symposium zu Ehren von H. STILLE, ed. F. LOTZE, p. 396. Stuttgart: Deutsche Geologische Gesellschaft 1956.

[4] WILSON, J. T.: Amer. Sci. 47, No. 1, 1 (1959).

[5] SONDER, R. A.: Mechanik der Erde. Stuttgart: Schweizerbart 1956.

where the shrinkage of the Earth's interior has been attributed to a nuclear reaction by which the silicates of the mantle are transformed into the much denser material of the core.

In spite of the apparent success of the contraction hypothesis, there remain several points in which it is not very satisfactory. The first of these is that it leads to a discrepancy if the extension factor is calculated, and the second is that the contraction theory cannot explain the observed transcurrent motion in earthquake foci. These difficulties will be discussed in Secs. 6.25 to 6.27. Furthermore, the contraction theory does not even attempt to face the mounting evidence that the stresses in the Earth's crust are not at all uniform (cf. Sec. 2.47). It appears, therefore, that the contraction hypothesis (in its present form at least) cannot be accepted as an entirely satisfactory explanation of orogenesis.

6.22. The Existence of a Level of No Strain. The contraction hypothesis stands or falls with the possibility of the existence of a level of no strain in the light of thermal considerations. The problem has been investigated by JEFFREYS[1] who considered a shell of internal radius r and thickness δr. During a rise of temperature by the amount dT, the density, originally equal to ϱ, becomes $\varrho(1-3n\,dT)$ where n denotes the coëfficient of linear expansion due to a temperature change. If it is assumed that the radius r becomes equal to $r(1+d\alpha)$, then the external radius becomes after the change of temperature

$$r(1+d\alpha)+\delta r\left\{1+\frac{d}{dr}(r\,d\alpha)\right\}.$$

Consequently, the mass of the shell will be (neglecting the squares and products of dT and $d\alpha$):

$$4\pi\varrho r^2\delta r\left\{1+2d\alpha+\frac{d}{dr}(r\,d\alpha)-3n\,dT\right\}.$$

Since the mass of the shell cannot alter, this leads to the following equation of continuity:

$$2d\alpha+\frac{d}{dr}(r\,d\alpha)-3n\,dT=0. \qquad (6.22\text{–}1)$$

This is a differential equation which may be used to determine $d\alpha$. (Note that it was assumed that dT is known throughout the Earth.)

If the shell could expand without straining, the radius would increase by $r\,n\,dT$ instead of $r\,d\alpha$, so that the amount of straining required to make it fit into its place is

$$r(d\alpha-n\,dT)=r\,dS. \qquad (6.22\text{–}2)$$

[1] JEFFREYS, H.: The Earth, 2nd ed. London: Cambridge Univ. Press 1929.

One thus obtains for dS, a quantity immediately connected with the strains:

$$\frac{d}{dr}(r^3 dS) = -r^3 \frac{d}{dr}(n\,dT) \tag{6.22–3}$$

or

$$dS = -\frac{1}{r^3}\int_0^r r^3 \frac{d}{dr}(n\,dT)\,dr. \tag{6.22–4}$$

By partial integration, this yields

$$dS = -n\,dT + \frac{1}{r^3}\int_0^r 3r^2 n\,dT\,dr, \tag{6.22–5}$$

and, since the differential dT may be understood with respect to time, one obtains:

$$\frac{\partial S}{\partial t} = -n\frac{\partial T}{\partial t} + \frac{1}{r^3}\int_0^r 3r^2 n \frac{\partial T}{\partial t}\,dr. \tag{6.22–6}$$

So far, we have been following JEFFREYS[1]. But now we note[2] that we can substitute $\partial T/\partial t$ from the heat conductivity equation:

$$\varrho c \frac{\partial T}{\partial t} = k \operatorname{lap} T. \tag{6.22–7}$$

In spherical co-ordinates this is (if T depends on r only):

$$\varrho c \frac{\partial T}{\partial t} = \frac{k}{r^2}\frac{\partial}{\partial r}\left(r^2 \frac{\partial T}{\partial r}\right). \tag{6.22–8}$$

Hence

$$\frac{\partial S}{\partial t} = -\frac{n}{\varrho c}\left[\frac{k}{r^2}\frac{\partial}{\partial r}\left(r^2\frac{\partial T}{\partial r}\right)\right] + \frac{3}{r^3}\int \frac{nk}{\varrho c}\frac{\partial}{\partial r}\left(r^2\frac{\partial T}{\partial r}\right)dr. \tag{6.22–9}$$

Now, the object is to evaluate the integrals in Eq. (6.22–9) from the center of the Earth up to a point r somewhere in the mantle. Thus, we have to integrate over the whole core, across the discontinuity between core and mantle, and finally over part of the mantle. Let us denote the radius of the core by R_0 ($=3470$ km), then we have a discontinuity in the integration for $r = R_0$. Before (undashed) and after (dashed) this level, n, k, ϱ and c are assumed to be constant. Hence, for $r > R_0$

$$\left.\begin{aligned}\frac{\partial S}{\partial t} &= -\frac{n'}{\varrho' c'}\left[k'\frac{1}{r^2}\frac{\partial}{\partial r}\left(r^2\frac{\partial T}{\partial r}\right)\right]\\ &\quad + \frac{3}{r^3}\left\{\frac{nk}{\varrho c}\left(r^2\frac{\partial T}{\partial r}\right)\Big|_0^{R_0} + \frac{n'k'}{\varrho' c'}\left(r^2\frac{\partial T}{\partial r}\right)\Big|_{R_0}^{r}\right\}.\end{aligned}\right\} \tag{6.22-10}$$

[1] JEFFREYS, H.: The Earth, 2nd ed., p. 280. London: Cambridge Univ. Press 1929.

[2] SCHEIDEGGER, A. E.: Canad. J. Phys. 30, 14 (1952).

Thus

$$\left.\begin{aligned}\frac{\partial S}{\partial t} &= -\frac{n'}{\varrho' c'}\left[k' \frac{1}{r^2}\frac{\partial}{\partial r}\left(r^2 \frac{\partial T}{\partial r}\right)\right] \\ &\quad + \frac{3}{r^3}\left\{\left(\frac{n\,k}{\varrho\,c} - \frac{n' k'}{\varrho' c'}\right) R_0^2 \frac{\partial T}{\partial r}\bigg|_{R_0} + \frac{n' k'}{\varrho' c'} r^2 \frac{\partial T}{\partial r}\right\}.\end{aligned}\right\} \qquad (6.22\text{–}11)$$

The term from the discontinuity vanishes if we assume that the temperature gradient for R_0 vanishes. Then we have for $r > R_0$:

$$\frac{\partial S}{\partial t} = \frac{n'}{\varrho' c'}\frac{k'}{r}\frac{\partial T}{\partial r} - \frac{n'}{\varrho' c'} k' \frac{\partial^2 T}{\partial r^2}. \qquad (6.22\text{–}12)$$

This is the equation connecting stress and temperature in the Earth. The condition for a level of no strain is (according to JEFFREYS)

$$\frac{\partial S}{\partial t} = 0. \qquad (6.22\text{–}13)$$

Hence

$$\frac{1}{r}\frac{\partial T}{\partial r} = \frac{\partial^2 T}{\partial r^2}. \qquad (6.22\text{–}14)$$

If we have any given temperature distribution within the Earth, then the layers in which the above equation is satisfied are levels of no strain. It is obvious that the position of the level of no strain has nothing to do with the coëfficients of heat conduction etc., but is determined wholly by the *shape* of the temperature curve with depth.

We are now in a position to compare Eq. (6.22–14) with the various estimates of temperature curves above 1000 km shown in Sec. 2.62. In particular, it is of interest whether any of the estimates mentioned in Sec. 2.62 are compatible with the assumption of a level of no strain at 70 km depth. The gradient at that depth is about 1 or $1\frac{1}{2}$ °/km. Then, formula (6.22–14) gives us

$$\partial^2 T/\partial r^2 = 0.00025°/\text{km}^2. \qquad (6.22\text{–}15)$$

The change of the gradient over a distance of 100 km in the level of no strain becomes

$$(\partial^2 T/\partial r^2)\, dr = 0.00025 \times 100 = 0.025°/\text{km}. \qquad (6.22\text{–}16)$$

This means that, over a distance of 100 km, the gradient should change in the level of no strain by about 1/40°/km, which is almost negligible. This indicates that in the level of no strain, the temperature curve must be almost straight. This would still be the case if the temperature gradient were assumed to be much larger.

It is obvious that JEFFREYS' temperature estimate (cf. Sec. 2.62) fits this condition fairly well. On the other hand, GUTENBERG's assumption of a sharp kink in the temperature-depth curve at about the depth of the level of no strain is certainly not in conformity with it. The conclusion would therefore be that a temperature curve of the type postulated by GUTENBERG would preclude the existence of a level of no strain, but that JEFFREYS' curve would be compatible with it. Since both these temperature curves are largely hypothetical, not much of a definite conclusion regarding the existence or nonexistence of a level of no strain can be derived in this manner.

The stress state in the various shells of the Earth as envisaged by the contraction theory has also been analyzed more closely by HALES[1]. Making various plausible assumptions and using essentially JEFFREYS' method indicated above, HALES estimated the level of no strain at 59 km depth. However, he also found that the stress differences increase much more rapidly in relation to the strength of the material below the level of no strain than above it. This would indicate that failure occurs below the level of no strain rather than above it so that an oceanic trough would be formed rather than mountains folded up. It is not clear whether this does not, in fact, constitute a severe difficulty in the commonly envisaged mechanism of the contraction theory.

6.23. The Thickness of the Earth's Crust and Mountain Building. The Available Contraction. The next problem to be investigated concerns the relative size of present and very ancient mountains. Presumably, the depth to which cooling has penetrated (according to the contraction hypothesis) must be assumed to increase during the Earth's history, which, in turn, might give an explanation of why very ancient mountains show a different pattern from recent ones, as has been outlined earlier.

We shall investigate now whether such a theory can be confirmed by physical considerations of the types of folding that must occur when the outermost shell above the level of no strain is supposed to become thicker[2]. Thus, the model of the mechanics of mountain building which we are to consider is as follows: Below a surface shell of constant density we assume a contracting spherical interior. If the contraction takes place at a constant rate, the question arises as to what extent folding depends on the thickness of the surface shell.

Let the volume of the shell be V and the inner radius of this shell be a and its thickness b. If the shell is a relatively thin and cool layer, it will retain its volume almost unchanged during any short period of contraction brought about by cooling in the larger and hotter interior. The situation is illustrated in Fig. 88. Then we have

$$V = 4\pi a^2 b. \tag{6.23-1}$$

[1] HALES, A. L.: Month. Not. Roy. Astr. Soc., Geophys. Suppl. **6**, 458, 486 (1953).

[2] SCHEIDEGGER, A. E.: Canad. J. Phys. **30**, 14 (1952).

If the radius of the interior sphere is changed by the amount da and the thickness of the shell by db, then we obtain

$$dV = 4\pi a^2 db + 8\pi a b da. \qquad (6.23\text{–}2)$$

However, dV must vanish, as stated above, hence

$$2da\, b + a\, db = 0, \qquad (6.23\text{–}3)$$

$$db = -2b\, da/a. \qquad (6.23\text{–}4)$$

This equation indicates that, for a shrinking of the interior by the amount da, the amount of material of the shell that has to be moved

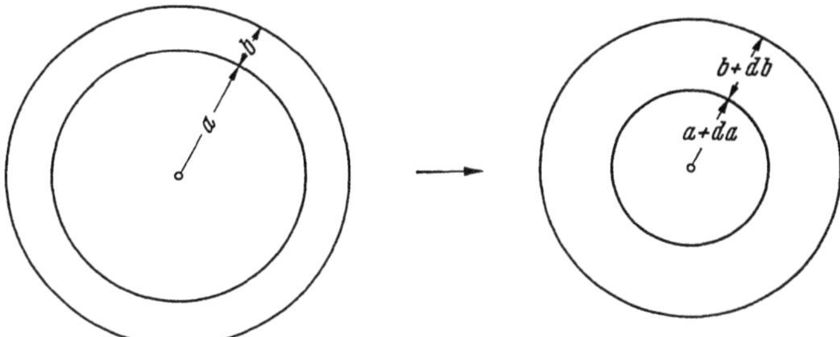

Fig. 88. The change of the Earth during contraction

in order to make it continue to fit the interior is proportional to its thickness.

If we assume that the shrinking of the interior occurs at a constant rate $\dot{a}$ in time, then (6.23–4) becomes

$$db/dt = -(2b/a)\,\dot{a}, \qquad (6.23\text{–}5)$$

which indicates that the material in the shell that has to be moved around per unit time is just proportional to the thickness of the shell. This shows indeed that a thin shell must form smaller mountain ranges than a thick one, if the interior shrinks by the same amount. Thus, if the ancient folds were formed in that manner, they would be expected to be closer together than later folds formed in a similar way.

We may corroborate this statement and estimate how much the mantle has to cool so as to cause a major world-wide orogenetic system. We have already calculated in (6.12) that the volume of an orogenetic system is equal to 32×10^6 km³. This corresponds to an over-all increase in thickness of the outermost shell of 7×10^{-2} km. It follows from Eq. (6.23–4) that in order to obtain an increase of thickness of $db = 7 \times 10^{-2}$ km, assuming $b = 70$ km, one has

$$da/a = -(1/2)\, db/b = -5 \times 10^{-4}. \qquad (6.23\text{–}6)$$

Thus, one needs an over-all relative linear contraction of 0.05% in the interior to fold up one orogenetic system. The radius of the interior of the Earth is thereby contracting by about 3 km.

If we would have taken the depth of the level of no strain at 140 km (i.e. $b=140$ km), the overall linear contraction would have come out as 0.095% and the radial contraction as 1.5 km.

We shall assume now that the cooling and contraction takes place in depths between 70 and 700 km only. Thus the thickness of the corresponding layer is changed from 630 to 627 km during one orogenetic cycle. We can calculate the relative volume contraction which has therefore to occur, it is 0.0106. The relative linear contraction is one third of this, and thus equal to 0.0035 or 0.35%. The relative linear expansion of granite equals[1] about 10^{-5} per degree centigrade, so that an over-all change in temperature of about 350° C in the cooling layer would cause sufficient contraction to give rise to the mountains of one orogenetic cycle.

This seems indeed quite reasonable. If the liquid iron core is assumed to be at a temperature of about 4500° C, then about 12 orogenetic cycles would be possible while a particular part of the mantle cooled down from the temperature of the core to that of outer space. This is the right order of magnitude in comparison with the actual number of observed orogenetic cycles which is usually given as 10.

For the smaller shortening required if the level of no strain is put at 140 km depth, it is even easier to account for the required contraction in terms of a cooling process.

As noted in Sec. 6.1, some people have held that the mid-ocean ridges are due to crustal shortening. For ocean ridges, the contraction required can be calculated[2] in the same fashion as is done for continental mountain ranges. The over-all increase, db, of the outermost shell of the Earth must be (using the values from Sec. 6.1)

$$db = \frac{V}{4\pi R^2} = \frac{96 \times 10^6\,\text{km}^3}{4\pi \times (6.3)^2 \times 10^6} = 0.2\,\text{km}. \qquad (6.23\text{–}7)$$

Furthermore, one has (according to 6.23–4)

$$db = -2b\, da/a. \qquad (6.23\text{–}8)$$

This yields with $b=70$ km

$$\frac{da}{a} \cong -0.14\%. \qquad (6.23\text{–}9)$$

---

[1] BIRCH, F.: Handbook of Physical Constants. Geol. Soc. Amer. Spec. Pap. No. 36 (1942).

[2] After SCHEIDEGGER, A. E.: J. Alberta Soc. Petrol. Geol. 6, 266 (1958).

The radius of the Earth must therefore contract by about 9 km to create a system of the size of the present day mid-ocean ridges. This contraction of the Earth's radius by 9 km seems somewhat high. According to the writer's estimates, this is roughly three times larger than the contraction required to create a continental orogenetic system, and hence represents three times the temperature decrease. The latter would now amount to about 1000° C. Whereas approximately 12 continental orogenetic cycles (each calling for a lowering of temperature by 350°) would be required to cool the contracting layer from the temperature of the core (some 4500°) to be equilibrium temperature with solar radiation, about four ocean-ridge cycles would produce the same effect. In computing the maximum number of cycles that could have occurred since the Earth solidified, each "marine" cycle counts the same as three continental cycles, which cuts down the number of the latter which would be possible. It would thus appear that the contraction theory requires that the "orogenetic" cycles producing ocean ridges must be much less frequent than cycles causing continental mountains.

The thermal effect of the cooling has been estimated above in a very simplified way. REITAN[1] has reported some calculations where the radius-change was estimated upon the basis of the models of the thermal history of the Earth referred to in Sec. 2.63. He showed that the change of the circumference of the Earth, based on these models, can lie only between −13 km (contraction) and +31 km (expansion). The maximum radial contraction, according to REITAN[1], is therefore only about 2 km. This could, at most, account for only one continental orogenetic cycle. However, since REITAN's calculations are tied up with such uncertainties as the amount of radioactivity in the Earth, it is not certain whether they constitute an unsurmountable difficulty to the acceptance of the contraction theory.

Apart from the creation of mountains, contraction has also been held responsible (by SONDER[2]) for small changes in elevation of the continents such as caused the inundation by the sea (transgression) of various parts of the former throughout geological history. Such changes in elevation, according to SONDER, are caused by a buckling mechanism induced by the varied tangential compression owing to the steady contraction of the Earth.

6.24. The Junctions of Island Arcs. A further instance in which the contraction hypothesis has lead to fairly reasonable results is the explanation of the types of junctions between orogenetic elements (i.e. island arcs) that have been observed in reality (cf. Sec. 1.43). In order

[1] REITAN, P. H.: J. Geol. **68**, 678 (1960).
[2] SONDER, R. A.: Mechanik der Erde. Stuttgart: Schweizerbart 1956.

to show this, one has to split the implications of the contraction hypothesis into "primary" effects taking place in the contracting shell and into "secondary" effects which are the surface expressions of the primary ones[1].

We shall turn first to the *primary effects*. During the formation of an island arc, the phenomenon thought to occur in the contracting shell is a conical fracture with slippage taking place along the cone of fracture. Within a cross section through the contracting layer parallel to the surface of the Earth, the motion is therefore as illustrated in Fig. 89. Within the cross section, the surface of slippage is seen as a line which is, in the mathematical sense, singular. The displacements of the material elements in the cross section of the contracting shell are everywhere continuous during the orogenetic diastrophism except on that singular line which is the trace of the surface of slippage on the cross section.

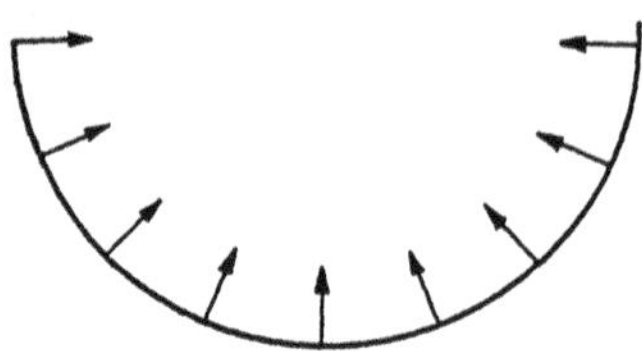

Fig. 89. Motion within a cross section parallel to the surface of the Earth through the contracting shell during the formation of an arc

It is quite obvious that the displacements of material within the contracting shell of the Earth during an orogenetic diastrophism must obey a continuity equation, since no material can be created or annihilated in the process. Whatever the displacements (denoted by ξ) may be, they must therefore be subject to the equation

$$\operatorname{div} \varrho \xi = 0. \tag{6.24-1}$$

If the density ϱ of the material is assumed as (more or less) constant, this means that the displacement field must be a divergence-free field. It is well known that such a divergence-free field cannot satisfy arbitrarily singular boundary conditions.

This fact can be illustrated by elementary geometry which is directly applicable to the theory of island arcs and mountain belts.

Thus, assume that two island arcs join, forming an angle α at the junction point. In a cross section of the contracting shell parallel to the surface of the Earth, within the neighborhood of the junction, one will have a singular line, having a kink of angle α at the junction point of the arcs. If it be borne in mind that *ab hypothesi* the motion has to be at right angles to the singular line, it is immediately clear that, for continuity reasons, at least one other singular line must join the first one at the kink. Furthermore, elementary geometry shows that this

[1] SCHEIDEGGER, A. E.: Canad. J. Phys. **31**, 1148 (1953). A Similar analysis, though less mathematical, has also been made by ROBINSON [ROBINSON, R. O. A.: Canad. J. Phys. **35**, 536 (1957)].

new singular line must lie within the sector formed by the two normals to the original singular line at the kink. Fig. 90 will clarify the situation: the permitted sector for the additional singular lines is shaded; a possible position for the latter has been drawn in and the arrows signify the motion of every element in the cross section of the contracting shell which is under consideration. It is seen that this arrangement of singularities corresponds to the junction termed "linkage" in Sec. 1.43

It has been stated above that there has to be *at least* one additional singular line joining the island arcs at their junction. There may be

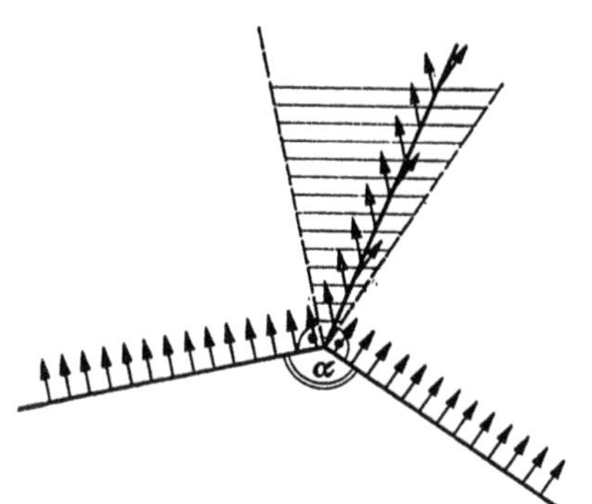

Fig. 90. Motion within a cross section parallel to the surface of the Earth though the contracting shell in the neighborhood of the junction of two arcs if only *one* additional singular line is present. The permitted sector for the additional singular line is shaded

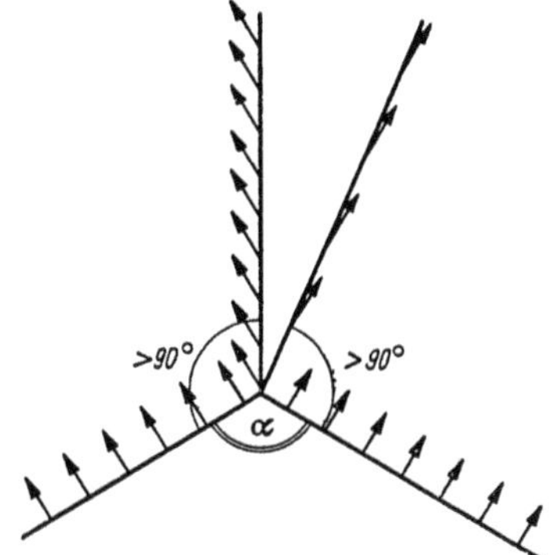

Fig. 91. Motion within a cross section parallel to the surface of the Earth through the contracting shell in the neighborhood of the junction of two arcs if *two* additional singular lines are present

more. In every case, the additional lines must lie within the shaded area of Fig. 90; a possible position of *two* singular lines together with the displacement field has been drawn in Fig. 91. It must be expected that this is the case which gives rise to a junction of arcs with cap range and lineaments. However, those two latter occurrences are phenomena of the surface and not of the contracting shell, and will be discussed below.

It is obviously not meaningful to have more than two additional singular lines joining the arcs at the junction, as the motion of the additional sector created would be entirely arbitrary and in no way connected with the formation of the arcs.

There is one additional possibility; that is when the junction of two arcs forms no kink at all. If the arcs are to be distinct, this suggests that they must be joined in reverse. It is anticipated that this arrangement will give rise to the reversed arcs as discussed in Sec. 1.43.

The primary effects discussed above are not immediately manifest on the surface of the Earth. Owing to *secondary effects* occurring within the surface layer of the Earth, features additional to those occurring in the contracting layer may be expected.

The surface shell of the Earth overlies the contracting layer. Any movement in the contracting layer must therefore transmit itself to the surface shell because the two layers act upon each other. It will therefore be expected that fractures within the contracting layer will also be reflected as fractures on the surface of the Earth. The types of fracture to be expected at the surface will depend on the relative motion of the fractured parts of the underlying (contracting) shell. There will be two extreme cases: viz. if the motion of the contracting shell is at right angles to the singular line, a hiatus will be the result on the surface; contrariwise, if the relative motion of the contracting shell is parallel to the singular line, a simple shearing fracture or shift will be the surface result. The two types of fractures are illustrated in Fig. 92.

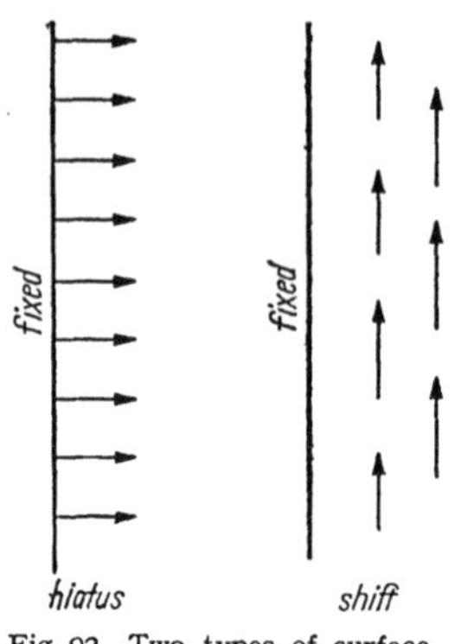

Fig. 92. Two types of surface fracture

There will be some additional effects, however. It is a salient feature of the contraction hypothesis that it assumes that the Earth as a whole is shrinking. The surface is assumed to be in a state of (planar) compression. Anticipating results to be presented in the sections on the theory of folding, we shall assume that *buckling* will be responsible for determining the position of mountain ridges to be expected due to the compression.

If we examine the phenomena occurring in the contracting layer with reference to their effect on the surface, it is seen that the case illustrated in Fig. 90 where there is only *one* singular line at the junction, will give rise essentially to a hiatus. Were it not for the compression which is assumed to be superimposed, the surface expression of this hiatus would always be a trench. Owing to the compression, however, this is not necessarily the case, as following the formation of a trench, the latter may play the rôle of a weak region in a shell about to fail, so that all that can be said is that the singular lines below are reflected as tectonic disturbances on the surface. These disturbances are arrayed in exactly the same manner as below, i.e. they correspond to "linkages".

More complex are the phenomena to be expected if there are *two* additional singular lines originating from the junction point of arcs. In this case, the relative motion of the contracting shell is essentially parallel to the singular lines so that a shearing fracture or shift would be expected to be the surface result. Geologically, this would cause lineaments on the Earth's surface.

Such lineaments have actually been observed in certain cases, but the superimposed (assumed) compression may be held to be able to change the picture considerably. The shearing may occur under any circumstances and the surface will therefore be cut up into a tooth-shaped triangular pattern which is pointed toward the junction points of the arcs. The maximum size of the triangles depends on the size of the arcs as illustrated in Fig. 93: a triangle cannot be bigger than that formed by the centers and the junction points of two adjoining arcs.

We have outlined above that we shall assume now that buckling will determine where these triangles will fold up under the assumed compression; for a discussion of this assumption one may refer to Sec. 7.32. The buckling of a triangle can easily be calculated. If we denote the angle at the vertex by β, die distance from the vertex of a line parallel to the base by x, and the deflection of that line from the original plane of the triangle by y, the condition of static equilibrium yields

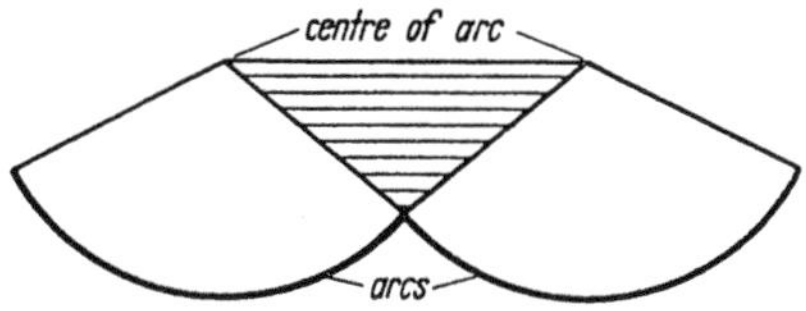

$$-Fy(x) = Mx \tan\beta \, d^2y/dx^2. \quad (6.24\text{–}2)$$

Fig. 93. Maximum size of triangle (shaded) into which the surface may be cut owing to a pure shearing motion of the contracting shell

The left-hand side of the equation is the moment of a force F parallel to the original plane of the triangle and normal to the base, applied at the vertex (that is the force causing buckling), and the right-hand side is the resistance to bending provided by the material of the triangle, M being a constant. The solution of this differential equation can be represented in terms of a Bessel function

$$y = x^{\frac{1}{2}} J_1[2x^{\frac{1}{2}}(M \tan\beta)^{-\frac{1}{2}} F^{\frac{1}{2}}] \quad (6.24\text{–}3)$$

whose zeros are at $X=0$, 3.83, ... with

$$X = 2(M \tan\beta/F)^{-\frac{1}{2}} x^{\frac{1}{2}}. \quad (6.24\text{–}4)$$

The maxima of y are at those points where $dy/dx=0$. Thus

$$dy/dx = \tfrac{1}{2} x^{-\frac{1}{2}} J_1(X) + (M \tan\beta/F)^{-\frac{1}{2}} J_2'(X). \quad (6.24\text{–}5)$$

However, one has

$$J_1' = -\tfrac{1}{2}(M \tan\beta/F)^{\frac{1}{2}} x^{-\frac{1}{2}} J_1(X) + J_0(X) \quad (6.24\text{–}6)$$

so that it is seen that the zeros of $y' = dy/dx$ are the same as those of $J_0(X)$. The latter are $X=2.40$, 5.52, ...

In order for buckling to be possible, the buckling force F must have such a value that y is zero for $x=0$ and $x=h$ (the latter being the height of the triangle); thus one has for first-order buckling

$$3.83 = 2(M \tan\beta/F)^{-\frac{1}{2}} h^{\frac{1}{2}}. \quad (6.24\text{–}7)$$

The maximum deflection (the "ridge") is obtained for $X=2.40$; thus

$$2.40=2(M\tan\beta/F)^{-\frac{1}{4}}x_{\max}^{\frac{1}{4}} \qquad (6.24\text{–}8)$$

whence

$$x_{\max}=0.393\,h. \qquad (6.24\text{–}9)$$

It is thus seen that the distance of the buckling ridge from the vertex of the triangle depends only on the height of the triangle and not on the angle at the vertex.

If we apply these results to the buckling of the surface of the Earth, the following may be seen.

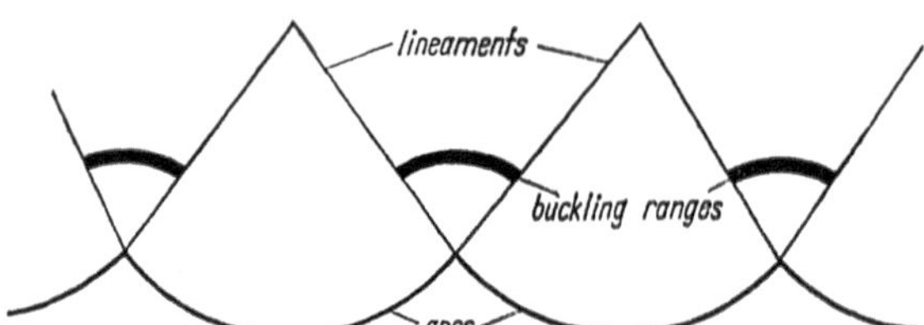

Fig. 94. Geographical appearance of buckling ranges

In the case of two singular lines, lineaments may be expected to start out from the junction of two arcs. Some distance away, a high ridge (the buckling ridge or the secondary arc of a common deflection) would be formed facing the junction. The foregoing discussions refer to a plane. On a sphere, conditions may be slightly modified so that the picture shown in Fig. 94 may be expected to result. In the case of pure shear, the lineaments are at 90° to the arcs. The larger the angle between the arcs, the smaller is the angle at the vertex of the triangle, and the larger the triangle may be. Therefore, the smaller the angle between the arcs, the closer to the junction point the buckling range will be found. If it is very close, it could correspond to the "cap range", if it is far away, it could correspond to the "front range".

Finally, one can also explain reversed arcs by the earlier remark that there is one possibility for which there may be no kink between arcs: when the arcs are joined in reverse. The surface features connected with reversed arcs may then be thought of as a direct manifestation of the arrangements in the contracting shell. The contraction theory, thus, yields an acceptable explanation of the observed junctions of island and mountain arcs.

6.25. The Extension Factor. After having outlined such features of the contraction hypothesis that lead to reasonable results, let us now examine some implications that are somewhat less satisfactory.

We have discussed a basic relationship (in Sec. 6.13) which every theory of orogenesis must fulfill. Let us investigate how well the contraction hypothesis as outlined above is compatible with this relationship. The depth h to which orogenesis is felt is given by the depth of the level of no strain; thus

$$h = 70\ \text{km}. \tag{6.25-1}$$

This yields at once, in virtue of (6.13-5):

$$\gamma' = 5.8. \tag{6.25-2}$$

According to earlier estimates (Sec. 6.23) an orogenetic system can be created by a shortening of the Earth's radius by 3 km. This yields a shortening of each meridian by (approximately) 9 km, or a true shortening of

$$s''_T = 4.5\ \text{km} \tag{6.25-3}$$

for one mountain range since a meridian intersects the orogenetic system twice. However, according to Eq. (6.13-2), the true shortening (with $s_A = 50$ km) is given by

$$s'_T = s_A/\gamma' = 8.6\ \text{km}. \tag{6.25-4}$$

The fact that the two estimates for the true shortening (viz. s'_T and s''_T) are not identical, shows that the usual form of the contraction theory is inconsistent in the light of the basic Eq. (6.13-4).

Finally, it may be remarked that the discrepancy cannot be removed by introducing a metamorphosis factor (cf. 6.13-6) ζ different from (i.e. greater than) 1. One simply needs less for s' and s'' to produce the required volume V, but the true shortening is reduced by the same factor in both cases. The discrepancy can also not be removed by choosing a different depth for the level of no strain since both estimates of s_T are proportional to the inverse of that depth (note that one must put $b = h$ in [6.23-6]).

As noted on earlier occasions, the contraction theory has also been advocated as an explanation of oceanic ridges. Thus let us repeat the above calculations for *marine* features[1]. Using the values found in Sec. 6.12, with the depth, (h), to which orogenesis is felt equal to 70 km, we have for the true shortening required

$$s'_T = \frac{A}{h} = \frac{2400}{70}\ \text{km} = 34\ \text{km}. \tag{6.25-5}$$

On the other hand, we have shown above (Sec. 6.23) that the radius of the Earth must contract by about 9 km to cause the bulge represented by the mid-ocean ridges. This corresponds to the shortening of a meridian by $2 \times 10^4 \times 1.4 \times 10^{-3} = 28$ km. Each meridian intersects the system of ridges once, or nearly so does, so that this figure is equal to the crustal shortening to be expected:

$$s''_T = 28\ \text{km}. \tag{6.25-6}$$

[1] Scheidegger, A. E.: J. Alberta Soc. Petrol. Geol. 6, 266 (1958).

It is seen that the crustal shortening obtained in the two ways is of the same order of magnitude. This is in contrast to what occurs in an investigation of continental mountains, where it has been shown that the depth of the level of no strain at any level is incompatible with the shape of these mountains. The shortening required in the contraction theory is thus not in such direct conflict with the shape of the oceanic ridges as it is with the shape of continental mountains, although here, too, is there a discrepancy.

6.26. Compatibility with Seismic Data. The chief results of statistical evaluations of seismic data, as had been explained in Sec. 2.24, are (a) that the great majority of earthquakes represents transcurrent faulting, (b) that there is some doubt whether the earthquake foci are really arranged on such a well-defined plane beneath recent island arcs as had been thought before, and (c) that, considering the dip-slip components of earthquakes, there is no indication of a level of no strain. All these results, if correct, constitute severe difficulties regarding the geological implications of the contraction theory.

First, if the earthquake foci are not arranged in planar zones beneath orogenetic systems, then there is obviously no support for the hypothetical cross-section of such systems as envisaged in Fig. 87. Since, in earthquakes below 70 km depth, thrust faulting is preponderant, the assumption of a level of no strain at that depth and a state of internal tension below that depth seem highly questionable. Second, from the drawing in Fig. 87, it is also apparent that there is no room for any transcurrent adjustments as have been found to take place during earthquake-faulting. It is therefore evident that there is not much support for the contraction hypothesis from recent seismological investigations.

6.27. Compatibility with Oceanic Features[1]. The general difficulties that one encounters in connection with continental orogenesis in the contraction theory do not become less if oceanic features are also taken into account. Thus, the general pattern of stress distribution in the contraction hypothesis does not allow for anything but thrust faulting to occur above the level of no strain (at 70 km depth) and tension faulting below that level. The graben-type topography of the crest of the Mid-Atlantic Ridge, therefore, remains totally unaccounted for. In addition, the mid-ocean ridges are assumed to be akin to folded mountains; no other type of origin could possibly be envisaged. The archipelagic aprons and volcanic islands must be chance-occurrences that do not fit at all into the general scheme of the contraction hypothesis.

Other marine features which seem to remain unexplained in any version of the contraction theory are: the gravity over oceans and,

[1] After SCHEIDEGGER, A. E.: J. Alberta Soc. Petrol. Geol. 6, 266 (1958).

especially, the geographical trend of the mid-ocean ridges. Whereas in the contraction theory there exists a reasonable explanation for the circular strike of off-shore island arcs and for the type of their junctions, etc., no such explanation can be obtained for the seemingly rather irregular geometry of the ridges.

6.3. Continental Drift Theory

6.31. Principles. After the contraction theory, we turn our attention to the continental drift theory of orogenesis. This theory assumes the whole Earth as practically liquid, with the land masses floating upon it in almost total local isostasy like ice on water. Forces of unknown origin are pushing the land masses around and their interaction with the substratum and with each other gives cause to orogenesis.

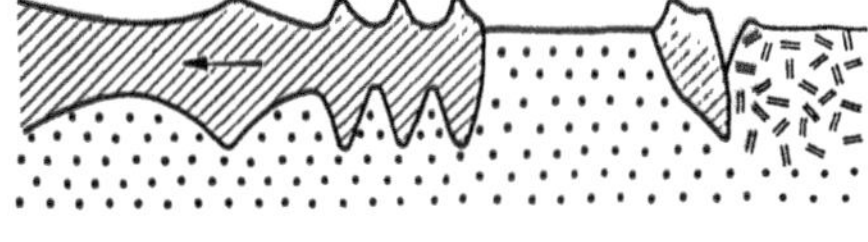

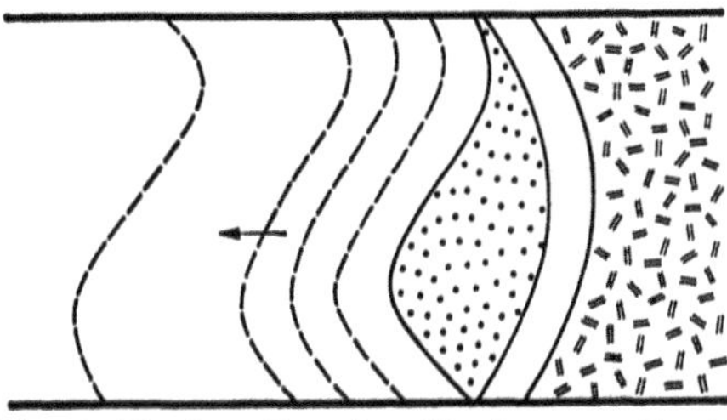

Fig. 95. The formation of an island arc according to the drift theory of WEGENER[1]

The continental drift theory has been supported chiefly by WEGENER[1]. The development of the theory as presented by that author is mainly based on physiographic evidence as mentioned in Sec. 1.31. This general physiographic evidence has then been (heuristically) followed up into much detail[2–7], even to the investigation of ancient wind directions[8, 9].

Thus, the crust of the Earth is assumed to have originally solidified to form a uniform, thin layer of the composition of the present continents. Due to some unknown cause, it broke up and started to drift around on the still liquid substratum. During this process, the "front" of any land mass (with respect to the drifting motion) drags against the substratum

---

[1] WEGENER, A.: Die Entstehung der Kontinente und Ozeane, 3. Aufl. Braunschweig: Vieweg & Sohn 1922.

[2] MA, T. Y. H.: Proc. 8th Pac. Sci. Cong. **2**A, 731 (1956). — Res. Past Climate and Cont. Drift **13**, 22pp. (1957). — Bull. Volc. **21**, 103 (1959).

[3] GUSSOW, W. C.: J. Alberta Soc. Petrol. Geol. **6**, 253 (1958).

[4] HAPGOOD, C. H., and J. H. CAMPBELL: The Earth's Shifting Crust. New York: Pantheon 1958.

[5] WILSON, D. W. R.: J. Alberta Soc. Petrol. Geol. **6**, 174 (1958).

[6] HEEZEN, B. C.: Coll. Int. Cent. Nat. Rech. Sci., Paris **83**, 295 (1959).

[7] HAVEMANN, H.: Geologie **10**, 185 (1961).

[8] RAASCH, G. O.: J. Alberta Soc. Petrol. Geol. **6**, 183 (1958).

[9] LAMING, D. J. C.: J. Alberta Soc. Petrol. Geol. **6**, 179 (1958).

and therefore becomes folded up. Behind, island arcs are broken off as demonstrated in Fig. 95. The net effect of this process is that the land masses became thicker vertically and smaller horizontally during geological evolution. This concept is exactly opposite to the idea of growing continents as exhibited in the discussion of the contraction hypothesis.

The history of the continents since the Carboniferous epoch was traced back by WEGENER from the evidence listed in Sec. 1.3. He arrived at the result that all the continents formed at that time a connected block which broke up subsequently, gradually yielding the distribution as we know it to-day. The evolution as postulated by WEGENER is shown in Fig. 96. Accordingly, the average drift of the southern continents would have been some 14000 km since the Carboniferous (cf. Sec. 1.31).

In contrast to WEGENER[1], DU TOIT[2] postulated the primeval existence of *two* continents at the poles of the Earth, called Laurasia and Gondwanaland, originally formed by convection. By a mechanism similar to that envisaged by WEGENER, these *two* original continents are then supposed to have broken up and to have gradually drifted into the present distribution of land masses. It is, in fact, quite possible to start with any hypothetical primeval layout of land and to arrive at the present distribution by a suitable shift. The particulars of the shifts are, in any case, developed from purely physiographic considerations.

With regard to marine orogenesis[3], drift could have caused mid-ocean ridges by inducing fissuring, through which material from below could have been extruded, in the wake of the moving continents. The present system of ridges is more or less located in such areas from which the continents would have moved away, according to the visualization of WEGENER. This would also explain the median rift on the ridges.

In this explanation of orogenesis, it is not necessary to assume that crustal shortening has taken place across mid-ocean ridges. The latter may be assumed to be built up entirely from volcanic material. Consequently, the estimates regarding the total availability of volcanism, rather than those regarding the values of crustal shortening, may be applicable. Since it has been shown that the ridges may have been built up in as little as 200 million years (using SAPPER'S[4] [cf. Sec. 5.24] estimate), there is no difficulty with such an explanation. However, if the lesser estimate of VERHOOGEN'S[5] is used, some 800 million years

[1] WEGENER, A.: Die Entstehung der Kontinente und Ozeane. 3. Aufl. Braunschweig: F. Vieweg & Sohn 1922.

[2] DU TOIT A.: Our Wandering Continents. Edinburgh: Oliver & Boyd 1937.

[3] After SCHEIDEGGER, A. E.: J. Alberta Soc. Petrol. Geol. **6**, 266 (1958).

[4] SAPPER, K.: Vulkankunde. Stuttgart 1927.

[5] VERHOOGEN, J.: Amer. J. Sci. **244**, 745 (1946).

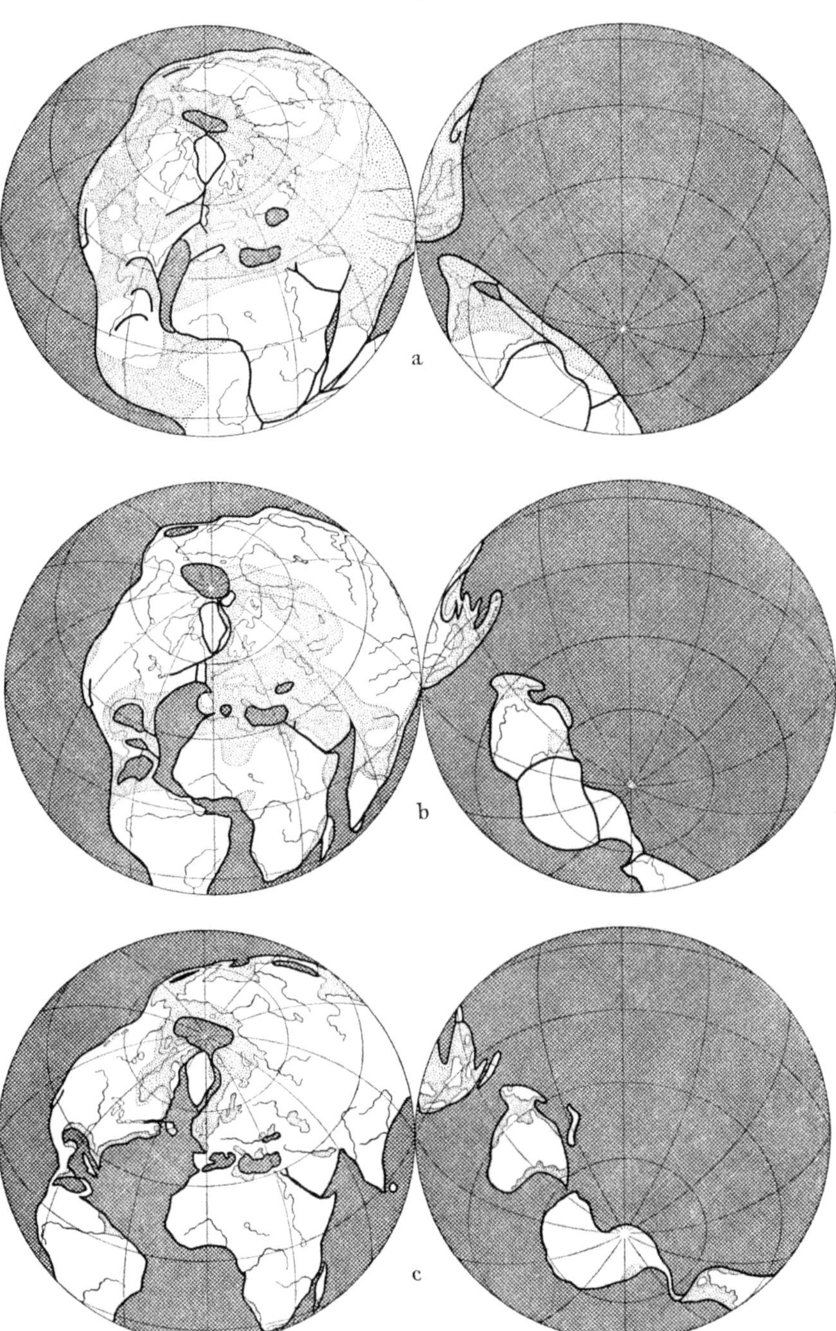

Fig. 96a–c. The evolution of continents as envisaged by WEGENER. a Carboniferous, b Eocene, c Pleistocene

would be required to build up the ridge system. This would severely curtail the possible movements of the continents, since they obviously could not have moved over an area where there are now ridges, at a time less than some 800 million years ago. If ridges are simply bulges in the crust, without crustal thickening, then they could have been built up very rapidly and hence the continents could have moved around very easily.

If SAPPER's estimate is adopted, then it is possible to assume that other ridges caused by the continual drifting of the continents have existed at other times. That such a phenonmenon might have taken place is indicated by the fact that the relief of the Pacific Ocean bottom shows quite a varied topography. The latter might be regarded as the remnant of old mid-ocean ridges, worn-down possibly by turbidity-current erosion. Difficulties, however, are encountered if lower rates of volcanic activity are assumed.

Similarly, if the ridges are crustal bulges (without thickening), systems which subsequently have again subsided could have existed at various times, in conformity with the wandering of the continents nearby. Furthermore, the hilly relief of the Pacific Ocean bottom could be explained as a "remnant" of the ridges and there would be no necessity to resort to the erosional properties of hypothetical turbidity currents.

The physiographic theories outlined above fit many of the observed facts rather well. One of the chief difficulties, however, is the explanation of the occurrence of earthquakes to a depth of 700 km. The existence of deep-focus earthquakes was, of course, not known at the time when WEGENER developed his theory. A further difficulty is that drift theories usually envisage that the "primeval" state of the Earth subsisted essentially to a time not much before the beginning of the traditional geological time scale. That the latter constitutes only about one-fifth of the total geological history of the Earth, was also not known before the development of the recent techniques for radioactive age determinations.

Nevertheless, because of its good aspects, it will be necessary to examine the continental drift theory somewhat further.

6.32. Extension Factor. Transcurrent Faulting. In Sec. 6.13 we have introduced the extension factor γ. In the continental drift theory, the value of this extension factor comes out as equal to 1.25. This can be shown by recalling Eq. (6.13–5)

$$h/\gamma = 12\ \text{km}. \qquad (6.32\text{–}1)$$

If we insert for h the weighted mean thickness of the crust as defined by the Mohorovičić discontinuity, viz.

$$h = 15\ \text{km}, \qquad (6.32\text{–}2)$$

we obtain

$$\gamma = 15/12 = 1.25 \tag{6.32-3}$$

as claimed above. This means that the continental drift theory permits the apparent shortening in mountains to be approximately equal to the true shortening. The fact that the assumption of an extension factor very nearly equal to 1 is consistent with the limitation of folding to precisely that depth which is indicated by the Mohorovičić discontinuity, is a strong argument in favor of a drift theory.

A similar strong argument in favor of the theory considered here follows from a consideration of the results from fault plane solutions of earthquakes. In Sec. 2.24 we have demonstrated that the average horizontal motion in earthquakes is large compared with the corresponding vertical one. This is well in accordance with the idea of large continental shifts, whatever may be their cause.

If one uses the GUTENBERG low velocity layer at 140 km depth as the base of the drifting continents rather than the Mohorovičić discontinuity, it would not be so easy to visualize the drifting motion.

It thus appears that the continental drift theory yields a basically very satisfactory explanation of geophysical facts. Of the criticisms that might be launched against it is chiefly the existence of deep-focus earthquakes. However, this can be countered by the remark that the drifting motion must of necessity belong to the "long" time range in the sense of Sec. 3.6, whereas the build-up of stresses leading to earthquakes belongs to the "intermediate" time range. Since it has been shown earlier (in Sec. 3.6) that the dynamical behavior in the various time ranges may be entirely different, it is quite possible that the drifting motion and the occurrence of earthquakes can take place concurrently at their own individual characteristic speeds.

6.33. Origin of the Forces Causing Drifting. No theory of orogenesis can be called such until it is put upon a sound mechanical basis. The continental drift theory, so far, is a purely heuristic *hypothesis* which has been rather successful in explaining a series of observations. The crux of this hypothesis, however, lies in whether it will be possible to find a reasonable explanation for the mechanical cause of the postulated continental drifts.

WEGENER himself thought that EÖTVÖS' polfluchtkraft would provide a suitable physical basis for his continental drift idea. However, since PREY questioned the existence of the polfluchtkraft altogether (cf. Sec. 4.23), WEGENER'S assumption does not seem to be very strongly supported. A similar conclusion must be reached with regard to tidal and Coriolis forces which are also much too small to have any significant effects in relation to orogenesis.

The question remains, then, to postulate a mechanical process which would provide for the forces necessary to produce the continental shifts. To-date, no such mechanical process is known. It is interesting to note, however, that certain observations can be made regarding some of the properties of this process.

The basic data that are to be explained are as follows:

(i) Drift of India since the Eocene (60 million years ago) has been 6000 km.

(ii) Drift of India since the Carboniferous (265 million years ago) has been 14000 km.

It is obvious that the above data cannot be explained by the assumption of uniform motion, since the "short-term velocity", v:

$$v = \frac{6000 \text{ km}}{60 \times 10^6 \text{ yrs.}} = 10 \text{ cm/year}$$

is much greater than the "long-term velocity", V:

$$V = \frac{14000 \text{ km}}{265 \times 10^6 \text{ yrs.}} = 5 \text{ cm/year}.$$

It turns out that the actual values of the continental drifts as suggested by the factual (although still somewhat doubtful) evidence can be explained by the assumption of a *random* force[1]. This implies that the continents are subject to random drifting.

Since geological investigations lead to a *continental path*, the logical type of analysis of continental drift is the Lagrangian analysis. We assume that the mean velocity $\bar{v}$ of all the continents is zero; for the analysis, the Earth can of course be considered as flat (co-ordinates of a continent are then Cartesian; they may be denoted by x and y). Thus we have

$$\bar{v}_x = \bar{v}_y = 0, \tag{6.33-1}$$

$$\overline{v_x^2} = \overline{v_y^2} = \tfrac{1}{2}\,\overline{v^2} = \text{const}. \tag{6.33-2}$$

A value for the velocity-square $\overline{v^2}$ can be obtained from the short-term velocity v calculated above; one obtains

$$\overline{v^2} = 10^2 \text{ cm}^2/\text{year}^2 = 100 \text{ [cm/year]}^2. \tag{6.33-3}$$

Then the total displacement of a continent evolving in time is

$$x = \int_0^t v_x(\tau)\, d\tau, \tag{6.33-4}$$

$$y = \int_0^t v_y(\tau)\, d\tau. \tag{6.33-5}$$

---

[1] SCHEIDEGGER, A. E.: Canad. J. Phys. **35**, 1380 (1957). — J. Alberta Soc. Petrol. Geol. **6**, 170 (1958).

This leads to[1]

$$\overline{x^2} = \overline{y^2} = \Phi(t) \tag{6.33-6}$$

where $\Phi(t)$ can be calculated as follows:

$$\overline{x^2} = \overline{\left[\int_0^t v_x(\tau)\,d\tau\right]^2} = \overline{\int_0^t \int_0^t v_x(\tau_1)\,v_x(\tau_2)\,d\tau_1\,d\tau_2}. \tag{6.33-7}$$

If we introduce the *Lagrangian correlation coëfficient*:

$$R_x(\tau) = \overline{v_x(t)\,v_x(t+\tau)}/\overline{v_x^2}, \tag{6.33-8}$$

relation (6.33–7) can be written as follows:

$$\overline{x^2} = \overline{v_x^2} \int_0^t \int_0^t R(\tau_1 - \tau_2)\,d\tau_1\,d\tau_2 \tag{6.33-9}$$

and, upon making a minor transformation[1]

$$\overline{x^2} = 2\overline{v_x^2} \int_0^t (t-\tau)\,R_x(\tau)\,d\tau. \tag{6.33-10}$$

The last relation is useful to investigate limit cases. Thus, introducing the Lagrangian autocorrelation time L_t

$$L_t = \int_0^\infty R(\tau)\,d\tau \tag{6.33-11}$$

one can for instance investigate the case

$$t \gg L_t. \tag{6.33-12}$$

One then obtains from (6.33–10)

$$\overline{x^2} \simeq 2\overline{v_x^2} L_t t. \tag{6.33-13}$$

With

$$\overline{r^2} = \overline{x^2} + \overline{y^2}, \tag{6.33-14}$$

this yields

$$\overline{r^2} = 2\overline{v^2} L_t t. \tag{6.33-15}$$

This formula for the average square of the displacement is valid for time intervals which are long compared with L_t. If we insert the estimates for the displacements and the velocity, we obtain a value for the autocorrelation time L_t. Recalling that the displacements of the pieces of "Gondwanaland" since the Carboniferous epoch ($t = 2.65 \times 10^8$ years) are on the average 14000 km, and using the earlier value of 100 (cm/year)² for $\overline{v^2}$, we obtain

$$L_t = \frac{\overline{r^2}}{2\overline{v^2} t} = \frac{t}{2} \cdot \frac{\overline{V^2}}{\overline{v^2}} \simeq 30 \times 10^6 \text{ years}. \tag{6.33-16}$$

[1] Kampé de Fériet, J.: Ann. Soc. Sci. Bruxelles. Sér. I 59, 145 (1939). See also Pai, S.: Viscous Flow Theory, vol. II, p. 174. New York: D. van Nostrand Inc. 1957.

It may be noted that this is indeed small compared with the time since the Carboniferous (as was supposed when making the calculation). On the other hand, for very short time intervals, one obtains

$$\overline{r^2} = \overline{v^2}\, t^2 . \tag{6.33-17}$$

Therefore, if the time since the Eocene (60 million years) can be regarded as "short", the time since the Carboniferous (265 million years) as "long", one can explain the order of magnitude of the drift of India and of the drift of the pieces of Gondwanaland simply by assuming a random drift with a velocity-square of, on the average, 100 cm^2/year^2 and an autocorrelation time of 30 million years. It is worthy of note that the autocorrelation time is precisely of the minimum order of magnitude to divide "short" from "long" time intervals to make the theory consistent.

The above kinematical discussion of continental drift does not elucidate the nature of the random force which is supposed to cause the drifting. One of the efforts to do this has been the assumption of subcrustal convection currents with which we shall deal in Sec. 6.4. However, this introduces further difficulties in that it requires a special mechanism below the crust. From factual evidence with regard to continental drift, there is no indication that anything special should happen below the continents: the deep-focus earthquakes could be treated as secondary effects. What one would like to have, therefore, is a universal force capable of shifting the "floating" masses around. Although this has not yet been found, it appears that the continental drift idea is one of the more hopeful alleys in which the cause of orogenesis might be further explored.

6.4. Convection Current Hypothesis of Orogenesis

6.41. General Principles. We have shown above that there are indications which make it conceivable that parts of the Earth's crust underwent large displacements. We have also shown that it is difficult to envisage forces which would be large enough to produce such shifts. It has also been seen that the contraction theory is unable to account for the various types of stress states found in the Earth; this has led people to the search for other mechanisms. One of these is represented by convection currents, the history of which has been reviewed e.g. by KRAUS[1]. The concept of convection currents seems to have been postulated first by AMPFERER[2], who assumed that currents exist in the mantle of the Earth whose drag would be sufficient to supply the forces of orogenesis.

[1] KRAUS, E. C.: Geologie **7**, 237 (1958).

[2] AMPFERER, O.: Jb. geol. Reichsanst. **56**, 539 (1906).

Subcrustal convection currents, as is implied by the name, are hypothetical currents which are supposed to exist in the mantle of the Earth. In general terms, they are thought to supply orogenetic forces such as shown in Fig. 97. The patterns of such convection currents that have been proposed are quite varied. A good summary has been given in the colloquium of the American Geophysical Union on plastic flow and deformation within the Earth under the chairmanship of ADAMS[1]. We shall discuss these proposals one by one below.

The question of the physical possibility of convection currents is tied up with the chemical and physical conditions in the Earth's mantle. In particular, if the chemical or physical phase-change at 900 km depth

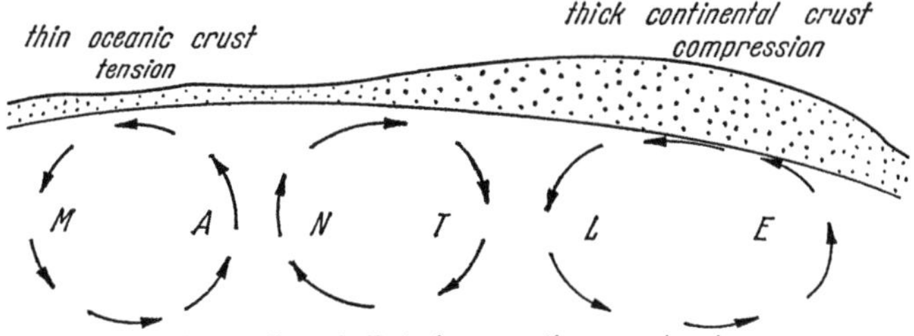

Fig. 97. General effect of a convection current system

which has been postulated by BIRCH (cf. Sec. 2.81) is real, it would presumably obviate any convection hypothesis. Convection currents are possible only if the mantle consists of homogeneous material. The possibility of currents under such conditions depends on the assumed rheological equation, on the assumed thermal gradient and on the assumed thermal properties of the interior of the Earth;—in other words, there are a great number of assumptions involved. It is quite obvious that these assumptions can only be tested by an elaborate discussion of the interior of the Earth. The chemistry and physics of the interior of the Earth, however, forms a subject distinct from geodynamics and is even more speculative than the latter. For the present purpose we shall therefore confine ourselves to the calculation of some of the critical parameters without going into too many details.

6.42. Steady-State Convection. The most straightforward way to establish a convection-current theory of orogenesis is by investigating large-scale patterns of thermo-mechanical flow systems that might become established in the mantle of the Earth. This leads to steady-state convection-currents in which the motion of the fluid particles would be continual. The established patterns would be world-wide; they have been discussed e.g. by PEKERIS[2] and by CHANDRASEKHAR[3].

[1] ADAMS, L. H.: Trans. Amer. Geophys. Un. **32**, 499 (1951).
[2] PEKERIS, C. L.: Month. Not. Roy. Astron. Soc., Geophys. Suppl. **3**, 343 (1935).
[3] CHANDRASEKHAR, S.: Phil. Mag. **43**, 1317 (1952).

Because of the world-encompassing scale of steady convection currents, one would expect that they would influence the distribution of continents rather than the mechanics of mountain-building. In this

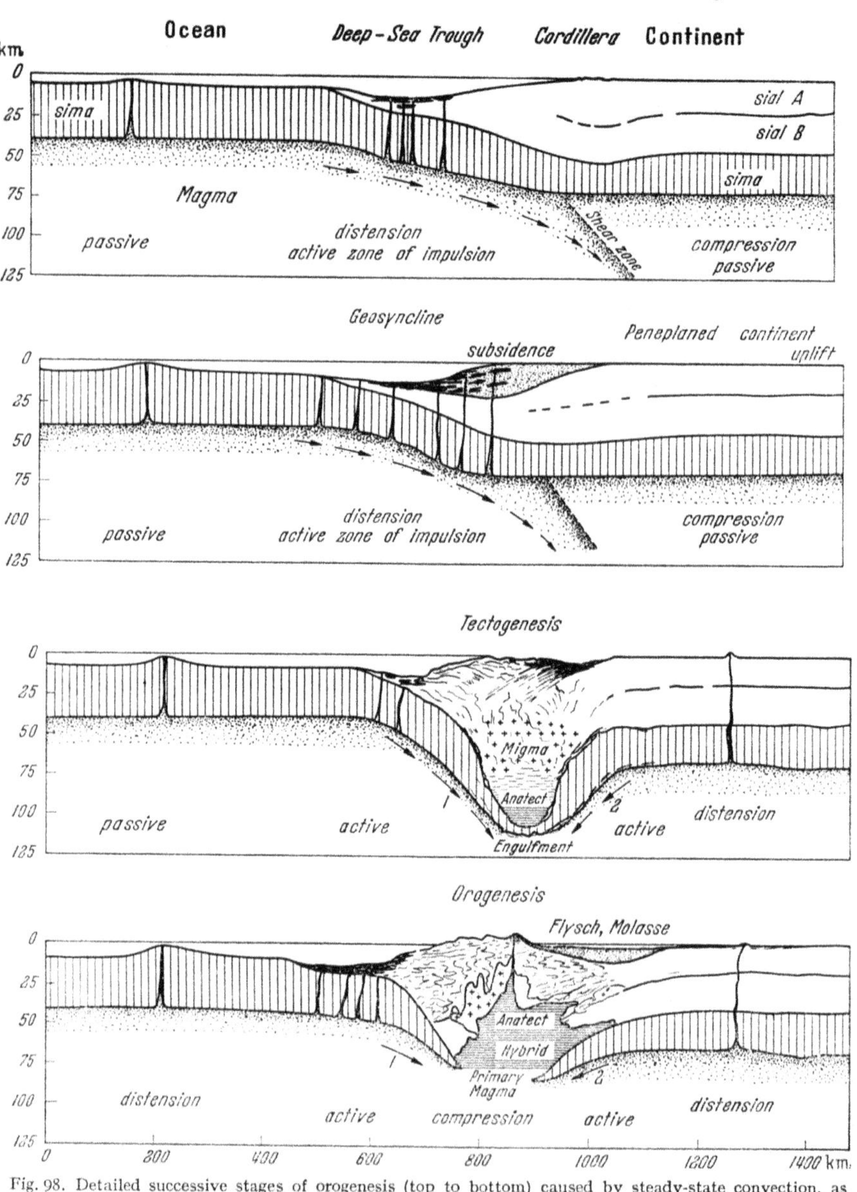

Fig. 98. Detailed successive stages of orogenesis (top to bottom) caused by steady-state convection, as envisaged by RITTMANN[1]

[1] RITTMANN, A.: Bull. Soc. Geogr. Égypte **29**, 111 (1956).

connection, they have been discussed in Sec. 5.53. Mountain-building would be induced by them only owing to secondary effects, particularly owing to drag-effects which would occur at the margins of continents. In this instance, the mechanics of mountain-building would be identical to that envisaged in the continental drift theory (cf. Fig. 95): the drag of the convection currents would simply replace the Eötvös force.

However, if the above theory of mountain-building be adopted, it does not seem possible that orogenesis could be a continuing process because of the intrinsic time-independence of steady-state flow. Thus, once the continents are in their proper places, all the drag forces are at equilibrium and mountain-building should cease. Furthermore, the geographical distribution of orogenesis does not fit into a simple geometrical pattern as postulated e.g. in the scheme of PEKERIS (cf. Fig. 82). Therefore, modifications of the simple theory discussed above have to be attempted.

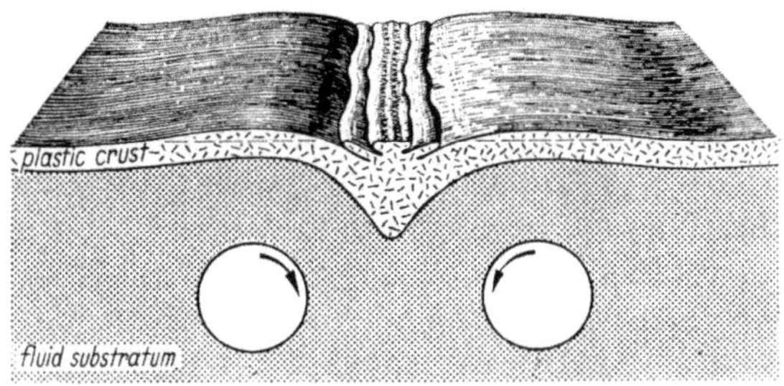

Fig. 99. Stereogram of GRIGGS'[1] model of convection currents showing the development of an orogenetic system

One of these modifications is to abandon the idea that there is a world-wide convection pattern, but to assume instead that, owing to a suitable temperature gradient in the mantle, many single convection currents would become established in various regions without any regularity in geometrical arrangement. Each convection current would have a closed axis so as to be in conformity with kinematical requirements, and may last only for a certain length of time. The geological significance of such little currents has been followed up in detail by RITTMANN[2]; the latter author's representation is shown in Fig. 98.

To substantiate the above ideas, GRIGGS[1] performed model experiments in which the currents were represented by rotating drums in a viscous liquid. A typical result, meant to represent the development of orogenetic systems and an accumulation of crustal material is shown in Fig. 99. Although the outcome of the experiments has a striking resemblance with certain orogenetic phenomena, there are, in fact, several difficulties to be overcome until GRIGGS' model studies can be taken as a basis for a theory of mountain building.

[1] GRIGGS, D.: Amer. J. Sci. **237**, 611 (1939).

[2] RITTMANN, A.: Bull. Soc. Géogr. Egypte **29**, 111 (1956); also: Arch. Sci., Genève **56**, 275 (1951).

First, the models deal only with the cross-section of mountain belts. The plan and the three-dimensional arrangement of convection currents in existing island and mountain arcs is completely neglected. It was mentioned above that the axes of convection currents have to be closed in themselves. The rotating drums correspond to these axes and the question arises, therefore, as to how they should be arranged if they are to represent orogenetic features. If they are connected with island arcs, they must be curved, too, and where the island arcs join each other, the axes of the convection currents must disappear. From a hydrokinematical standpoint this is, to say the least, very difficult to explain.

Second, although GRIGGS has been very careful to construct his models in such a fashion as to be in conformity with the laws of mechanical similarity, the thermal similarity has ben completely neglected. According to Sec. 3.33, the product λ of Grashoff and Prandtl numbers must be at least 1709 in order to start convection currents. A little calculation shows that this is possible in the *Earth* (for approximately $D=500$ km, $\varrho=3$ g/cm^3, $\beta=5\times10^{-6}$ deg^{-1}, $c=0.25$ cal/g deg, $\eta=10^{23}$ cgs and $k=5\times10^{-3}$ cal/sec cm deg, one obtains $\lambda=3000$), but the *thermal* facts have no analogues whatsoever in the *model*. It is questionable, therefore, whether the convection currents would occur in the supposed way. Several authors have attempted to determine whether a thermal instability is, in fact, possible in the mantle[1-3]. At best, the results of these investigations are inconclusive, at worst, downright negative. Thus the possibility of the very existence of convection currents remains neither affimed nor obviated.

Third, the convection currents described above are laminar and steady and thus do not reflect the fact that orogenetic events occur in single diastrophisms and not steadily.

Fourth, the concept of depressions, basic to GRIGGS' idea of mountain-building, has recently been questioned by the recent seismic work (cf. Sec. 2.13) which failed to detect a depression of the Mohorovičić discontinuity below deep-sea troughs.

Last but not least there is the ever-present possibility mentioned in Sec. 2.81 that a chemical or physical phase-change exists at about 900 km depth. If this is true, convection currents as envisaged above are impossible.

6.43. Intermittent Convection Currents. GRIGGS was fully aware of the fact that orogenetic cycles cannot easily be explained by steady-state convection currents. He therefore postulated that the substratum be

[1] JEFFREYS, H.: Geophys. J. Roy. Astr. Soc. **1**, 162 (1958).

[2] LATYNINA, L. A.: Izv. Akad. Nauk SSSR., Ser. Geofiz. **1958**, No. 3, 391 (1958).

[3] LYUSTIKH, E. N.: Izv. Akad. Nauk SSSR., Ser. Geofiz. **1960**, No. 1, 3 (1960).

pseudoviscous instead of viscous. It is not quite clear what he meant by the term "pseudoviscous". In the few calculations which he made, a rheological equation was used which resembles a creep condition of a Maxwell liquid, but the model which he constructed appears to show the characteristics of a St. Venant yield stress. At any rate, the idea was to introduce a substratum of such properties that it behaved like a solid until a sufficiently high temperature gradient was reached, and then executed a half revolution in laminar viscous motion. It then should stop dead, because the temperature gradient had been abolished by this process and not until a sufficient gradient had again been established was another cycle to start. A convection cycle should therefore have distinct phases which would be connected with corresponding phases of the mountain-building cycle. This connection is shown in Fig. 100. A modification of the above ideas has been suggested by REITAN[1] who proposed a "rock-back" in a convection cell.

The above ideas have been formulated somewhat more precisely by VENING MEINESZ[2] who stated clearly that he expected that a plastic substratum with a finite yield strength would be capable of sustaining intermittent convection currents. VENING MEINESZ supposed that a direct consequence of this finite yield strength would be the counteraction of the instability which otherwise the cooling of the Earth would bring about. He considered that this cooling causes a lowering of the temperature at the surface of the convective layer, the upper part of this layer thus acquiring a greater density than the lower half. VENING MEINESZ then proceeds to state that "in the case of a Newtonian liquid, this would mean instability, as the slightest disturbance would start a convection current. If, however, the layer possesses a certain strength, the disturbance must be powerful enough to overcome this strength before the current can be started."

In this connection, the reader is reminded of the investigations discussed in Sec. 3.33 into the thermohydrodynamics of viscous fluids and of the results reported in Sec. 3.44 regarding heat convection in general rheological substances. In view of these results, the statement that an arbitrarily small temperature gradient would mean instability in a viscous medium does not seem to be correct. The temperature gradient has to reach a definite value even in viscous media before instability occurs. The additional assumption of a finite yield strength, therefore, does not seem so tremendously important except to increase the temperature gradient necessary for instability.

Furthermore, VENING MEINESZ believes that some other starting phenomenon is also necessary to get the currents moving and that a

[1] REITAN, P. H.: J. Geol. **67**, 129 (1959).
[2] VENING MEINESZ, F. A.: Quart. J. Geol. Soc. Lond. **103**, 191 (1948).

plastic substance would cause the currents to be intermittent. He proceeds with the following statement:

"Because of the initial temperature difference needed to overcome the strength limit and start the current, we may assume that this will

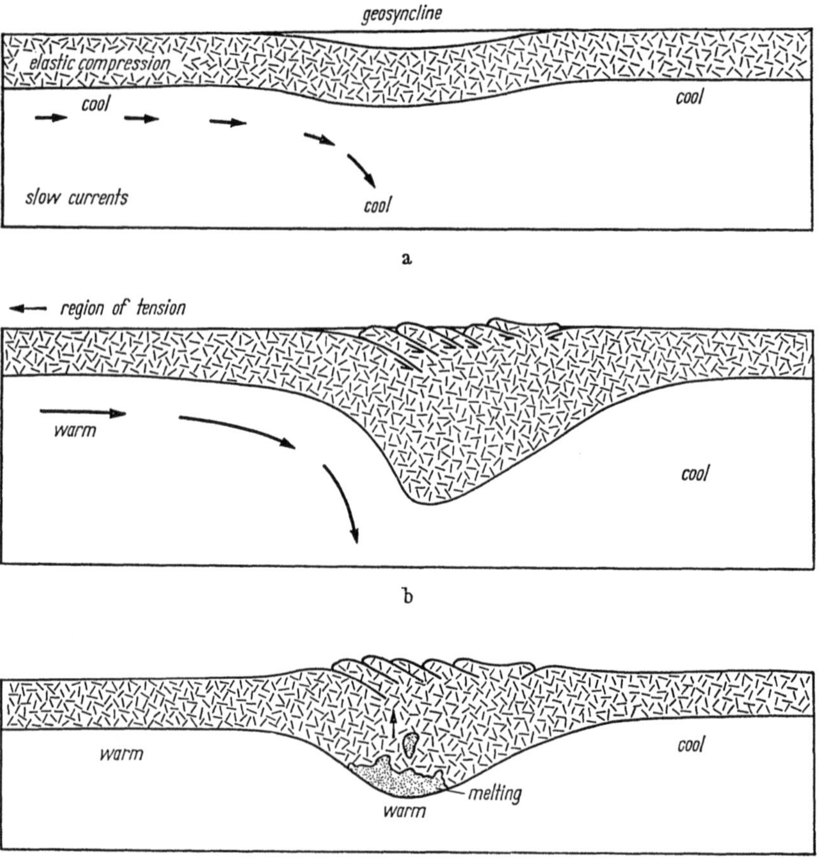

Fig. 100a—c. Hypothetical correlations between a convection-current cycle and a corresponding mountain-building cycle as envisaged by GRIGGS[1]; (a) first stage: period of accelerating current; (b) period of fast currents; (c) end of convection current: period of emergence of the mountains

quickly attain a fairly high speed in comparison with that which a steady convection in a Newtonian liquid would assume. Moreover, during the first quarter of a complete revolution it will no doubt further accelerate, since the sinking column will receive more and more low temperature matter from the surface while in the rising column the higher temperature of the deeper layers penetrates further towards the surface. So

[1] GRIGGS, D.: Amer. J. Sci. **237**, 611 (1939).

the difference of mass in the two columns caused by their different temperatures will increase till a maximum is reached after about a quarter revolution. This increasing mass difference will produce an increasing pressure difference, with a resultant acceleration of the current during this period. (...) After about a quarter revolution the speed will gradually decrease as higher temperature matter penetrates from above into the sinking column and lower temperature matter from below into the rising column. The mass difference between the two columns will be thereby diminished and the current brought gradually to a stop. This must occur before a half revolution is made, since at that point, roughly speaking, the higher temperature matter has arrived in the upper layer and the cooler matter at the bottom; the situation has become stable. No new current can come into being before first, the cooling of the Earth has re-established a vertical gradient of the temperature, and second, a starting phenomenon of sufficient intensity has occurred."

The assumption of a plastic substratum is thus believed to give rise to intermittent currents. The picture created by VENING MEINESZ is, however, very phenomenological and not based upon sound rheological reasoning. BROOKS[1] has made a mathematical analysis of convection currents in a plastic substratum and came up with the result that such currents as envisaged by VENING MEINESZ can actually exist. However, no full account of BROOKS' investigations seems to have been published and from the short communication mentioned above it is not possible to check the validity of the claim.

We have tried to obtain an idea of what happens when a plastic material is heated from below (in Sec. 3.44). Although an exact solution of the problem admittedly does not exist, the evidence seems to point toward a steady phenomenon, not too much different from convection currents in a viscous material. Therefore, VENING MEINESZ' suppositions need, to say the least, a lot more physical corroboration.

6.44. Roller Cell Theory. Another modification of the convection-current hypothesis of orogenesis seems to have been made by HESS[2]. In it is assumed that convection currents are *stacked* in the mantle of the Earth. The topmost convection cells are called *roller cells*, they reach from just beneath the crust to a depth of about 500 km. Below the roller cells are *mega-cells*. The arrangement is shown in Fig. 101.

The system of convection currents envisaged by HESS has been postulated purely upon physiographic grounds in order to accommodate a suspected lack of deep-focus earthquakes at a depth of approximately 475 km. Since this lack is by no means definitely established, and since

[1] BROOKS, H.: Trans. Amer. Geophys. Un. **27**, 548 (1941).
[2] HESS, H. H.: Trans. Amer. Geophys. Un. **32**, 528 (1951).

no investigations into the mechanism of roller cells have been made, the postulated existence of the latter is still largely a speculation.

From the roller cell theory of HESS, it is only a small step to the assumption that convection currents of various shapes and sizes are distributed randomly below the crust[1] dragging the continents hither and yon.

A support for this assumption may be obtained from the observation (cf. Sec. 6.33) that the alledged drifting motion of the continents can be described adequately by assuming that it is random. This naturally suggests that the forces causing the drift are also random, and in particular, that they are due to the drag of random convection currents. If the continents are assumed to be *large* with regard to the convection currents, then it is obvious that the effect of the latter on the former will, on the average, be zero, since the various convection currents are just as likely to pull in any one direction as in any other. However, if the continents are *small* compared with the convection cells, then the former will be totally immersed in the latter and therefore assume the surface speed of the convection current that happens to be underneath. The net effect of this argument is that the larger the continents, the slower should be their motion. It cannot yet be said whether this is actually the case. Evidence rather seems to indicate that all continents, regardless of size, move on the average equally fast. Therefore, if convection currents are the cause of the motion of the continents, they should be larger than the largest continents, which would make them of some 6000 km radius. The same 6000 km is then also the distance over which a continent may move at uniform speed; the velocity of the current just below the surface should be equal to the average velocity of the continents, viz. 5 cm/year. The currents, thus, would have to be of gigantic proportions indeed which, in turn, makes their reality somewhat doubtful.

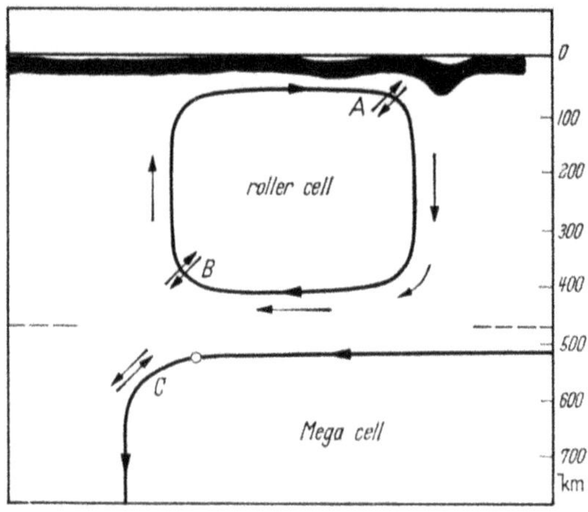

Fig. 101. Roller cell of HESS[2]

[1] MATSCHINSKI, M.: Ann. Geofis. 7, 1 (1954).

[2] HESS, H. H.: Trans. Amer. Geophys. Un. 32, 528 (1951).

The roller cell idea has been followed up by KRAUS[1] into very great geological detail. His view of the effect of roller cells is shown in Fig. 102.

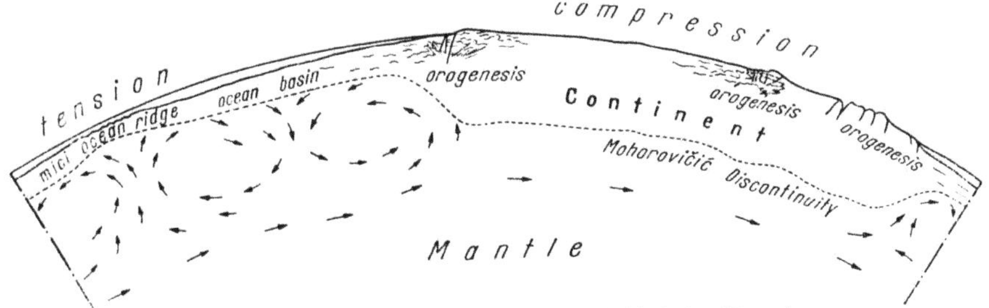

Fig. 102. Geological effects of roller-cells. Modified after KRAUS[1]

6.45. Rotation Fields. The latest modification of the convection-current theory seems to have been suggested by PAVONI[2] who assumes

Fig. 103. Principal fault slip pattern of the circum-Pacific margins, showing a rotation of that ocean. After BENIOFF[3]

that the most recent tectonic movements can be described in terms of rotation-fields. A rotation had already been suggested by BENIOFF[3] for

[1] KRAUS, E. C.: Geologie **7**, 261 (1958).
[2] PAVONI, N.: Vjschr. naturforsch. Ges. Zürich **105**, Sept. (1960).
[3] BENIOFF, H.: Publ. Dom. Obs. Ottawa **20**, 395 (1958).

the Pacific Ocean from seismic evidence (see Fig. 103) and PAVONI assumes that a similar picture can be postulated for the world as a whole. Thus, one would have three rotation fields, one associated with the Pacific Ocean, one with Laurasia and one with Gondwanaland. The idealized scheme of the rotations, then, is as shown in Fig. 104. Such

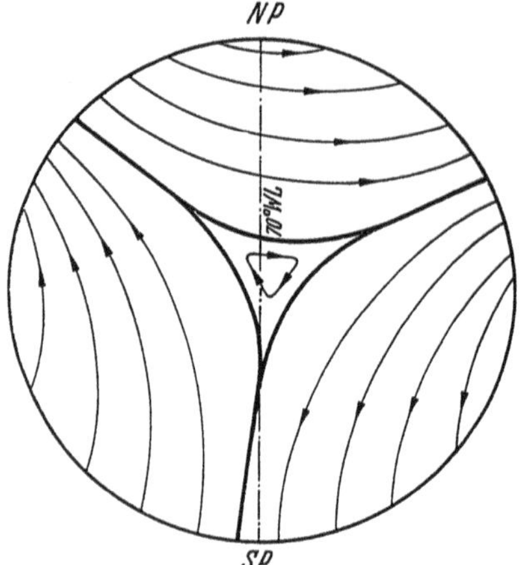

Fig. 104. Schematic drawing of the three rotation fields of the Earth. After PAVONI[1]

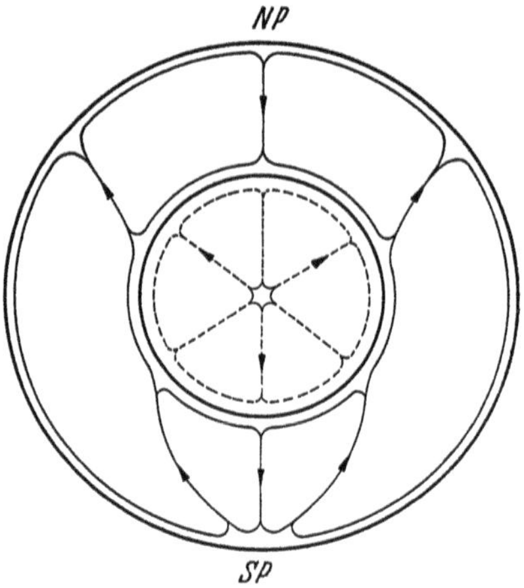

Fig. 105. World wide convection currents as an explanation of rotation fields. After PAVONI[1]

a scheme can be explained in terms of a world-wide system of convection currents if the latter are assumed as shown in Fig. 105.

6.5. The Hypothesis of Zonal Rotation

6.51. Principles. We now turn our attention to the next attempt at explaining orogenesis: to the hypothesis of zonal rotation. This hypothesis grew out of the observation that the Sun as well as the larger planets do not rotate uniformly like a solid body. In fact, the angular velocity ω increases in these celestial bodies towards the equator. For the Sun, the law of rotation has been determined empirically as follows

$$\omega = a + b \sin^2 \Theta, \tag{6.51-1}$$

where Θ is the heliocentric latitude and a and b are constants[2], b being negative. Setting $\cos \Theta = s/R$ (with s distance from axis of rotation, R radius of the Sun) this law can also be written:

$$\omega = \omega_0 + \lambda s^2 \tag{6.51-2}$$

where $\omega_0 = a + b$ and $\lambda = -b/R^2$. Then, λ is positive. No similarly definite law can be given for the larger planets, but there, too, it is quite clear that the angular velocity of rotation increases towards the equator.

[1] PAVONI, N.: Vjschr. naturforsch. Ges. Zürich **105**, Sep. (1960).

[2] SOTOME, K.: Proc. Imp. Acad. Tokyo **3**, 317 (1927).

It thus appears as possible to postulate that all bodies of the solar system, at least as long as they are in a fluid state, would exhibit the phenomenon of zonal rotation with increasing angular velocity towards the equator. The implications of such a hypothesis with regard to the Earth and orogenesis have been investigated by JARDETZKY[1-3] and by others[4-6]. In this connection, it must be imagined that at depth, the material is still fluid enough to show differential rotation, or else that there was a time when this was the case.

6.52. The Origin of the Atlantic Ocean. JARDETZKY assumes that a primeval continent formed at one time while the Earth was still essentially fluid, as is the case in many other theories of the origin of continents and oceans. This primeval continent was a large mass, roughly equivalent to the present land masses combined. Owing to the zonal rotation of the substratum, stresses would act upon this large mass which would crack up in big rifts.

Thus, let us assume that the strains in the primeval continent are due to the zonally differentiated rotation of the substratum. The parts near the equator would be dragged faster than the parts near the pole. One can therefore calculate the displacements in the substratum, hence the stresses in the primeval continent and hence those curves along which fracture (envisaged as occurring by a tensile mechanism) would most likely occur. The shape of the latter curves depends on the law or rotation that is oroginally assumed; for the particular case represented by Eq. (6.51–2) JARDETZKY obtained

$$\cos\varphi = \text{const}\; e^{-\psi} \qquad (6.52\text{–}1)$$

where ψ is the longitude and φ the latitude. This is the equation of an S-shaped curve which has been referred to as the "equation of the Atlantic Ocean"; the corresponding curve is shown (denoted by $e\,f\,g$) in Fig. 106. If the primeval continent is assumed as somewhat square, then the line represented by the equation of the Atlantic Ocean will conveniently cut off two pieces resembling North and South America (A and B in Fig. 106); the rest, then, would form the Eurasian block.

The forces of the differential rotation would give rise to mountains and mid-oceans ridges through the interaction of the continents with the

[1] JARDETZKY (ŽARDECKI), W.: Recherches mathématiques sur l'évolution de la terre. Éd. spéc. Acad. Roy. Serbe, tome **107** (1935).

[2] JARDETZKY, W.: Denkschr. Österr. Akad. Wiss. **108**, No. 3 (1948).

[3] JARDETZKY, W.: Science **119**, 361 (1954).

[4] DINGEMANS, G.: Formation et transformation des continents. Paris: Colin 1956.

[5] SCHOENBERG, E.: Geofis. Pura Appl. **43**, 1 (1959).

[6] STOVAS, M. V.: Vest. Leningr. Un-ta, Ser. Mat. Mekh. Astr. **1959**, No. 1, 119 (1959).

material upon which they float, through the pressure of one continent against another and through shearing motions. In the above theory one assumes that all continents (and islands etc.) on the northern hemisphere are rotating in an anticlockwise direction and on the southern hemisphere in a clockwise direction; this would cause a piling up of substratum effecting this rotation in the equatorial regions, whereas in the polar regions rifts would occur that could give rise to island arcs.

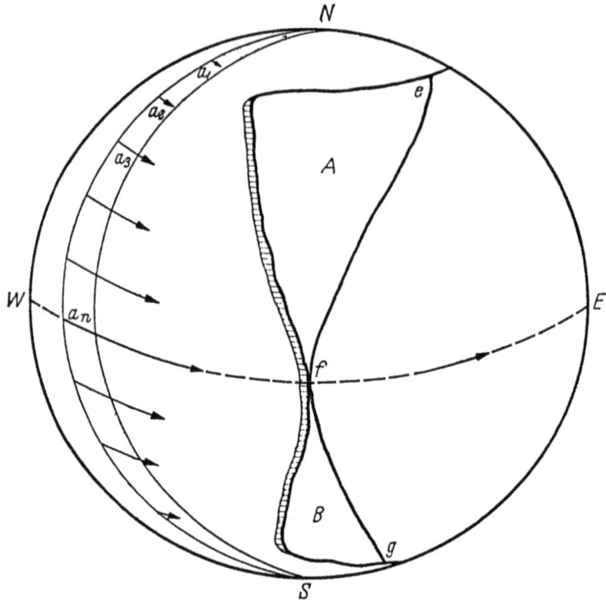

Fig. 106. JARDETZKY's hypothesis of zonal rotation. $\alpha_1\ldots$ are the zonal velocities, the line $e\,f\,g$ represents the equation of the Atlantic Ocean, A and B are the pieces left (i.e. the Americas) if the big block breaks along $e\,f\,g$

A connection with zonal rotation has also been postulated by CAREY for an explanation of his "*oroclines*" (cf. Sec. 1.45).

6.53. Persistence of Zonal Rotation. The above scheme rests on the possibility of zonal rotation persisting through the ages. It is of course quite obvious that zonal rotation cannot represent an equilibrium state of rotation in a viscous Earth since the viscosity has the ultimate effect that the Earth, given an initial state of zonal rotation, will finally rotate *en bloc* (i.e. like a solid). What is of importance, therefore, is the decay constant of the zonal rotation.

There is a way by which zonal rotation can be treated in a simplified manner. Thus, let us think of the Earth as represented by co-axial cylinders of equal length, the common axis being the axis of rotation of the Earth, and assume that each cylindrical shell rotates *en bloc* with an angular velocity characteristic of its radius s; the latter being given by the rotation-equation (6.51–2), with $\omega_0 = 0$ since we are only interested in the deviation from the mean angular velocity. The model described here corresponds essentially to a rotation viscosimeter.

The equation of motion for each cylindrical shell of thickness ds is

$$-sF'(s)\,ds = dI\,\dot{\omega} \tag{6.53–1}$$

where the left hand side represents the moment exerted by the frictional forces, and the right hand side the deceleration (dI being the moment of inertia of the shell) thereby effected. In a viscous fluid, the force F exerted on one side of the cylindrical shell (so that $sF'\,ds$ is the moment originating from the different velocities on both sides) is

$$F = \eta \frac{dv}{ds} \times \text{surface} \tag{6.53–2}$$

with v being the linear velocity of motion. With JARDETZKY's assumption for the distribution of angular velocity, this yields (note that $v = s\omega$)

$$F(s) = 6\lambda\,\pi\,\eta\,s^3 R. \tag{6.53–3}$$

Noting that for a cylindrical shell

$$dI = s^2\,2\pi\,sR\,\varrho\,ds, \tag{6.53–4}$$

(ϱ being the density) one obtains by straightforward differentiation from the equation of motion (6.53–1)

$$-\lambda\,\eta = \tfrac{1}{9}\,\varrho\,s^2\,\dot{\lambda}. \tag{6.53–5}$$

Integrating, we obtain:

$$\lambda = \exp[-9\eta\,t/(\varrho\,s^2)] \tag{6.53–6}$$

which shows that the decay constant T of the feature is

$$T = \varrho\,s^2/(9\eta). \tag{6.53–7}$$

The decay constant depends on s which reflects the oversimplification employed in the model, but the order of magnitude can still be calculated. With $\varrho = 3$, $\eta = 10^{22}$, $s = 1000$ km, one obtains

$$T = 3 \times 10^{-7}\ \text{sec}. \tag{6.53–8}$$

This shows that the decay time for zonal rotation is extremely short. It seems therefore quite unlikely that a "primeval" zonal rotation could subsist in the Earth for very long, as the assumed viscosity (10^{22} cgs units) simply makes this impossible. It is true enough that the analogy with the model of a rotation viscosimeter is not very accurate, but it seems unlikely that the order of magnitude could be changed very much by a better model. The decay-time of zonal rotation, in order for the latter to have any orogenetic significance, should be of the order of hundreds of millions of years, i.e. by about a factor of 10^{20} larger than what was found above. It is extremely doubtful whether any small adjustments in the model employed could effect this, even if the "viscosity of the Earth" were lowered by a substantial amount. A possible way out of the difficulty might be found by assuming that the differential zonal rotation is upheld by some thermal process, i.e. for instance by convection currents. However, this sounds very much like an *ad hoc* hypothesis.

There are other difficulties with the hypothesis of zonal rotation. One is the questionability of the fact that a primeval continent could have existed until the Carboniferous epoch. The orogenesis which is evident at the West Coasts of the Americas and the Alpide-Himalayan orogenesis are only the most recent ones, and only they are explained by the zonal hypothesis. The theory completely ignores the fact that the geological history of the Earth is much longer than that evidenced by the most recent orogeneses. The earlier orogeneses (about 9 of them) and the marine features are therefore totally unaccounted for.

6.6. Undation Theory

6.61. Principles. The next theory of orogenesis to be discussed here is the undation theory. This is an attempt at explaining orogenesis in terms of rhythmic oscillations (undations) of the Earth's surface. The theory seems to have been proposed for the first time by HAARMANN[1] from purely physiographic reasoning; its mechanical aspects have later been analyzed mathematically by VAN BEMMELEN and BERLAGE[2] and recent accounts of it have been given by BELOUSOV[3].

As indicated above, the undation theory assumes that orogenesis is due to oscillations of certain parts of the globe. These oscillations constitute the *primary* cause of orogenesis. During their occurrence, certain parts of the world become elevated and others become lowered with regard to their surroundings. The elevated parts have been called *geotumors*, the lowered ones *geodepressions*. As soon as the limit of stability is exceeded, it is usually assumed that material from the top of the geotumors is supposed to slide down into the depressions. This

[1] HAARMANN, E.: Die Oszillationstheorie. Stuttgart: Ferdinand Enke 1930.

[2] BEMMELEN, R. W. VAN, u. H. P. BERLAGE: Gerlands Beitr. Geophys. **43**, 19 (1935); see also BEMMELEN, R. W. VAN: Proc. 21st Int. Geol. Congr. (Norden) **18**, 99 (1960).

[3] BELOUSOV, V. V.: Trudy geofiz. In-ta Akad. Nauk SSSR., No. **26** (**153**), 51 (1955). — Endeavour **17**, No. 68, 173 (1958). — Publ. Bur. Centr. Seism., Ser. A **20**, 369 (1959).

phase of the process is called *secondary* orogenesis. By it are produced contortions and folds in the strata as the depressions are rapidly being filled in (see Fig. 107). Finally, after the depressions have become geotumors in their own turn during the continuation of the undations, the familiar picture of folded mountain ranges is created. Thus, primary orogenesis is represented by slow irregular oscillations; their periods are of the order of millions of years. Secondary orogenesis is usually assumed to be caused by gliding and sliding of material from the top of the elevations; this produces the detailed patterns of tectonic elements ("gravity tectonics").

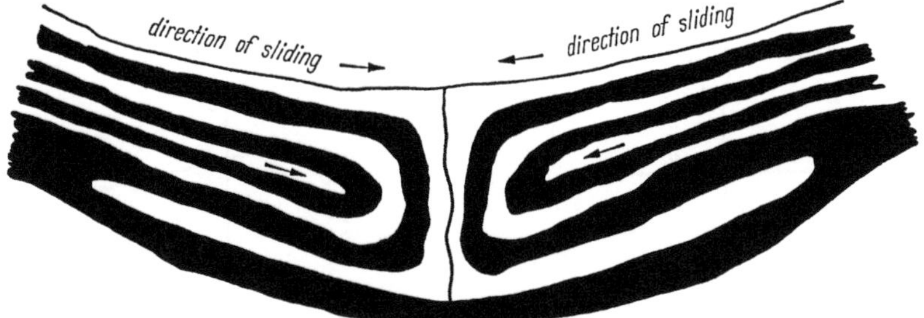

Fig. 107. Fold-formation through sliding

The undation theory was postulated by HAARMANN to account for certain physiographic features of the Earth. The uplift of Fennoscandia, the existence of submarine canyons and the fact that various parts of the world seem to have been flooded by the sea at various times, gain indeed a very natural explanation in this way. A further peculiar feature of the theory is that it does not have to assume any crustal shortening at all, i.e. all the folds are produced by extension of the original strata.

HAARMANN gives in his book a great number of examples of his interpretation of the physiographic aspects of orogenesis. He maintains, for instance, that the craters on the Moon represent the remnants of primary orogenesis[1],—no secondary orogenesis exists on the Moon. The arcuate strike of most mountain ranges and island chains is explained by the remark that most of the sliding into the depressions would presumably occur in the form of giant tongues (like avalanches); the front of these tongues would then yield the mountain-arcs.

HAARMANN did not proceed to a detailed investigation into the mechanics of his undation theory. He stated, however, that he expected that the primary undations would be caused by differentiation of magma

[1] This is, of course, completely at variance with the current view according to which the craters on the Moon have been formed by meteorite impact.

in the mantle of the Earth into two components, and that the action of gravity would cause the tops of the geotumors to slide down into the depressions. The action of a load (such as ice, mountains etc.) upon the Earth's surface is by HAARMANN entirely discounted.

There have been other speculations regarding the origin of forces which could make the Earth pulsate, such as that of HAVEMANN[1] postulating that this behavior be caused by periodic accumulation and escape of radioactive heat. Since this is much less specific than the theory outlined above, it is wide open to criticism.

6.62. Forces in the Undation Theory. We shall now investigate what the possible causes are that might give rise to the *primary* motions of the Earth's surface which were assumed in the undation theory.

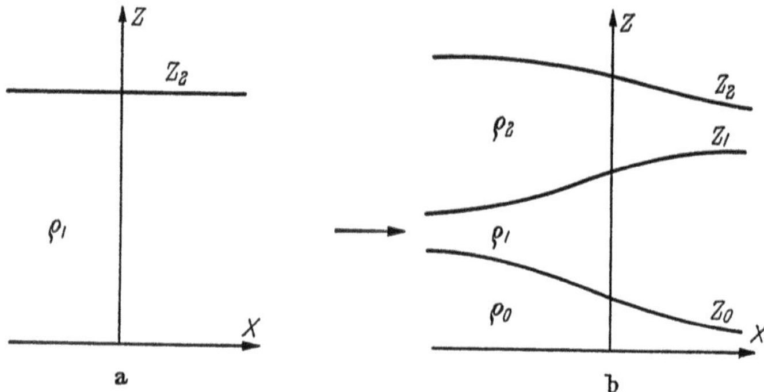

Fig. 108a and b. Magma differentiation from one into 3 layers as envisaged in the undation theory

Following HAARMANN, BEMMELEN and BERLAGE[2] postulated that the undations could be brought about by the uneven differentiation of "fundamental magma" into heavier and lighter material.

Accordingly, the proposed mechanism would be as follows. Originally, the surface of the Earth was flat and a homogeneous layer of "fundamental magma" (envisaged as a mixture of Si-Al and Si-Mg compounds) was resting upon some "basement layer" (possibly identified with the doubtful Birch discontinuity at 900 km depth). In a cross-section through the upper part of the Earth, the situation was therefore as shown in Fig. 108a. Herein, the curvature of the Earth has been neglected, the x-axis represents the bottom boundary of the fundamental magma, the z-axis is the vertical.

After the postulated differentiation of the magma into Si-Al and Si-Mg has proceeded for a while, it may be assumed that the picture shown in Fig. 108b results. This picture implies that the speed of differ-

[1] HAVEMANN, H.: Trans. Amer. Geophys. Un. **33**, 749 (1952).

[2] BEMMELEN, R. W. VAN, u. H. P. BERLAGE: Gerlands Beitr. Geophys. **43**, 19 (1935).

entiation may vary with the geographical location. Furthermore, it is assumed that the density of the fundamental magma is larger than the mean density of the end products, i.e.

$$\frac{1}{a+b}(a\varrho_0+b\varrho_2)<\varrho, \tag{6.62-1}$$

if the differentiation into material 0 and 2 takes place in the ratio $a:b$. Thus, with continuing differentiation, the surface z_2 is pushed to a higher level than was the case originally. During the differentiation, isostasy is maintained with regard to the material below $z=0$, but there will of course be a tendency to smooth out the differences in height of the surface z_2 by lateral displacements.

The heights z_0, z_1, z_2 of the corresponding interfaces above the datum-level $z=0$ are functions of x and t. The speed of differentiation of the fundamental magma depends on the pressure and on the temperature; if this speed is set equal to an undetermined function $F(z_0, z_1, z_2)$, one obtains for the change of thickness of each of the layers under consideration:

$$\left.\begin{aligned}\varrho_2\frac{\partial(z_2-z_1)}{\partial t}&=c_1F,\\ \varrho_1\frac{\partial(z_1-z_0)}{\partial t}&=-(c_0+c_1)F,\\ \varrho_0\frac{\partial z_0}{\partial t}&=c_0F.\end{aligned}\right\} \tag{6.62-2}$$

Here, c_0 and c_1 are constants. One thus obtains for the vertical component of the change of each interface

$$\left.\begin{aligned}\frac{\partial z_2}{\partial t}&=\left(\frac{c_1}{\varrho_2}-\frac{c_0+c_1}{\varrho_1}+\frac{c_0}{\varrho_0}\right)F,\\ \frac{\partial z_1}{\partial t}&=\left(\frac{c_0}{\varrho_0}-\frac{c_0+c_1}{\varrho_1}\right)F,\\ \frac{\partial z_0}{\partial t}&=\frac{c_0}{\varrho_0}F.\end{aligned}\right\} \tag{6.62-3}$$

The next task is to calculate the disappearance of the differences in level in the surface z_2 through lateral flow of the rocks. It is quite hopeless to attempt this by recurrence to the fundamental equations of a Maxwell body, and hence BEMMELEN and BERLAGE made the following intuitive assumption

$$\left.\begin{aligned}\frac{\partial z_2}{\partial t}&=\alpha_2\varrho_2\frac{\partial^2 z_2}{\partial x^2}+\beta_2(\varrho_1-\varrho_2)\frac{\partial^2 z_1}{\partial x^2}+\gamma_2(\varrho_0-\varrho_1)\frac{\partial^2 z_0}{\partial x^2},\\ \frac{\partial z_1}{\partial t}&=\alpha_1\varrho_2\frac{\partial^2 z_2}{\partial x^2}+\beta_1(\varrho_1-\varrho_2)\frac{\partial^2 z_1}{\partial x^2}+\gamma_1(\varrho_0-\varrho_1)\frac{\partial^2 z_0}{\partial x^2},\\ \frac{\partial z_0}{\partial t}&=\alpha_0\varrho_2\frac{\partial^2 z_2}{\partial x^2}+\beta_0(\varrho_1-\varrho_2)\frac{\partial^2 z_1}{\partial x^2}+\gamma_0(\varrho_0-\varrho_1)\frac{\partial^2 z_0}{\partial x^2}.\end{aligned}\right\} \tag{6.62-4}$$

Here, α, β, γ are all positive constants indicative of the "viscosity" of the substance under consideration. The set of Eq. (6.62–4) has been chosen because it has essentially the form of a set of diffusivity equations. It is well known that, if in a system the microscopic resistance to motion is everywhere proportional to the velocity, one ends up, macroscopically, with a diffusivity equation. Proportionality between resistance and flow velocity is, however, characteristic for bodies of a Maxwell type and hence the basic structure of the system (6.62–4) of equations is justified. The factors ϱ_2, $(\varrho_1 - \varrho_2)$ etc. are conditioned by the requirement that the corresponding terms must vanish if there is no density difference above and below the interface.

If one combines the motion resulting from the differentiation with that resulting from the smoothing-out of the surface, one obtains finally

$$\left.\begin{aligned}\frac{\partial z_2}{\partial t} &= \left(\frac{c_1}{\varrho_2} - \frac{c_0 + c_1}{\varrho_1} + \frac{c_0}{\varrho_0}\right) F + \alpha_2 \varrho_2 \frac{\partial^2 z_2}{\partial x^2} + \beta_2 (\varrho_1 - \varrho_2) \frac{\partial^2 z_1}{\partial x^2} \\ &\qquad + \gamma_2 (\varrho_0 - \varrho_1) \frac{\partial^2 z_0}{\partial x^2}, \\ \frac{\partial z_1}{\partial t} &= \left(\frac{c_0}{\varrho_0} - \frac{c_0 + c_1}{\varrho_1}\right) F + \alpha_1 \varrho_2 \frac{\partial^2 z_2}{\partial x^2} + \beta_1 (\varrho_1 - \varrho_2) \frac{\partial^2 z_1}{\partial x^2} \\ &\qquad + \gamma_1 (\varrho_0 - \varrho_1) \frac{\partial^2 z_0}{\partial x^2}, \\ \frac{\partial z_0}{\partial t} &= \frac{c_0}{\varrho_0} F + \alpha_0 \varrho_2 \frac{\partial^2 z_2}{\partial x^2} + \beta_0 (\varrho_1 - \varrho_2) \frac{\partial^2 z_1}{\partial x^2} + \gamma_0 (\varrho_0 - \varrho_1) \frac{\partial^2 z_0}{\partial x^2}.\end{aligned}\right\} \tag{6.62–5}$$

It is possible to integrate this for some special cases. First, let us consider the case where the upper layers of the Earth are homogeneous so that $\varrho_0 = \varrho_1 = \varrho_2 = \varrho$, and where there is no differentiation of magma. Then one has

$$\frac{\partial z}{\partial t} = \alpha \varrho \frac{\partial^2 z}{\partial x^2}. \tag{6.62–6}$$

This is the diffusivity equation which describes, for instance, the disappearance of an initial trough. Let the trough be represented by

$$z = -h\, e^{-\mu^2 x^2}, \tag{6.62–7}$$

then the solution of (6.62–6) yields

$$z = \frac{-h}{\sqrt{1 + 4\alpha \varrho \mu^2 t}} \exp\left\{-\frac{\mu^2 x^2}{1 + 4\alpha \varrho \mu^2 t}\right\}. \tag{6.62–8}$$

This confirms the earlier inference that the rheological equation is such that the material behaves like a Maxwell liquid. Thus, if a trough is impressed on the Earth by some mechanism, it will disappear asymptotically with a characteristic time interval determined by the parameter α. If such a trough, i.e. a geosyncline, should be filled-in (partly or wholly) with light sediments, then we have here a mechanism whereby a mountain-range can eventually be created.

The implications of the basic Eq. (6.62–5) can be analyzed more fully by neglecting fewer terms than in the example just discussed. A characteristic case is obtained by setting $\varrho_0 = \varrho_1$ in the system (6.62–5), but retaining F. Writing as an abbreviation

$$\delta = c_1/(\varrho_1 \varrho_2) \tag{6.62–9}$$

we obtain

$$\left.\begin{aligned} \frac{\partial z_2}{\partial t} &= \delta(\varrho_1 - \varrho_2) F + \alpha_2 \varrho_2 \frac{\partial^2 z_2}{\partial x^2} + \beta_2 (\varrho_1 - \varrho_2) \frac{\partial^2 z_1}{\partial x^2}, \\ \frac{\partial z_1}{\partial t} &= - \delta \varrho_2 F + \alpha_1 \varrho_2 \frac{\partial^2 z_2}{\partial x^2} + \beta_1 (\varrho_1 - \varrho_2) \frac{\partial^2 z_1}{\partial x^2}. \end{aligned}\right\} \tag{6.62–10}$$

The next task is to make a reasonable assumption for F. BEMMELEN and BERLAGE assume

$$F = D - (z_2 - z_1) \tag{6.62–11}$$

mainly because it makes the calculations easy. If this be substituted into (6.12–10), one obtains

$$\left.\begin{aligned} \frac{\partial z_2}{\partial t} &= \delta(\varrho_1 - \varrho_2)\{D - (z_2 - z_1)\} + \alpha_2 \varrho_2 \frac{\partial^2 z_2}{\partial x^2} + \beta_2 (\varrho_1 - \varrho_2) \frac{\partial^2 z_1}{\partial x^2}, \\ \frac{\partial z_1}{\partial t} &= - \delta \varrho_2 \{D - (z_2 - z_1)\} + \alpha_1 \varrho_2 \frac{\partial^2 z_2}{\partial x^2} + \beta_1 (\varrho_1 - \varrho_2) \frac{\partial^2 z_1}{\partial x^2}. \end{aligned}\right\} \tag{6.62–12}$$

The general character of the solution of this system can be recognized if, for the sake of simplicity, only the terms dependent on F are calculated (i.e. the α's and β's are set equal to zero). One obtains:

$$\left.\begin{aligned} \frac{\partial z_2}{\partial t} &= \delta(\varrho_1 - \varrho_2)\{D - (z_2 - z_1)\}, \\ \frac{\partial z_1}{\partial t} &= - \delta \varrho_2 \{D - (z_2 - z_1)\}. \end{aligned}\right\} \tag{6.62–13}$$

Hence by subtraction

$$\frac{\partial}{\partial t}\{D - (z_2 - z_1)\} = - \delta \varrho_1 \{D - (z_2 - z_1)\} \tag{6.62–14}$$

and finally

$$z_2 - z_1 = D(1 - e^{-\delta \varrho_1 t}). \tag{6.62–15}$$

This shows that the upper layer (ϱ_2), if zero thickness is assumed at $t = 0$, will asymptotically grow to having a thickness D.

The general solution of (6.62–12) is therefore a combination of the growth of the upper layer plus a smoothing-out of any differences in level created. It can be obtained by a series-expansion of the form

$$z_i = a_{i0} + a_{i1} t + a_{i2} t^2 + \cdots \tag{6.62–16}$$

where the coëfficients a_{ij} will be functions of x. The calculations become very involved, but the result can be stated briefly as follows: The general solution $z_i(x, t)$ leads to a Fourier series which, for reasonable parameters, allows only for aperiodic undations of the same type as that discussed in connection with the filling-in of a trough presented earlier. Oscillations in the proper sense or progressive waves do not seem to be possible.

The calculations in this Section (6.62) rest upon the validity of the assumption [used especially in writing down Eq. (6.62–2)] that, once primary differentiation of the magma into its two components (whatever they be) has started, the latter accumulate *immediately* into their respective layers. This implies that the diffusion of the two components through the original magmatic layer is very rapid. This, however, is very doubtful. The speed of diffusion can be estimated as follows. The viscosity η of the layer in question is of the order of 10^{22} to 10^{23} cgs. If we assume a spherical droplet of radius a with a density $\varrho_1 = 1.1\,\varrho_0$ (i.e. 10% denser than the surrounding fluid), one can calculate the speed v with which it would drop. The latter should be indicative of the order of magnitude of the speed of diffusion of the heavier fluid through the original one. The resistance R of the fluid to the droplet is given by STOKES' law [see Eq. (3.32–5)]:

$$R = 6\pi\, a\, \eta\, v. \qquad (6.62\text{–}17)$$

The force F acting on the droplet effecting its sinking is

$$F = \frac{4}{3}\pi\, a^3\, 0.1\, \varrho_0\, g \qquad (6.62\text{–}18)$$

where g is the gravity acceleration. Thus, from $F = R$

$$v = 0.1\,\frac{2}{9}\,\frac{a^2 \varrho_0}{\eta}\, g; \qquad (6.62\text{–}19)$$

thus, with $a = 1$ cm (as an upper limit; v increases with a), $\varrho_0 = 3$, $\eta = 10^{23}$, one has

$$v \cong 10^{-21}\ \text{cm/sec}. \qquad (6.62\text{–}20)$$

The time T for the droplet to fall through, say, 100 km (which may be taken as about one-half of the thickness of the original layer) is

$$T = 10^{28}\ \text{sec} = 3 \times 10^{20}\ \text{years}. \qquad (6.62\text{–}21)$$

This is about 10^{11} times the total estimated life span of the Earth. It seems therefore extremely doubtful wheter the postulated mechanism could have any real significance, quite apart from any chemical considerations.

The above mechanism is that most commonly quoted in connection with the undation theory. The above theory is somewhat oversimplified; it has been elaborated upon by SHIMAZU[1], ASLANYAN[2], SUBBOTIN[3, 4] and others. In all these attempts, some kind of phase or chemical change with subsequent "magma differentiation" is assumed. It is difficult to see, however, how any attempt along these lines could be upheld in view of the slowness of the diffusive processes in the Earth.

In addition to the above investigations, causes completely different from those envisaged above have also been proposed in connection with the "primary" undations. Thus, BEMMELEN[5] envisaged some kind of a magmatic diapirism and LYUSTIKH[6] assumed that magma rises along planetary fractures. TRECHMANN[7] assumed a cosmic upward pull and BÜLOW[8] assumed vertical currents in the mantle so that the motion of the crust would be like that of ice on troubled water. Some authors thought to put the energy source for the primary undations entirely into the crust. Thus, CONTANT[9] attributed it to distilled gases from organic matter in sediments, and ODHNER[10, 11], followed by MALAISE[12, 13], to the constriction caused by climatic temperature variations (*"constriction theory"*). Needless to say, these various theories are mostly speculations as a true analysis of their quantitative aspects has yet to be supplied.

6.63. Gravity Tectonics. We are turning now to the secondary orogenesis in the undation theory. For this, one generally resorts to a gravitational gliding process although different mechanisms have also been proposed. We first review this gravitational gliding process leading to the notion of "gravity tectonics"; this idea seems to have been initiated by HAARMANN[14] in his book.

According to HAARMANN, gliding seems to be a rather rapid event which is compared with turbidity currents known to occur on inclined

[1] SHIMAZU, Y.: J. Earth. Sci., Nagoya **7**, 91 (1959).
[2] ASLANYAN, A. T.: Trudy Uprav. Geol. i Okhran. Nedr **2**, 141 (1959).
[3] SUBBOTIN, S. I.: Soobshch. Akad. Litovsk. SSSR., Inst. geol. i geog. **5**, 5 (1957).
[4] SUBBOTIN, S. I.: Heol. Zhur. Akad. Nauk Ukr. SSSR. **22**, No. 5, 3 (1960).
[5] BEMMELEN, R. W. VAN: Madj. Ilmu Alam untuk Indones. **113**, 1 (1957).
[6] LYUSTIKH, E. N.: Izv. Akad. Nauk SSSR., Ser. Geofiz. **1960**, No. 3, 402 (1960).
[7] TRECHMANN, C. T.: Geol. Mag. **95**, 426 (1958).
[8] BÜLOW, K. v.: Geotektonisches Symposium zu Ehren von HANS STILLE, ed. LOTZE, publ. by Dtsch. Geol. Ges., p. 45 (1956).
[9] CONTANT, H.: C. R. Cong. Soc. Sav. Paris, 79e Congr., Alger, Sec. Sci. 144 (1954).
[10] ODHNER, N. H.: Geogr. Ann., Stockh. **16**, 109 (1934).
[11] ODHNER, N. H.: Ark. Mineral. Geol. **2**, No. 24, 353 (1958).
[12] MALAISE, R.: Atlantis. Stockholm: Kalmar 1951.
[13] MALAISE, R.: Geol. Fören. Förh. **79**, 195 (1957).
[14] HAARMANN, E.: Die Oszillationstheorie. Stuttgart: Ferdinand Enke 1930.

slopes of the ocean bottom. Consequently, the front of a gliding tongue might even detach itself from the main part (presumably owing to the momentum inherent in it) to form a detached mountain chain.

The process of gliding depends according to HAARMANN on three factors. First, there is a requirement for the existence of suitable strata such as water-logged sediments. The absence of such strata on the Moon would, in turn, explain the absence of folded mountains on that celestial body. Second, an adequate inclination of the gliding surface is of equal importance. The difference in elevation between geotumors

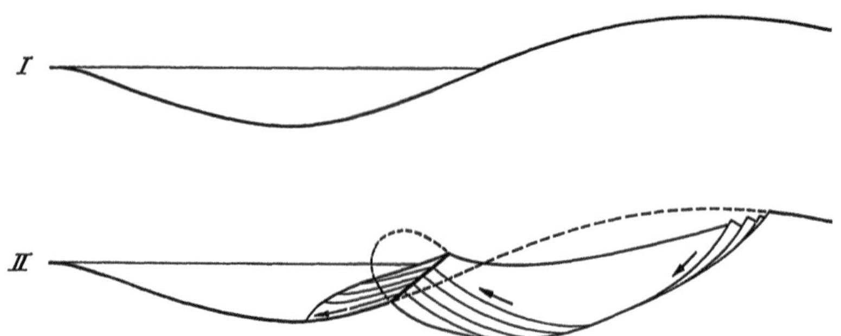

Fig. 109. Scheme of gravitational sliding. After BEMMELEN[1]

and geodepressions may be envisaged as up to about 3000 feet (1000 meters), the distance between the maxima and minima of elevation perhaps as of the order of several tens of miles. The available inclination would therefore be of the order of 1:10 (6°). According to HAARMANN, this is sufficient to cause downhill slides. Third, the time required for the sliding process to take place has been estimated to reach from that corresponding to a "sudden" catastrophe (an avalanche) to hundreds of thousands of years. The latter would correspond to slow creep.

The above ideas have been taken up by a variety of authors, notably BEMMELEN[1] whose view of the processes occurring is shown in Fig. 109.

HAARMANN arrived at his picture of the sliding process by purely intuitive arguments. Testing some of his statements somewhat more closely reveals that the possibility of the envisaged process is in reality rather doubtful. The fact that high mountains with a slope angle of much more than 6° can persist for a long time (at least for several millions of years) pretty well obviates the possibility of much sliding due to gravity occurring over slopes with such small inclinations. There is no doubt, of course, that land slides *can* and *do* occur, but in order to produce anything like, say, the islands of Japan by a slide off the mainland of China, such land-slides would have to be of fantastic dimensions indeed. It is true enough that there remains the possibility of gliding by

[1] BEMMELEN, R. W. VAN: Bull. Soc. Belge Géol. **64**, No. 1, 95 (1955).

slow creep. Indeed, if the rheological properties of the Earth's surface would come close to those of a Maxwell liquid (cf. Sec. 3.42), then there would be no doubt that slow creep could occur no matter how small the inclination of the gliding plane would be. However, there are no indications whatsoever that the topmost part of the crust does behave like a Maxwell body. It was shown in Sec. 3.43 that there seems to be a finite yield-strength; if the material is stressed below that strength, no permanent deformations can occur. It is precisely the topmost kilometers of the crust which HAARMANN supposed to undergo deformations, in spite of the fact that this is the only part of the Earth for which some trustworthy information is available. Unfortunately, the latter runs contrary to the assumption of Maxwell behavior.

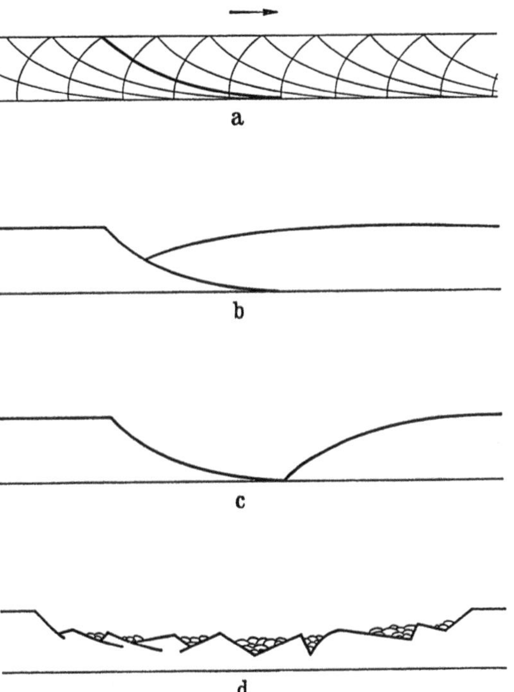

Fig. 110. Creation of a rift valley by plastic slip caused by gravity. After EVISON[1]

The above remarks do not intend to say that there is no such thing as gravity tectonics. The nappes of the Alps may well have been acted upon by gravity, and EVISON[1] has proposed that the rift valleys of Africa might have been caused in this fashion (plastic slip in a continental mass; cf. Fig. 110). However, it seems impossible to invoke gravity tectonics as an agent on a grand scale so as to create, for instance, the islands of Japan by a slide off Tibet.

6.64. DALLMUS'[2] Secondary Orogenesis[3]. We are turning now to some views which base the secondary orogenesis in the undation theory upon mechanisms different from that of gravity tectonics.

[1] EVISON, F. F.: Geophys. J. Roy. Astr. Soc. **3**, 155 (1960).

[2] DALLMUS, K. F.: In: Habitat of Oil, ed. WEEKS, p. 883. Tulsa: Am. Ass. Pet. Geol. 1958.

[3] This Section after SCHEIDEGGER, A. E.: J. Alberta Soc. Petrol. Geol. **6**, 266 (1958).

In pursuit of the ideas of the undation theory, DALLMUS[1] noted that a piece of the Earth's surface which is subsiding must of necessity undergo a compression. This is simply due to the fact that the Earth's surface is a sphere, not a plane. The situation that may be expected in a subsiding basin is shown in Fig. 111.

This compression due to subsidence provides DALLMUS with an alternative possibility of explaining secondary orogenesis in accordance with the undation theory. According to DALLMUS, the crustal shortening taking place in subsidence is sufficient to cause such features as mountain ranges and mid-ocean ridges by a regular folding process.

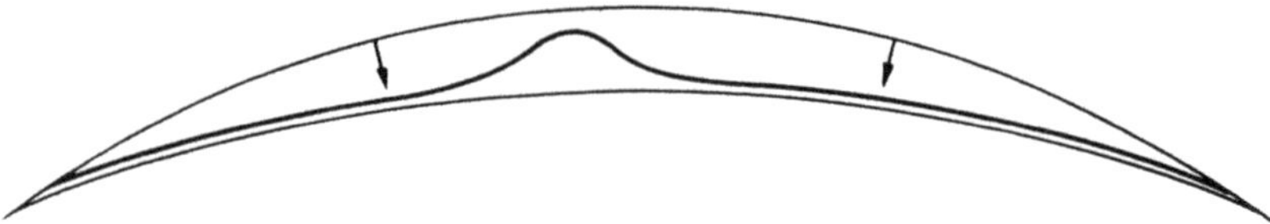

Fig. 111. Compression due to subsidence of a spherical shell

The claim can easily be checked. For a subsidence of dr, the circumference of the Earth is shortened by the amount dc equal to:

$$dc = 2\pi\, dr. \tag{6.64-1}$$

With $dr = 200$ meters (the maximum to be expected from a change of level of 100 m around a mean) this yields:

$$dc = 1.256 \text{ km}. \tag{6.64-2}$$

This is the shortening for the *whole* circumference of the Earth. Since the largest diameter of a subsiding block (corresponding to, say, an abyssal plain) could be at most equal to one-quarter of the Earth's circumference, the shortening that could be expected, would be equal to:

$$s_T = 314 \text{ meters}. \tag{6.64-3}$$

This is obviously much too little to cause mid-ocean ridges by a folding process.

DALLMUS obtained much higher values by postulating that the total subsidence is of the order of 8 km rather than of a few hundred meters. He assumed that there is no difference between continents and oceans and that a continent can subside to become an abyssal ocean-plain and vice versa. Under these circumstances, it is possible to get shortening up to 12 to 15 km. This is still very little. Furthermore, a postulated transformation of continents into oceans and vice versa runs completely contrary to well established results from geophysical exploration.

[1] DALLMUS, K. F.: In: Habitat of Oil, ed. WEEKS, p. 883. Tulsa: Am. Ass. Pet. Geol. 1958.

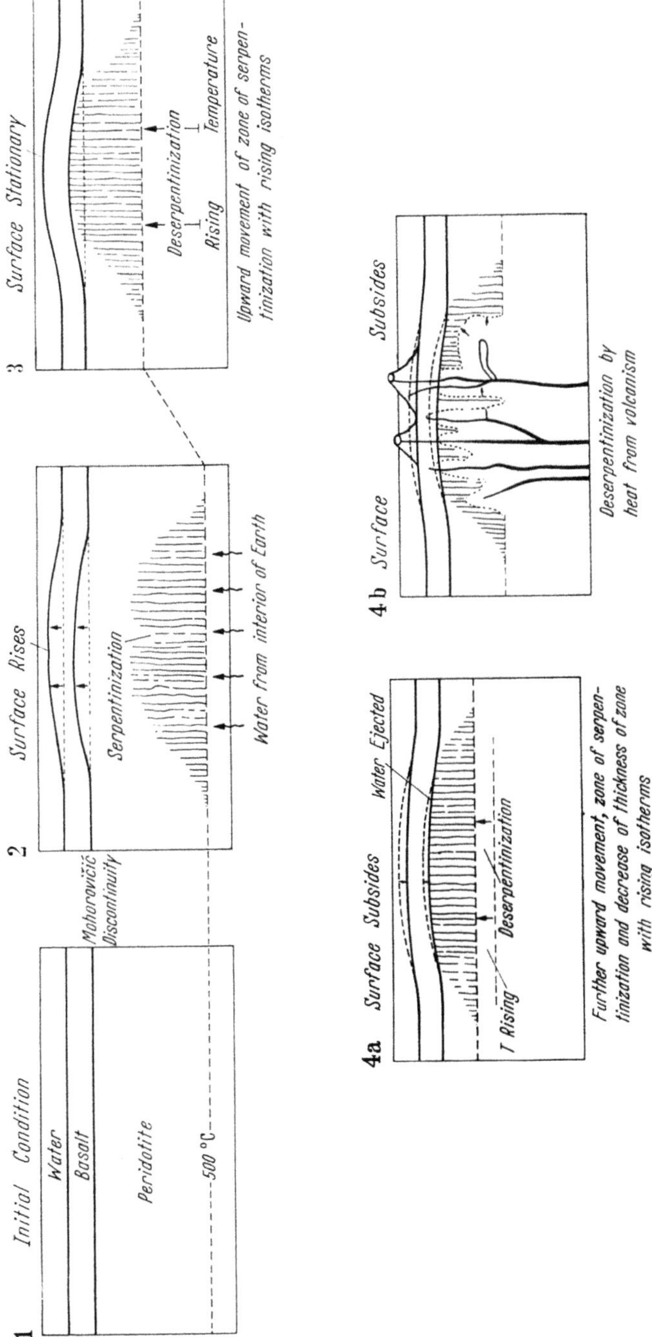

Fig. 112. The operation of the serpentinization cycle in the case of oceanic islands. (After HESS[1])

[1] HESS, H. H.: J. Marine Res. **14**, 423 (1955).

6.65. Serpentinization[1]. A new idea which might be regarded as a modification of the undation theory has recently been suggested by HESS[2].

It is based on the assumption that water would be leaking from the interior of the Earth and, upon passing through the 500° C isotherm below the Mohorovičić discontinuity, would react with the material there to form serpentine. This reaction would produce an increase of volume of about 25%, which in turn would produce an elevation of the crust above. However, the reaction is reversible, so that there is also a mechanism available to explain subsidence by assuming that water may be lost from the elevated areas to the atmosphere. With regard to oceanic islands, the process would then operate in a manner as shown in Fig. 112.

The feasibility of the process depends on whether enough water can be passed into the atmosphere to produce the proper size islands (and oceanic ridges). Such calculations do not seem to have been undertaken.

6.66. Evaluation of the Undation Theory. An inspection of the Sections above shows that the primary as well as the secondary orogenesis envisaged by the undation theory is still in want of much substantiation.

In spite of these difficulties, BELOUSOV[3] has followed the undation theory into even greater detail than HAARMANN. The formation of geosynclines, the building of mountains, in fact all geodynamic effects are attributed to up-and-down movements of larger or smaller portions of the Earth's crust. Thrust and tangential motions are entirely discounted. Stretching is supposed to occur during the up-movements, folding during the down-movements.

Summarizing, all that can be said about the undation theory is that it presents some interesting ideas about mountain bilding. There is little hope, however, that it will have much chance to be proven acceptable, unless some facts would happen to be uncovered which are completely at variance with all present ideas about the Earth's rheological properties.

6.7. Expansion Hypothesis of Orogenesis

6.71. Principal Outlines. It is a most notable feature of most theories of fracture of materials that the predicted failure-patterns are independent of the sign of the applied stresses. With the one exception of the possible occurrence of chasms under tension, the expectable failure pat-

[1] This Section after SCHEIDEGGER, A. E.: J. Alta. Soc. Petrol. Geol. **6**, 266 (1958).

[2] HESS, H. H.: J. Marine Res. **14**, 423 (1955).

[3] BELOUSOV, V. V.: loc cit. in footnote 3 on p. 258 of this book, also: Tez. Dokl. Mezhd. Ass. Seismol., pp. 5 and 9. Moscow 1957.

terns in an expanding Earth[1] would therefore be the same as those in a contracting Earth.

In an expansion theory, one would assume that the Earth consists of several layers, the topmost representing the crust, and the next one down representing an expanding layer. The orogenetic effects would originate in this expanding layer; the crust, in turn, would yield under the stresses which are created. According to the above remarks all that has been said regarding plastic and brittle failure patterns in the contraction theory of orogenesis, can also be said with regard to an expansion theory of orogenesis. The formation of island arcs, their position on the Earth's surface, the junctions between orogenetic elements etc. could equally be explained by assuming a slight expansion in a layer beneath the crust. In addition, there is the possibility of deep chasms occurring in an expansion theory which could be thought of as an explanation of deep ocean trenches.

In this connection, it may be interesting to note that most oceanic features can indeed be explained by the assumption of tensional forces. Thus, if only oceanic orogenesis be considered, some support might be gained for an expansion hypothesis of orogenesis. Because much knowledge about oceanic features has been collected rather recently, this has contributed to a revival of the expansion theory[2].

However, there is one fundamental difficulty. This is that in an expansion theory, it is no longer easy to account for the observed crustal shortening as there is no reason for the "skin" of an expanding sphere to become crumpled up. It would therefore appear that all the expansion could create, is a pattern of fissures through which the liquid "magma" below could rise to cause mountains. There seems to be no possibility of explaining nappes and similar phenomena.

In spite of these difficulties, EGYED[3] followed the hypothesis of an expanding Earth into some detail, claiming that it could account for many observed facts. According to EGYED, the tension in the crust would affect the modulus of rigidity which, in turn, would produce warping of parts of the crust. In this fashion, one again arrives basically at an undation theory with all its drawbacks. In an expansion-undation theory, however, it seems doubly impossible to obtain the crustal shortening necessary for mountain formation, because of the tensional character of the stresses involved. If a slight expansion of the Earth's interior did occur, it may be responsible for such features as rift valleys, the mid-Atlantic rift etc., but certainly not for the folding-up of mountains.

---

[1] Cf. Sec. 5.5.

[2] EGYED, L. wrote many papers on this subject, some of which have already been cited in Sec. 5.5. For other authors, also cf. Sec. 5.5.

[3] EGYED, L.: Acta Geolog. Magyar Todom. Akad. Föld. Közl. **4**, 43 (1956).

6.72. MATSCHINSKI's Buckling Theory. There have been several attempts at overcoming the difficulty of the seeming impossibility of producing crustal shortening or nappe structures in any type of expansion theory. A particularly notable one is that proposed by MATSCHINSKI[1].

Accordingly, when the Earth is expanding, the (solid) crust gets torn to pieces. Due to the expansion underneath, the pieces of the crust will soon exhibit a smaller radius of curvature than the substratum which forces them to "stand up" (cf. Fig. 113a). It stands to reason that such a state could not persist for very long and that a bulging would occur in the middle. This bulging would finally develop into nappes (cf. Fig. 113b) and thus explain orogenesis.

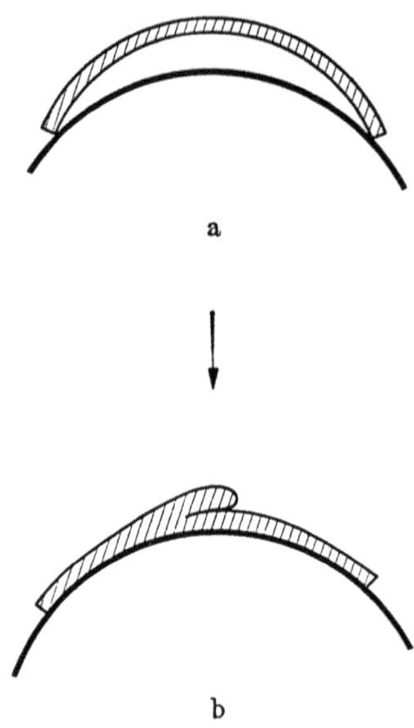

Fig. 113a and b. MATSCHINSKI's explanation of orogenesis

From the rheological properties of rocks it seems quite impossible that a state as envisaged in Fig. 113a could exist at all. Seen in a section, it would have to be expected that the crust conforms at all times to more or less an equilibrium position,—this, in reality, is implied in the observation that isostasy (cf. Sec. 2.32) is maintained over large areas. In two dimensions, of course, it is thinkable that a certain amount of bulging in the middle of, say, a triangular plate could occur, but it would appear as more reasonable to adjust the plate to the expanding sphere by assuming that tension cracks develop running radially from the center on outward, particularly in view of the low tensile strength of rock strata. It would appear, thus, that the preponderant phenomenon in any expansion theory would be tensional fissures scattered over the Earth and not crustal shortening. It is therefore hard to see how orogenesis could originate in any expansion theory.

6.73. Expansion by Rock Metamorphism. A further theory which might be classed as an "expansion hypothesis" has been proposed by LEBEDEV[2]. In this, however, it is not assumed that the Earth as such is expanding, with the crust being torn in consequence, but that the crust itself in spots is intrinsecally expanding owing to metamorphism

[1] MATSCHINSKI, M.: Rend. Accad. Naz. Lincei, Cl. sci. fis., mat. e nat., Ser. VIII, **16**, 54 (1953).

[2] LEBEDEV, V. I.: Dokl. Akad. Nauk SSSR., **90**, 217 (1953).

taking place. In other words, the metamorphosis factor ζ introduced in Sec. 6.13 is assumed to be, in spots at least, significantly greater than 1.

There is no doubt that metamorphism of rocks could account for very considerable stresses in the Earth's crust. However, the chemical factors involved are still largely hypothetical although LEBEDEV cites various examples that could have produced the enormous pressures necessary to cause orogenetic activity. It is generally believed that

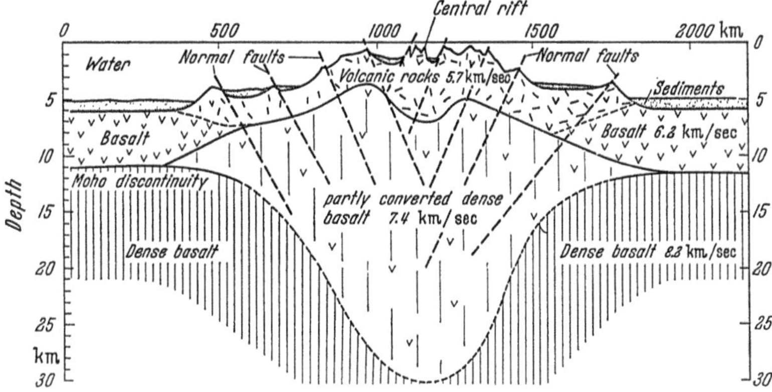

Fig. 114. DE SITTER's[1] interpretation of the formation of a mid-ocean ridge by local expansion (dense basalt into ordinary basalt)

metamorphism causes an *increase* in density, not a *decrease*, and LEBEDEV's theory needs, therefore, a lot more substantiation.

Into the present category also falls an attempt of DE SITTER[1] who tried to explain orogenesis by the assumption of a phase transition in basalt. In this fashion, he explained, for instance, the occurrence of mid-ocean ridges by assuming a local expansion, as shown in Fig. 114. DE SITTER[1] also allows the reverse phase transition to occur which, then, would explain crustal shortening. With this idea, however, one leaves again the domain of expansion theories and one returns to a type of undation theory exemplified, perhaps, in Sec. 6.65.

6.8. Orogenesis and Polar Wandering

6.81. The Problem. It has been outlined in Sec. 4.3 that there is a real possibility that polar wandering took place during the Earth's geological history, possibly on a very large scale. As stated earlier, polar wandering does not actually imply a change in the direction of the axis of rotation in space, but rather a change of position of the Earth with respect to its axis of rotation. It must be assumed that, for every position of the axis with respect to the Earth, the latter will take on an

[1] DE SITTER, L. U.: Geol. Rdsch. **50**, 219 (1960).

equilibrium figure which can be represented approximately by an ellipsoid. Thus, if the polar axis moves through an angle Θ, i.e. from the position represented by PQ to the position $P'Q'$ in Fig. 115, the crust of the Earth must adjust itself to the new shape. It is a reasonable conjecture that this adjustment will cause stresses that might have orogenetic significance. In the most extreme case the pole would move from its original position on the former equator through 90° normal to the latter. That would mean that the former equator becomes a meridional circle which would imply a shortening due to the ellipticity of the Earth of

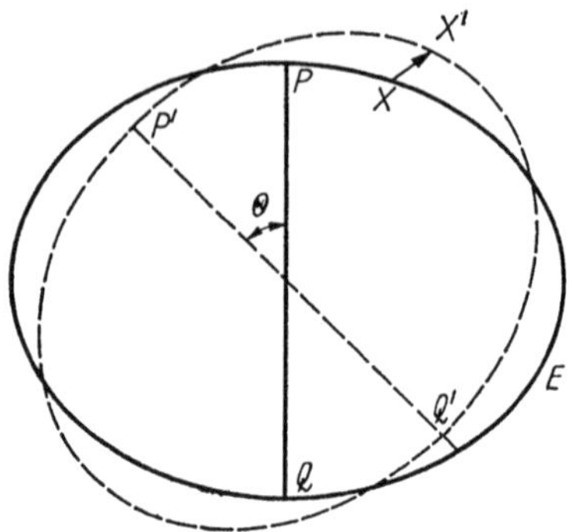

Fig. 115. Geometry of the shift of the Earth's axis of rotation

$$\frac{10019148.441}{10002288.299} = 1.0017 \quad (6.81\text{–}1)$$

or of 17 km for a quadrant. Since an orogenetic system consists of worldwide mountain ranges of the shape of roughly two great circles, it will presumably be intersected 4 times by any other great circle so that the above-mentioned shortening (17 km) would be that available for each mountain range. From various sources of evidence, it must be assumed that two orogenetic cycles took place in the time interval since the Devonian in which the pole moved, maybe, through 80°. This leaves a shortening of about 8 km per mountain range. However, it should be noted that the above value is for *maximum* shortening, valid only for that meridian which goes through the new pole at right angles to the direction of the polar shift. If we consider *any* quadrant, the available shortening is on the average again about halved which leaves approximately 4 km for a mountain range:

$$s_T = 4 \text{ km}. \quad (6.81\text{–}2)$$

This seems very little as it is somewhat smaller in magnitude than what is observed in the contraction theory. Correspondingly, we obtain the following parameters:

$$\gamma = s_A/s_T = 50/4 = 12.5, \quad (6.81\text{–}3)$$

$$h = 12\gamma = 150 \text{ km}. \quad (6.81\text{–}4)$$

These figures are even somewhat less reasonable than in the contraction theory. They constitute a strong argument against the possibility of polar wandering having been a cause of mountain building. Furthermore, polar wandering does not only induce crustal shortening as derived above, but also crustal lengthening in a different part of the world. Thus, if the shortening s_T is prevalent in one area, then a corresponding lengthening should prevail in another and one would expect trenches to develop. The observation that deep ocean trenches are associated with orogenetic developments and not one quarter of the way around the world from orogenetic developments, is a further difficulty in assuming that polar wandering causes mountains. Nevertheless, although it has thus been shown that polar wandering is inadequate to cause orogenesis, it might still give rise to shearing of the crust on a continental scale. This question has been investigated by various people[1-3]. A discussion will be given below.

6.82. General Theory. Referring to Fig. 115, let the original state of the Earth be represented by the ellipse drawn in solid lines, and the state corresponding to that after a shift of the polar axis through the angle Θ, by dotted lines. This picture implies that the axis of rotation and the geoidal axis always coincide corresponding to an infinitely great adaptability of the Earth to the prevalent dynamic forces. During the displacement, each point of the Earth shifts to a new position; such a shift is indicated for a point on the surface by an arrow pointing from X to X'.

What are the conditions governing the displacement? It is evident that there are two types of conditions necessary which can be designated as "kinematic" and as "dynamic". The kinematic condition requires that the change of shape of the total Earth be such that the original ellipsoid changes into another ellipsoid, the dynamic condition requires that the state before and after the displacement be an equilibrium state.

In order to investigate the possible effects of polar wandering upon stresses within the crust, it is necessary to consider such Earth models in which polar wandering is possible. This confines one, according to earlier remarks, to models in which the Earth is assumed as essentially fluid (at least with regard to the time intervals involved) with the crust being a thin skin covering the latter. What is of particular interest, is the displacement pattern of points on the surface as the displacement of the interior is not directly observable. Skin and interior will therefore be considered separately as this is presumably permissible in all models which exhibit polar wandering.

The kinematic condition can be taken care of as follows. In the original state we describe each point on the Earth's surface by its

[1] Vening Meinesz, F. A.: Trans. Amer. Geophys. Un. **28**, 1 (1947).

[2] Milankovitch, M.: Kanon der Erdbestrahlung und seine Anwendung auf das Eiszeitenproblem. Éd. spéc. Acad. Roy. Serbe, Tome **133**, Belgrade 1941.

[3] Schmidt, E. R.: Földtani Közlöny **18**, 94 (1948).

(geocentric) longitude α and corresponding latitude φ (or polar distance δ instead). It is to be understood that these co-ordinates represent an ellipsoid of revolution of a given ellipticity. The position of the pole is given by $\delta=0°$. A similar set of co-ordinates α', φ' (or δ'), then, corresponds to the deformed state with the new polar axis $P'Q'$ instead of PQ (cf. Fig. 115). Again, these co-ordinates are presumed to describe an ellipsoid of revolution of the same ellipticity as that above. The new position of the pole is given by $\delta'=0°$. In terms of the present scheme, the deformation is thus given by the following transformation equations

$$\alpha'=\alpha'(\alpha, \delta), \qquad (6.82\text{–}1\,a)$$

$$\delta'=\delta'(\alpha, \delta). \qquad (6.82\text{–}1\,b)$$

Thus, if the co-ordinates α, δ are taken as variables over the surface of a unit sphere, the shift of the polar axis represents a mapping of that sphere upon itself.

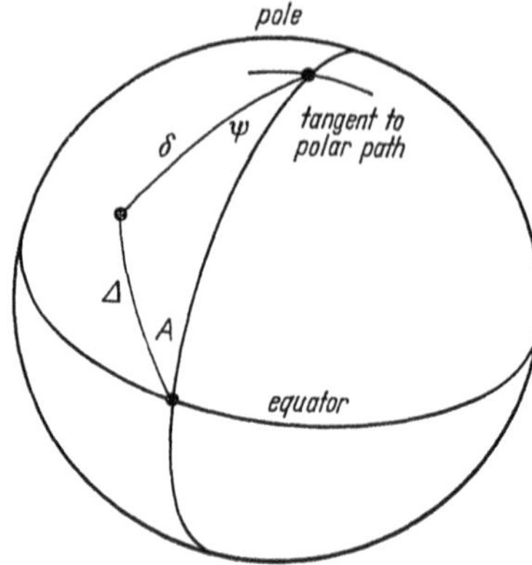

Fig. 116. The co-ordinate system A, Δ

A very convenient way to describe the *displacements* occurring during a shift of the polar axis is obtained by the following remark. A shift of the polar axis through the angle Θ corresponds to a rotation with regard to a certain axis. The pole of this rotation lies on a line vertical to the plane in which the polar shift occurs, i.e. it is on the intersection of the new and old equator of the Earth. It is convenient to introduce a co-ordinate system (A, Δ) where A and Δ are the longitude and polar distance, respectively, with regard to this pole, zero longitude corresponding to the position of the North Pole before the polar shift occurred. The use of this co-ordinate system enables one to describe the shift of the surface of the Earth by giving the displacement vector as a function of A, Δ (see Fig. 116).

The dynamic condition requires that the transformation expressed by (6.82–1 a/b) leads from an equilibrium state to another equilibrium state. In general, any possible set of equations of the above type will prescribe a set of *strains* in the Earth. However, only such strains are permissible whose associated *stresses* satisfy equilibrium conditions. The equilibrium conditions for the additional stresses (additional with regard to the hydrostatic state) have been investigated by VENING MEINESZ[1]

[1] VENING MEINESZ, F. A.: Trans. Amer. Geophys. Un. **28**, 1 (1947).

who found (refer to Fig. 117 for an explanation of symbols):

$$(\sigma_\alpha - \sigma_\delta)\cos\delta - (\partial\sigma_\delta/\partial\delta)\sin\delta + (\partial\tau/\partial\alpha) = 0, \qquad (6.82\text{–}2\text{a})$$

$$(\partial\sigma_\alpha/\partial\alpha) - (\partial\tau/\partial\delta)\sin\delta - 2\tau\cos\delta = 0, \qquad (6.82\text{–}2\text{b})$$

$$\sigma_r + \frac{T}{R}(\sigma_\delta + \sigma_\alpha) = 0 \quad \text{and} \quad \sigma_r = -\frac{T}{R}(\sigma_\delta - \sigma_\alpha). \qquad (6.82\text{–}2\text{c})$$

These formulas hold for a thin crust of thickness T and radius R in which $T \ll R$ so that bending stresses can be neglected and σ_ϱ increases linearly with depth from a value of zero at the surface to the value σ_r at the bottom of the crust.

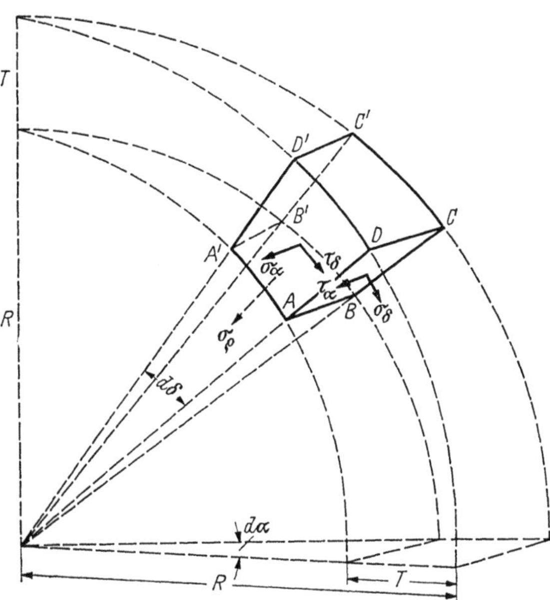

Fig. 117. Forces acting upon an element of a spherical shell. After VENING MEINESZ[1]

Upon inspection of the above discussion, it is evident that one is faced with (a) a set of constraints for the strains and (b) with a set of constraints for the stresses. It is therefore obvious that the problem of determining the deformation of the Earth's crust due to polar wandering cannot possibly be solved unless some stress-strain relations are assumed. These, however, constitute the "rheological condition" for the material in question and the difficulties for estimating the latter for the various layers of the Earth are well known.

From the above considerations it is clear that the orogenetic significance of a shift of the polar axis can be determined only if a certain rheological condition is assumed for the crust of the Earth. The two extreme conditions that have been investigated are (a) the assumption that the crust of the Earth is ideally elastic (VENING MEINESZ) and (b) the assumption that the crust of the Earth has no strength at all (MILANKOVITCH). These are, of course, the two logical extremes which one might attempt to investigate.

[1] VENING MEINESZ, F. A.: Trans. Amer. Geophys. Un. 28, 1 (1947).

However, one can try to obtain an idea as to the likelihood of applicability of either of these assumptions. If polar wandering is the cause of an orogenetic cycle, then its characteristic time must be of the order of 100 million years, i.e. "long" in the sense of Sec. 3.6. We have stated earlier that under "long"-duration stresses, *creep* is the significant behavior-pattern (once the strength-limit has been exceeded). This renders it extremely doubtful whether any model assuming an elastic Earth will be useful as the relaxation time for stresses that could produce significant elastic deformation is much shorter than the above-assumed time for their build-up. The opposite assumption of complete weakness in the Earth's crust is therefore much more realistic. A further difficulty with an elastic model of the Earth is that it is almost unavoidable to assume the crust as a homogeneous shell of uniform thickness. This is nowhere near reality. We shall therefore only briefly sketch the theory of elastic deformations due to polar wandering and then proceed to a more elaborate discussion of a weak Earth.

6.83. The Elastic Model. Thus we turn first to the assumption of ideal elasticity (POISSON's ratio $m=\frac{1}{4}$, YOUNG's modulus $E=10^{12}$ cgs) considered by VENING MEINESZ[1]. For a shift of the axis through an angle Θ, one finds for the additional stresses by a straightforward calculation (employing the co-ordinate system A, Δ introduced above)

$$\sigma_\Delta=[1/(5+m)]\,\nu E\sin\Theta\,(\sin^2\Delta+2)\sin(2A-\Theta)\,, \tag{6.83–1 a}$$

$$\sigma_A=[1/(5+m)]\,\nu E\sin\Theta\,(3\sin^2\Delta-2)\sin(2A-\Theta)\,, \tag{6.83–1 b}$$

$$\sigma_r=-[4/(5+m)]\,\nu E(T/R)\sin\Theta\sin^2\Delta\sin(2A-\Theta)\,, \tag{6.83–1 c}$$

$$\tau=-[2/(5+m)]\,\nu E\sin\Theta\cos\Delta\cos(2A-\Theta)\,, \tag{6.83–1 d}$$

where ν is as usual the ellipticity of the Earth ($=1/297$). The corresponding displacements s are

$$s_r=\nu R\sin\Theta\sin^2\Delta\sin(2A-\Theta)\,, \tag{6.83–2 a}$$

$$s_\Delta=[2(1+m)/(5+m)]\,\nu R\sin\Theta\sin\Delta\cos\Delta\sin(2A-\Theta)\,, \tag{6.83–2 b}$$

$$s_A=[2(1+m)/(5+m)]\,\nu R\sin\Theta\sin\Delta\cos(2A-\Theta)\,. \tag{6.83–2 c}$$

This gives the complete solution of the problem. VENING MEINESZ then proceeds to calculate the maximum shearing stress due to a polar shift of 90°, which is the maximum possible value. He obtains approximately

$$\tau_{\max}=1.2\times10^9\ \text{cgs} \tag{6.83–3}$$

using the present-day value for the ellipticity. The above value is just about equal to the lower limit of the strength of rocks at the surface of the Earth.

[1] VENING MEINESZ, F. A.: Trans. Amer. Geophys. Un. **28**, 1 (1947).

As outlined earlier, the poles of the Earth moved maybe about 40° during one orogenetic cycle, so that about half of the shearing stress calculated above in (6.83–3) would be available for the deformation of the rocks. This is well below the strength of the rocks at the surface, and it is therefore doubtful whether it could be sufficient to have any significance with regard to mountain-building. Entirely apart from the question of the applicability of the elastic model, which is uncertain in the first place, it does not appear that such a model would feature sufficient stresses to cause any orogenetic effects.

We therefore turn to the possibility of large-scale shearing effects being induced by polar wandering in an elastic Earth. In fact, VENING MEINESZ had this possibility in mind when he undertook the calculations mentioned above. In this instance, the low value for the shearing stress would serve as a justification of the use of an elastic model, as it might be argued that the latter is appropriate as long as the strength of the material is not exceeded.

Thus, from the displacements one can calculate the strains. One obtains

$$\varepsilon_A = \frac{1}{R \sin \varDelta} \frac{\partial s_A}{\partial A} = -\frac{4(1+m)}{5+m} \nu \sin \Theta \sin(2A - \Theta), \qquad (6.83\text{–}4\text{a})$$

$$\varepsilon_\varDelta = \frac{1}{R} \frac{\partial s_\varDelta}{\partial \varDelta} = \frac{2(1+m)}{5+m} \nu \sin \Theta \cos 2\varDelta \sin(2A - \Theta), \qquad (6.83\text{–}4\text{b})$$

$$\left.\begin{aligned} \varepsilon_{A\varDelta} &= \frac{1}{2}\left\{\frac{1}{R \sin \varDelta} \frac{\partial s_\varDelta}{\partial A} + \frac{1}{R} \frac{\partial s_A}{\partial \varDelta}\right\} \\ &= 3\, \frac{1+m}{5+m} \nu \sin \Theta \cos \varDelta \cos(2A - \Theta). \end{aligned}\right\} \qquad (6.83\text{–}4\text{c})$$

From these equations it is possible to calculate lines of maximum shear. The tangent to the line of maximum shear is the bisectrix of the principal strain axes at each point. The angle which the principal axes form with the given co-ordinate system is

$$\tan 2\varphi = -\frac{3 \cos \varDelta \cot(2A - \Theta)}{\cos 2\varDelta + 2} = \frac{2\varepsilon_{A\varDelta}}{\varepsilon_A - \varepsilon_\varDelta} \qquad (6.83\text{–}5)$$

where the angle φ is measured in the direction opposite to the sense of turning from the A-axis to the $\varDelta$-axis.

VENING MEINESZ did not calculate such lines of maximum shear. He used BIJLAARD's theory (cf. p. 127) in which the tangent to the line does not bisect the principal strain axes, but forms a slightly different angle which varies with the strain. He thus arrived at the picture shown in Fig. 118. If it is assumed that the maximum shear direction bisects the principal strain direction, the conjugate curves must intersect each other at right angles. One can obtain an idea of the pattern by noting that Eq. (6.83–5) can be written as follows (for $\Theta \sim 0$):

$$\tan 2\varphi = -f(\Delta) \cot 2A . \tag{6.83–6}$$

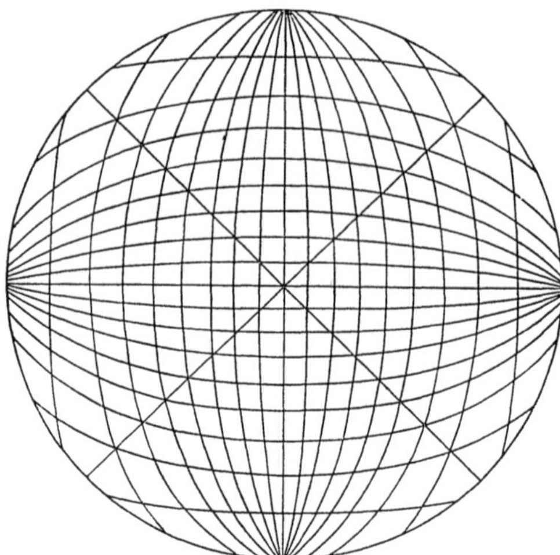

Fig. 118. VENING MEINESZ'[1] shear pattern

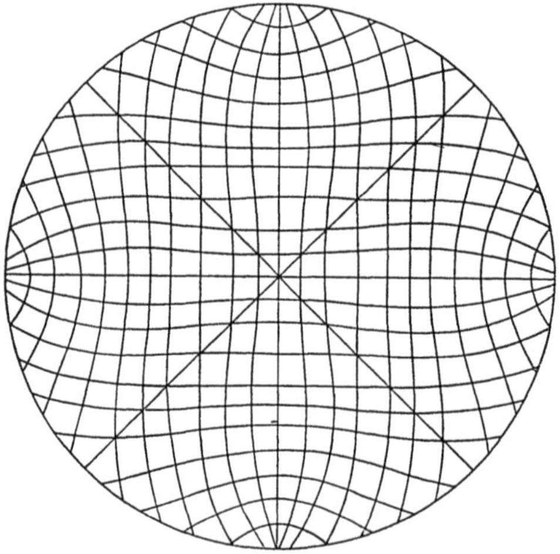

Fig. 119. Modified shear pattern

Here, f is some positive function with an upper limit for $0 \leqq \Delta < 90°$. It does not affect, therefore, the zeros and infinities of the tangents and cotangents, as long as $\Delta < 90°$. It is therefore possible to draw the directions of the shear lines (note that this direction is $\varphi \pm 45°$) for $A = 0°, 45°, 90°$ etc.; similarly, they can be drawn for $\Delta = 90°$. The shear lines, thus, are the orthogonal curves that fit into this pattern of tangents. The picture obtained by VENING MEINESZ must therefore be somewhat modified to look as shown in Fig. 119.

6.84. Model of a Weak Earth. Let us now assume that the crust of the Earth is so weak that it cannot offer any significant resistance to the adaptation of the Earth as a whole to the polar wandering. Then each point on the surface of the Earth will move in a direction corresponding to the change of the equilibrium surface. This problem has been studied by MILANKOVITCH but, unfortunately, the latter author only calculated the displacements in that meridional section of the Earth which is tangent to the direction of the motion of the pole. This is of little use for the investigation of possible shear zones and we shall

[1] VENING MEINESZ, F. A.: Trans. Amer. Geophys. Un. 28, 1 (1947).

therefore extend the calculations of MILANKOVITCH to the whole surface of the Earth.

The potential function W producing the equilibrium surface of the Earth has been given in Eq. (4.12–10). The calculations will be greatly simplified if the co-ordinate system A, Δ explained above is again introduced (cf. Fig. 116). One has then for instance

$$\cos\delta = \sin\Delta\,\sin A\,. \tag{6.84–1}$$

If we only take account of the rotational part of the potential function W (neglecting the small change in gravity due to the changing *shape* of the Earth), the change dW of the potential W during a polar shift $d\Theta$ can be found as follows:

$$\left.\begin{aligned} dW &= \frac{\partial}{\partial A}\,W\,d\Theta = \frac{\partial}{\partial A}\left[\frac{\omega^2 R^2}{2}\,(1-\sin^2\Delta\,\sin^2 A)\right]d\Theta \\ &= -\frac{1}{2}\,\omega^2 R^2 \sin^2\Delta\,\sin 2A\;d\Theta \end{aligned}\right\} \tag{6.84–2}$$

where now R is used for the radius. The force $d\boldsymbol{F}$ originated by this change is simply the gradient of the potential; one thus obtains

$$dF_R = \frac{\partial}{\partial R}\,dW = -\frac{\omega^2 R}{2}\sin 2A\,(1-\cos 2\Delta)\;d\Theta\,, \tag{6.84–3 a}$$

$$dF_A = \frac{1}{R\sin\Delta}\,\frac{\partial\,dW}{\partial A} = -\,\omega^2 R\sin\Delta\,\cos 2A\;d\Theta\,, \tag{6.84–3 b}$$

$$dF_\Delta = \frac{1}{R}\,\frac{\partial\,dW}{\partial\Delta} = -\frac{\omega^2 R}{2}\sin 2A\,\sin 2\Delta\;d\Theta\,. \tag{6.84–3 c}$$

One observes at once that the direction of the force $d\boldsymbol{F}$ always falls into that plane which contains the circle of latitude $\Delta = \text{const}$ of the co-ordinate system A, Δ. This becomes obvious if one calculates the angle α which the resultant $d\boldsymbol{F}_r$

$$d\boldsymbol{F}_r = d\boldsymbol{F}_R + d\boldsymbol{F}_\Delta \tag{6.84–4}$$

of the components that do not fall into that plane, forms with the radius vector:

$$\tan\alpha = dF_\Delta/dF_R = \cot\Delta = \tan(90^\circ-\Delta)\,. \tag{6.84–5}$$

This is the same angle which the radius vector forms with that plane. Furthermore, since the resultant $d\boldsymbol{F}_r$ must lie in a plane which contains the radius vector and is normal to the circle of latitude, it must also lie in the plane of the circle of latitude. This simplifies the calculations tremendously.

Let us therefore choose such a plane containing a circle of latitude $\Delta = \text{const}$. A particle on the surface of the Earth will follow the direction of the force until it arrives at the new equilibrium position, corresponding to the new ellipsoidal shape of the Earth. Thus, a particle will move only in a plane containing a circle of latitude $\Delta = \text{const}$; referring to Fig. 120 it will move from M to M_1 along dL while an (infinitesimal) polar shift through the angle $d\Theta$ occurs.

Furthermore, the resultant of $d\boldsymbol{F}_R$ and $d\boldsymbol{F}_\Delta$ also lies in the direction of r (the radius of the circle of latitude) and may therefore conveniently be denoted by dF_r. Its magnitude is

$$dF_r = (dF_R^2 + dF_\Delta^2)^{\frac{1}{2}} = \omega^2 R \sin 2A \sin \Delta \, d\Theta . \qquad (6.84\text{–}6)$$

Hence we obtain for the angle β shown in Fig. 120:

$$\tan \beta = \frac{dF_r}{dF_A} = \frac{\omega^2 R \sin 2A \sin \Delta \, d\Theta}{\omega^2 R \sin \Delta \cos 2A \, d\Theta} = \tan 2A \qquad (6.84\text{–}7)$$

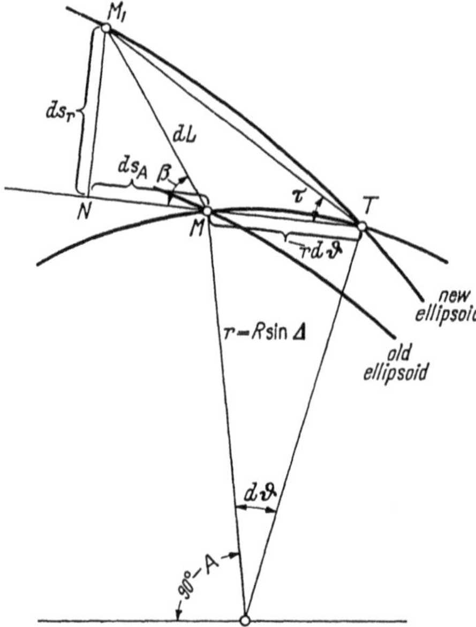

Fig. 120. Displacement in a weak Earth

or

$$\beta = 2A . \qquad (6.84\text{–}8)$$

Furthermore, since τ is the angle between the tangent to the ellipse and the normal to the radius-vector, it is given by (the first relation is valid for any plane curve in polar co-ordinates r, A; the second is obtained by using the equation of an ellipse $r = r_0 - r_0 \nu \cos^2 A$):

$$\left.\begin{aligned} \tan \tau &= \frac{1}{r} \frac{dr}{dA} \\ &= \frac{\nu \sin 2A}{1 - \nu \cos^2 A} \end{aligned}\right\} \qquad (6.84\text{–}9)$$

where ν is the ellipticity of the ellipse under consideration; this is the same as the ellipticity of the Earth. Since the latter is only 1/297, it is possible to neglect the term containing ν against unity, and one finally obtains for the components of the displacement denoted by ds_r, ds_A:

$$ds_A = \nu \cos 2A \, r \, d\Theta , \qquad (6.84\text{–}10\text{a})$$

$$ds_r = \nu \sin 2A \, r \, d\Theta . \qquad (6.84\text{–}10\text{b})$$

Splitting ds_r into the directions of R and Δ, and substituting for r, one finally has

$$ds_A = \nu \cos 2A \sin \Delta \, R \, d\Theta , \qquad (6.84\text{–}11\text{a})$$

$$ds_R = \nu \sin 2A \sin^2 \Delta \, R \, d\Theta , \qquad (6.84\text{–}11\text{b})$$

$$ds_\Delta = \nu \sin 2A \sin \Delta \cos \Delta \, R \, d\Theta . \qquad (6.84\text{–}11\text{c})$$

Of interest are the strains in the crust of the Earth:

$$\varepsilon_A = \frac{1}{R\sin\varDelta}\frac{\partial ds_A}{\partial A} = -2\nu\sin 2A\,d\Theta, \tag{6.84-12a}$$

$$\varepsilon_\varDelta = \frac{1}{R}\frac{\partial ds_\varDelta}{\partial\varDelta} = \nu\sin 2A\cos 2\varDelta\,d\Theta, \tag{6.84-12b}$$

$$\varepsilon_{A\varDelta} = \frac{1}{2}\left[\frac{1}{R\sin\varDelta}\frac{\partial ds_\varDelta}{\partial\varDelta} + \frac{1}{R}\frac{\partial s_A}{\partial\varDelta}\right] = \frac{3}{2}\nu\cos\varDelta\cos 2A\,d\Theta. \tag{6.84-12c}$$

It is most significant to note that these strains are the same as those found in the elastic model of the Earth for an infinitesimal polar shift, except for a constant factor of proportionality. Therefore, the lines of maximum shear turn out to be identical to those found in an elastic Earth. The tangent to the lines of maximum shear bisects at every point the principal strain axes. The angle which the principal strain axes form with the given co-ordinate system is given by Eq. (6.83-5) since the factor of proportionality is the same in numerator and denominator and therefore cancels out. The picture of the shear lines is therefore that shown in Fig. 119.

It may be of some interest to calculate the absolute maximum of strain and its location on the Earth. The maximum for the shear $\varepsilon_{\text{shear}}$ is at any one point

$$\varepsilon_{\text{shear max}} = \sqrt{\tfrac{1}{4}T^2 - K} \tag{6.84-13}$$

with

$$T = \varepsilon_1 + \varepsilon_2, \tag{6.84-14a}$$

$$K = \varepsilon_1\varepsilon_2 - \varepsilon_{12}^2. \tag{6.84-14b}$$

One obtains

$$\left.\begin{aligned}\varepsilon^2_{\text{shear max}} = \nu^2\,d\Theta^2\,\{&\tfrac{1}{4}\sin^2 2A\cos^2 2\varDelta\\ &+\sin^2 2A\cos 2\varDelta + \sin^2 2A + \tfrac{9}{4}\cos^2\varDelta\cos^2 2A\}.\end{aligned}\right\} \tag{6.84-15}$$

In order to calculate the absolute maximum of shear, one differentiates $\varepsilon^2_{\text{shear max}}$ first with respect to A and then with respect to $\varDelta$ and sets the result equal to zero. This yields the conditions

$$\left.\begin{aligned}&\sin 2A\,[\cos 2A\cos^2 2\varDelta\\ &\quad + 4\cos 2A\cos 2\varDelta + 4\cos 2A - 9\cos 2A\cos^2\varDelta] = 0,\end{aligned}\right\} \tag{6.84-16a}$$

$$\sin 2\varDelta\,[\sin^2 2A\cos 2\varDelta + 2\sin^2 A + \tfrac{9}{4}\cos^2 2A] = 0. \tag{6.84-16b}$$

One obtains from the second condition

$$\sin 2\varDelta = 0;\quad \varDelta = 90^\circ \tag{6.84-17}$$

if the bracket is not assumed zero. Inserting $\varDelta = 90^\circ$ into the first condition yields

$$\sin 2A\cos 2A = 0. \tag{6.84-18}$$

Thus one has the following solutions for A

$$A = 0°, 45°, 90° \dots \quad (6.84\text{–}19)$$

Hence

$$\varepsilon_{\text{shear max}} = \tfrac{1}{2}\, \nu\, d\Theta . \quad (6.84\text{–}20)$$

Thus, the maximum shearing strain occurs on that great circle which contains the direction of the polar shift at a distance of 45° from the pole.

The above considerations refer to an infinitesimal polar shift only. This is in accordance with the notion of a weak Earth; adjustment of the Earth to the prevailing forces is assumed to be instantaneous and no integrated form of the equations would be applicable.

6.85. Tectonic Significance. We are now in a position to evaluate the possible geological significance of polar wandering. As stated earlier, it cannot be expected that polar wandering can give rise to mountains; the only possible effect is therefore one of producing shear patterns.

The most important result is that, whether the Earth be considered as consisting of an elastic shell covering the interior or as possessing a crust that has no strength at all, the resulting shear pattern is the same. The only difference between the two models lies in the fact that, if the shell is assumed as elastic, the shear forces can be integrated over a finite shift of the pole, i.e. over some hundreds of millions of years, whereas in a weak Earth this is not possible. However, if one seeks to explain tectonic movements in this fashion, it is obviously implied that adjustments to the prevailing stresses do take place; this, in turn, obviates the principles of elasticity. It therefore appears that one ought to expect that the displacements to be explained correspond to the instantaneous strains; i.e. to the strains set up during an infinitesimal shift of the pole;—which are the same in either model.

If at every point the local adjustments take place in the plane of maximum shear, then one obtains the shear patterns shown in Fig. 119. If some other criterion is used (such as BIJLAARD's condition for bands of plastic instability considered by VENING MEINESZ) some other pattern is obtained. Again, it should be remarked that the VENING MEINESZ pattern does not really differ significantly from a maximum shear pattern; most of the deviations concern minor details. The fact that the result depends significantly neither on the model of the Earth nor on the yield criterion applied, lends considerable support to the argument presented above.

It therefore remains to correlate tectonic features on the Earth's surface with the possible shear patterns. The position of the shear pattern with regard to the Earth's surface is given firstly by the present position of the pole, and secondly by the tangent to the polar path at the present time. Whereas the present position of the pole is quite definite, the direction of the tangent to the path is not. Inspecting the various proposals for a polar path, we note that the tangent is according to KREICHGAUER along the 40° E meridian, according to KÖPPEN and WEGENER along 50° W meridian (see Fig. 1), according to MILANKOVITCH along the 140° W meridian (see Fig. 74), and according to RUN-

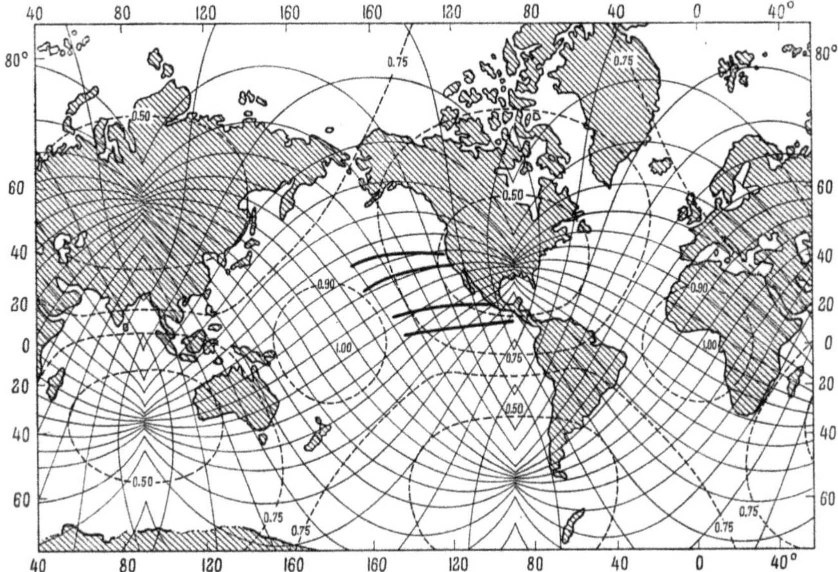

Fig. 121. Shear net in a strong Earth, using BIJLAARD's criterion, if the pole moved from Calcutta to its present position. Comparison with MENARD's fracture zones (after MENARD[1])

CORN (see Fig. 52) along the 120° E meridian. This completes pretty well the whole circle of the compass.

It is therefore evident that any correlations with known features on the surface of the Earth that one might be tempted to set up, are somewhat doubtful. The only instance where such correlations might be justified, is with regard to MENARD's fracture zones (cf. Sec. 1.54).

In the case of a strong Earth[2], the effect of infinitesimal polar shifts can be integrated over arbitrary time intervals. MENARD[1] has considered this case, testing a hypothesis advanced earlier by VENING MEINESZ. He assumed that the pole of the Earth moved from somewhere near Calcutta to its present position. The resulting shear net, using BIJLAARD's criterion, is that shown in Fig. 121.

[1] MENARD, H. W.: Bull. Geol. Soc. Amer. **66**, 1149 (1955).

[2] The remainder of this Section is after SCHEIDEGGER, A. E.: J. Alberta Soc. Petrol. Geol. **6**, 266 (1958).

In the opinion of the writer, it is entirely inconceivable that the Earth could be so strong as to maintain elasticity over the period during which the pole might have moved from Calcutta to its present position. The discussion of the rheology of the Earth seems to indicate that the Earth will yield, once the strength limit is exceeded. The maximum shearing strain, ε, for a shift of some 80° can be calculated [cf. formula (6.84–20)], and one then obtains (with ν=ellipticity, $d\Theta$=angle of shift)

$$\varepsilon = \frac{1}{2} \nu \, d\Theta = \frac{1}{2} \frac{1}{297} \frac{80}{360} 2\pi = 2.34 \times 10^{-3}. \qquad (6.85\text{–}1)$$

With the short-term value for the rigidity μ [see Table 18], this corresponds to a shear stress of τ

$$\tau = 2\mu\varepsilon = 2 \times 10^{12} \times 2.34 \times 10^{-3} = 4.68 \times 10^{9} \text{ cgs} \qquad (6.85\text{–}2)$$

which is of the order of the short-term strength, ϑ. The above calculation thus would appear to render reasonable the assumption of a strong Earth. However, this calculation has been made with the short term values for μ and ϑ. The writer has pointed out that, for "long" time intervals (which would be applicable in the present case), μ might be very much larger (in fact it might be possible to assume it as infinite in the Maxwell equation) and ϑ might be substantially smaller. The use of BIJLAARD's criterion, which is one of plastic instability, for determining the shear net seems to be at variance with the very basis of treating the Earth as elastic. The correctness of the assumption that the Earth behaves as an elastic body over such long-time intervals under stresses as produced by polar wandering is therefore open to question.

Furthermore, the assumption that the pole of the Earth ever was in the vicinity of Calcutta, does not seem to be very well supported. It is generally assumed that the pole travelled from somewhere near Hawaii to its present position during the past 300 million years. The position of the "old" pole of the Earth has been selected by MENARD so as to obtain a "statistically significant correlation" with the fracture zones determined by him physiographically. Furthermore, the coincidence of the shear lines and the fracture zones is not particularly good.

We shall therefore look at the possibility of explaining MENARD's fracture zones by assuming the Earth to be weak. In this instance, the all-important parameter is the direction of the *tangent* to the polar path by which the North Pole reached its present position. There is considerable ambiguity with regard to this direction, but it would appear to be reasonable to move the pole northwards from the Pacific say, along the 180° meridian. The shear net (*not* using BIJLAARD's criterion) for a weak Earth can then be redrawn, to represent the result of such a movement, on a Mercator chart of the world and this is shown in Fig. 122.

It is really quite amazing how well MENARD's fracture zones fit into this net. In spite of this success, the explanation of the zones as being caused by polar wandering should not be regarded as more than a reasonable possibility. One of the chief drawbacks of such an explanation is that any shear net produced by a polar shift would be perfectly symmetrical. Fracture zones similar to those found by MENARD should

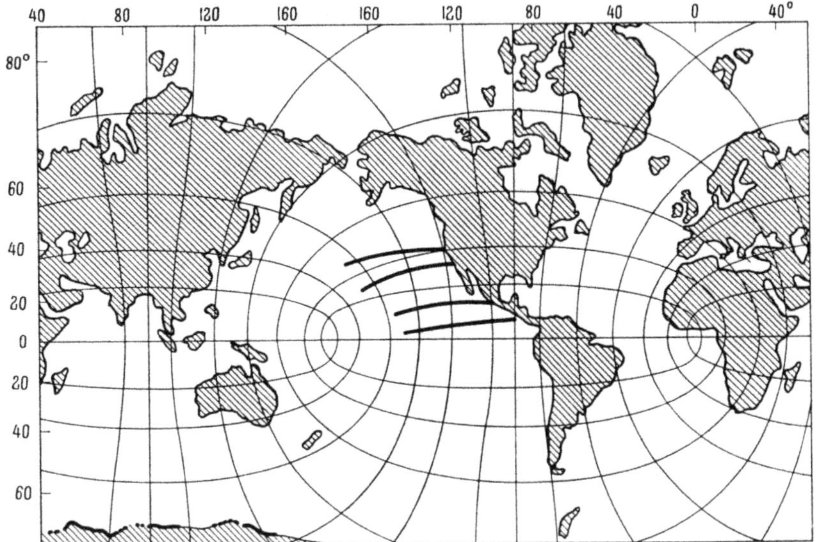

Fig. 122. Shear net in a weak Earth, assuming the polar path to be tangent to the 180° meridian. Comparison with MENARD's fracture zones

therefore be found in other parts of the world. Some zones found by HESS[1] fit the pattern very well. It is not known whether others exist.

6.9. Conclusion

6.91. General Remarks. In the concluding section on orogenesis we shall first present some views on the subject which do not fall into any of the schemes discussed above. Mostly, this concerns possible cosmic effects.

Finally, we shall attempt to give an evaluation of the various theories.

6.92. Orogenesis and Rotation of the Earth. We have already discussed some indirect effects of the rotation of the Earth in the Section on polar wandering (6.8). However, there may also be a much more direct effect: The ellipticity of the Earth is essentially conditioned by the angular velocity of the rotation. Thus, if the rotational speed were to

[1] HESS, H. H.: J. Marine Res. **14**, 423 (1955).

change, the ellipticity must change and this, in fact could cause the crust to adjust itself to the changed shape beneath, thereby initiating orogenesis.

There are, in fact, some indications that a decrease in the rotational speed might have occurred in geological times, although this is entirely speculative. Nevertheless, such effects as tidal friction or an increase of the moment of inertia in an expanding Earth could cause an increase in the length of the day. The above ideas have been elaborated by VUILLE[1].

A modification of the above ideas has been proposed by SCHMIDT[2]. Accordingly, the length of the day was not only increasing during geo-

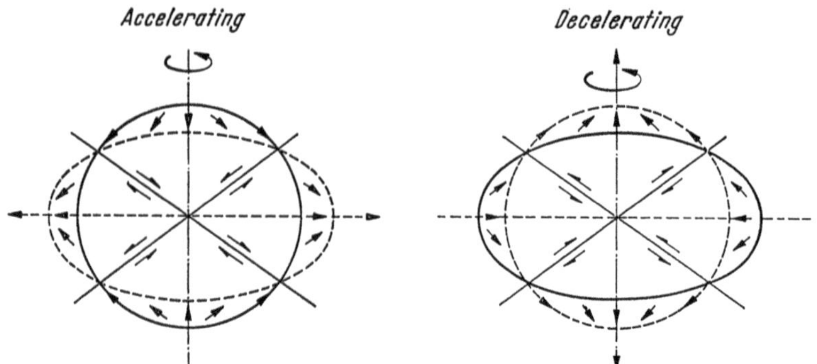

Fig. 123. Stresses in Earth due to accelerating and decelerating rotation. After SCHMIDT[2]

logical history, but was subject to a pulsation; during the accelerating and decelerating phases of the rotation, the stresses induced in the Earth are then as shown in Fig. 123. In this fashion, SCHMIDT attempts to account for many aspects of the orogenetic cycles. Unfortunately, it is not quite clear how the pulsation should come about; SCHMIDT assumes that in addition to the well-known Newtonian attraction, there is also a repulsion in cosmic bodies. Needless to say, this is highly speculative.

6.93. Cosmological Speculations. Finally, we may mention some cosmological speculations regarding the origin of orogenesis. Thus TAMRAZYAN[3] correlated geological diastrophisms with the passage of the Earth through the plane of the galaxy. It is thought that the latter occurrence, taking place every 200 million years, would trigger orogenesis. It is difficult, however, to envisage the suggested connection between the galaxy and mountain building.

[1] VUILLE, P.: Personal communication.

[2] SCHMIDT, E. R.: Rel. Ann. Inst. Geol. Publ. Hung. B **10**, 157 (1948).

[3] TAMRAZYAN, G. P.: Izv. Akad. Azerbaidzh. SSSR., No. 12, 85 (1957).

Another interesting attempt along these lines is due to HAITES[1] who noted that there are approximately equal perspective ratios between the times when orogenesis occurred and the distance of the planets from the Sun. Again, this observation is offered only as a correlation, but perhaps one might be tempted to speculate that orogenesis was connected with the creation of the planets.

6.94. Evaluation of Theories of Orogenesis. Of the many attempts at explaining the cause of orogenesis, only two have been considered (until very recently) as the chief contenders to be the correct ones: The contraction theory and the convection current hypothesis. Both seemed to fit the known data equally well. In the case of the contraction theory, specific calculations had been made yielding a corroboration thereof by showing that, quantitatively, the required amounts of cooling, contraction, upthrust and the resulting fault-patterns are entirely reasonable;—especially if the theory is combined with the notion of continental growth.

Similarly, the convection current hypothesis yielded a reasonable explanation of geosynclines with a subsequent upthrust of mountains. This theory, however, has not been developed to such a high quantitative refinement as the contraction theory. The size of the convection cells, the dynamical condition under which the latter are to appear and to disappear, their distribution over the surface of the Earth, are therefore wide open to speculation.

However, the outcome of recent geophysical investigations casts grave doubts upon the contraction theory as well as upon the convection current hypothesis of orogenesis. The first of these geophysical investigation is the oceanographic work (see Sec. 2.13) showing that the Mohorovičić discontinuity is not depressed beneath ocean trenches and thus dealing a severe blow to the notion of downwarping. The second is the work on faulting in earthquake foci (cf. Sec. 2.24) showing that the latter is, in the vast majority of cases, strike-slip and not dip-slip. This runs contrary to any idea sustained in either the contraction or the convection current theory. The third is the work on stresses (Sec. 2.4) showing that the Earth is subject to all types of stresses, not just one type as proposed by most theories of orogenesis.

It is therefore necessary to re-examine *all* theories of orogenesis that have ever been invented, in order to determine what can be saved of them in the light of the presently available facts. If this is done, it becomes immediately obvious that something fundamental is wrong with each and every of the theories: The difficulties in the contraction and convection theories have just been discussed; in the continental drift theory the forces producing the shifts are a mystery; the hypothesis

[1] HAITES, T. B.: J. Alberta Soc. Petrol. Geol. **8**, 345 (1960).

of zonal rotation is rendered impossible by the rapid decay of any differential rotation; in the undation theory it seems to be inconceivable that the small inclinations which are postulated, could induce orogenesis to take place on the required large scale; expansion theories could presumably produce only a few fissures here and there,—and the same is also true of hypotheses basing orogenesis upon effects of the rotation of the Earth.

If one tries to extract something constructive out of the above discussion, one notes first of all that the value of the extension factor γ introduced in Sec. 6.13, the fact that earthquake faulting is transcurrent, the paleoclimatic and paleomagnetic evidence all point towards the likelihood that at least parts of the Earth's crust have undergone considerable horizontal displacements. This points towards a continental drift hypothesis of orogenesis. The latter need not necessarily be opposed to the idea of continental growth which might well be superposed upon drift. It is difficult to see, however, where the forces causing the drift could have come from. Judging from present knowledge, forces due to the rotation of the Earth are presumably entirely ruled out owing to their minuteness so that convection currents would be the only agents left. Convection currents might not have to be thought of as forming part of a world-wide system, but they might exist in random patterns within the mantle of the Earth. The net drag upon each continent might therefore be the outcome of a composition of the various individual drags so that the continental path might correspond to that of a particle subject to Brownian motion. Such a theory, however, is not entirely satisfactory either because, first of all, it is uncertain whether the temperature distribution in the mantle and its rheological properties are such that convection currents can exist and second, because it is not known whether there is a significant discontinuity (Birch discontinuity) at 900 km depth (cf. Sec. 2.81) which would make convection cells reaching across that depth impossible. One way out of the last difficulty would be to assume that convection does not take place above the level of the Birch discontinuity but only below it. Then the mantle above that discontinuity would move in unison with the crust proper. This would also account for the fact that there appears to be no difference in focal mechanisms between deep and shallow focus earthquakes, transcurrency being preponderant at all depths. Assuming that primary orogenesis is felt to a depth of 700—900 km, however, would imply that the calculation of the extension factor (Eq. 6.13–5) has to be made with $h = 900$ km, yielding $\gamma = 75$, which is an unreasonable value. This, in turn, obviates one of the arguments in favor of random convection which stood at the beginning of the present train of thought. Thus, one has arrived at a vicious circle.

To sum up, it is almost inevitable to admit that the physiographic evidence regarding features on the Earth's surface is such that it must be assumed that some parts of the Earth's crust have been subject to compression (cf. crustal shortening), others have been subject to tension (mid-ocean rifts) and yet others to shearing (fracture zones). Most theories have been proposed for the purpose of explaining only one of these features and are therefore inadequate with regard to the others. A theory of orogenesis, to be acceptable, must be flexible enough to allow for compression, tension, *and* shear of sufficient magnitude to be produced in the Earth's crust. Setting aside those hypotheses which are inadequate for lack of self-consistency, the only acceptable concepts at present remaining are the continental drift hypothesis and the convection current hypothesis. The convection hypothesis has its own unsatisfactory features if it is used alone, but if the continental drift hypothesis is adopted, particularly with the assumption that the drift is (to a degree at least) random, a somewhat more satisfactory explanation for the various oceanic features is obtained. However, the reality of continental drift has by no means been established beyond reasonable doubt and, furthermore, the forces that could cause such drift are still somewhat of a mystery, although convection currents might be advocated as a possibility.

It appears, therefore, that the problem of finding the causes of the various geodynamic features must be regarded as still unsolved.

VII. Dynamics of Faulting and Folding

7.1. Dynamics of Faulting

7.11. Principles. We are turning our attention now to some smaller features of the Earth's surface as compared with those which have been considered in the preceding parts of this book. Such features are relatively easily amenable to an investigation of their physical causes because of the very smallness of the scale of the phenomena.

We shall start with *faults*. The principal features and classes of faults have already been described in Sec. 1.61; we shall proceed here with their physical explanation.

From the physiographic appearance of the faults it seems likely that they are simply the expression of localized mechanical failure of the material of the Earth's crust. Using one of the theories of such mechanical failure, it should therefore be possible to reconstruct the field of stresses which must have been in existence when a fault was created. If the prevalence of such a field of stresses can be rendered plausible from other considerations about the Earth's crust, the phenomenon of faulting will have been "explained".

The subject of mechanical failure has been discussed in Sec. 3.5 of this book. Accordingly, MOHR's "engineering" theory gives at least a qualitative description of how fractures occur. Based upon it, ANDERSON has evolved a theory of fault formation which will be discussed in Sec. 7.12. Following this, in Sec. 7.13, we shall add a few remarks regarding the possibility of devising an actual analytical theory of faulting. Finally, in Sec. 7.14 and 7.15 we shall discuss some special theories.

7.12. ANDERSON's Theory. In this paragraph we shall review some theories of faulting which are based upon "engineering theories" of (brittle) fracture. The idea to explain faulting in such terms seems to have been due originally to ANDERSON[1] who observed that an undisturbed or "standard" state of stress in the Earth's crust cannot be entirely arbitrary. First, it must be near the breaking point as is evidenced by the frequency of earthquakes. Second, there can be no pressure or tension perpendicular to the surface and no shearing force parallel to it.

[1] ANDERSON, E. M.: Trans. Edinb. Geol. Soc. **8**, part 3, 387 (1905). — ANDERSON, E. M.: The Dynamics of Faulting and Dyke Formation with Applications to Britain. Edinburgh: Oliver & Boyd Ltd. 1942.

The latter condition implies that the normal to the surface is one of the principal directions of stress at or near the surface. Thus, except in strongly folded areas, one principal direction of stress is nearly vertical, the other two are nearly horizontal.

Now, assuming comparatively small disturbances of this standard stress state in the Earth's crust, ANDERSON arrived, based on MOHR's engineering theory of fracture (cf. Sec. 3.52), at the following explanation of the three types of faults observed by geologists:

(i) *Normal faults.* Assuming that there is relief of pressure in all horizontal directions, the greatest pressure is the vertical pressure which is due to gravity. In general, the horizontal stresses will not be equal so that the greatest tension will prevail in a certain direction. Thus, the intermediate stress will also be horizontal, but at right angles to the direction of greatest tension (whatever this may be) and, if fracture occurs, this will happen along a plane containing the intermediate stress and inclined at an angle $\Phi \leqq 45°$ toward the vertical, which is the direction of smallest principal stress. One thus obtains the characteristics of a normal fault. From the geometrical pattern of the stresses it is obvious that the motion of the two parts must be such that the horizontal extent is increased.

(ii) *Transcurrent faults.* Assume that there is an increase of pressure in one horizontal direction and a relief of pressure in a horizontal direction at right angles to it. The smallest principal stress is then horizontal and the intermediate one is vertical. Now, if fracture occurs, this must happen according to MOHR's theory in a vertical plane inclined at an angle $\Phi \leqq 45°$ toward the greatest pressure. One obtains thus a fault with a vertical dip, the motion of the two parts being essentially horizontal. This is the characteristic pattern of a transcurrent fault.

(iii) *Reversed faults.* Assume that there is an increase of pressure in all horizontal directions. In general, one horizontal direction will be characterized by the fact that along it the pressure will be greatest. Thus the minimum pressure will be vertical and the intermediate principal stress will be horizontal, at right angles to the greatest pressure. If conditions are such that fracture occurs, this will happen according to MOHR's theory along a plane inclined at an angle $\Phi \leqq 45°$ toward one horizontal direction, the motion of the two parts being toward each other. One thus obtains the characteristics of a reversed fault: the dip is shallow, and the motion is such that the horizontal extent is shortened.

It may be noted that ANDERSON's theory has been confirmed in the field e.g. by LENSEN[1].

---

[1] LENSEN, G. J.: New Zeal. J. Geol. Geophys. **1**, 533 (1958).

Apart from the faulting phenomena discussed above, ANDERSON's theory also provides for an explanation of *dykes*. Dykes are in the main, nearly vertical fissures between 3 and 30 meters wide that have been infilled with some intrusive material. The two sides of a dyke appear to have moved apart in a direction normal to the fissure such that there is neither a lateral nor a vertical dislocation. ANDERSON explained dykes by the remark that MOHR's theory of fracture does not apply in the case of a tensile stress state. In the latter case, fracture is, according to engineering theories, normal to the "tensile stress". Thus, dykes may be explained by considering them as evidence of "tension fracture". The same explanation would hold for *joints*.

ANDERSON's theory, as outlined in the previous paragraphs, seems to account for the types of faults and related phenomena that have been observed by geologists. It might be desirable, however, to seek a more analytical description of its, after all, very qualitative statements. In this connection, we note that HAFNER[1] gave an analytical representation of ANDERSON's standard state. He then proceeded to calculate analytically such deviations from this standard state as would seem reasonable and which could produce faulting. The faulting patterns to be expected were then also calculated upon the assumption of MOHR's criterion of fracture. The results were compared with geologically observed facts.

Thus, ANDERSON's standard state can be expressed as a two-dimensional stress state. Let x and y be two Cartesian co-ordinates, x horizontal, y downward, in the direction of gravity ($y=0$ surface), and let the components of the stress tensor be $\sigma_x, \sigma_y, \tau_{xy}$. It is then convenient to express the stresses by means of AIRY's stress function Φ so that [cf. (3.21–8)]

$$\sigma_x = \partial^2\Phi/\partial y^2, \quad \sigma_y = \partial^2\Phi/\partial x^2 - \varrho g y, \quad \tau_{xy} = -\partial^2\Phi/\partial x\,\partial y \qquad (7.12\text{–}1)$$

if gravity is the only body force and ϱ = density. The stress function, furthermore, has to satisfy the following equation

$$\frac{\partial^4\Phi}{\partial x^4} + 2\frac{\partial^2\Phi}{\partial x^2\,\partial y^2} + \frac{\partial^4\Phi}{\partial y^4} = 0. \qquad (7.12\text{–}2)$$

Then, ANDERSON's standard state can be expressed by the following choice of the stress function:

$$\Phi = -\tfrac{1}{6}\varrho g y^3 \qquad (7.12\text{–}3)$$

which yields

$$\left.\begin{aligned}\sigma_x &= \sigma_y = -\varrho g y\\ \tau_{xy} &= 0.\end{aligned}\right\} \qquad (7.12\text{–}4)$$

[1] HAFNER, W.: Bull. Geol. Soc. Amer. **62**, 373 (1951).

As an example of a practical stress state, one can assume the presence of an "additional" horizontal component in addition to the standard stress state, but the absence of an associated additional vertical component. This is expressed by: (no body force for additional stress)

$$\sigma_y = \frac{\partial^2 \Phi}{\partial x^2} = 0 \quad \text{for all values of } y. \tag{7.12–5}$$

Integrating, one obtains the stress function:

$$\Phi = c\, f_1(y)\, x + a x + b\, f_2(y) + d. \tag{7.12–6}$$

To satisfy Eq. (7.12–2), the fourth order derivatives of f_1 and f_2 must vanish. Hence the second order derivatives must be either linear functions of y, constants, or zero. The stress components then are

$$\sigma_x = c\, f_1''(y)\, x + b\, f_2''(y); \quad \sigma_y = 0; \quad \tau_{xy} = -c\, f_1'(y). \tag{7.12–7}$$

The boundary conditions at the surface require that $f_1' = 0$ for $y = 0$. Keeping within the limits of the above restrictions, one can set up the following subgroups:

$$\text{(a)} \qquad f_1'(y) = 0; \quad f_2''(y) = y + d, \tag{7.12–8a}$$

$$\sigma_x = b(y + d); \quad \sigma_y = 0; \quad \tau_{xy} = 0. \tag{7.12–8b}$$

$$\text{(b)} \qquad f_1' = y; \quad f_2''(y) = 0, \tag{7.12–9a}$$

$$\sigma_x = c x; \quad \sigma_y = 0; \quad \tau_{xy} = -c y. \tag{7.12–9b}$$

$$\text{(c)} \qquad f_1'(y) = \frac{1}{2} y^2; \quad f_2''(y) = 0, \tag{7.12–10a}$$

$$\sigma_x = c x y; \quad \sigma_y = 0; \quad \tau_{xy} = -\frac{c}{2} y^2. \tag{7.12–10b}$$

The most general expression for the stress systems satisfying the assumption of absence of a vertical stress component is given by the superposition of Eqs. (7.12–8) to (7.12–10). It is seen that the stipulation $\sigma_y = 0$ is associated with two additional general properties of the internal stress system: (i) that the shearing stress is a function of y only, i.e. constant in all horizontal planes, and (ii) that σ_x has linear gradients in both the horizontal and vertical directions.

Of practical importance are the stress systems of the first two subgroups. The combination of (7.12–8), (7.12–9) with the standard stress state yields (with $d = 0$)

$$\sigma_x = c x + b y - a y; \quad \sigma_y = -a y; \quad \tau_{xy} = -c y. \tag{7.12–11}$$

where $a = \varrho g$. An analysis of these expressions yields that the trajectories of maximum principal pressure are curved lines dipping downwards away from the area of maximum compression. The curvature is strong if the vertical gradient of σ_x is small. From these trajectories, the potential fault surfaces can be calculated according to MOHR's criterion. The potential faults obviously belong into the class of thrust faults. The set dipping towards the area of maximum pressure is slightly concave upwards, the complementary set concave downwards. Thrust faults of the former type are very common in nature and the theoretically deduced curvature is frequently observed. The latter type appears to occur only rarely.

The above theory, which is essentially that of HAFNER, has been elaborated upon by SANFORD[1]. Calculating the elastic response of a rock layer to a two-dimensional stress distribution, SANFORD assumed a Mohr-type fracture criterion for the actual occurrence of the faults. His results were compared with scale-model experiments and a reasonable substantiation of the theory was obtained.

7.13. Analytical Theories. ANDERSON's theory gives a good and satisfactory qualitative explanation of the various types of faults that have been observed by geologists. However, it does not tell us how the breakage actually occurs, i.e. what the conditions are under which the material fails and what the speed is with which a fault slips or extends.

In order to obtain information on the above questions, it would be necessary to devise a complete analytical theory of faulting. Unfortunately, one runs in this connection into the same difficulties as those encountered in a discussion of the dynamics of fracture (cf. Sec. 3.54): a formulation of the appropriate dynamical laws has not yet been achieved.

Nevertheless, attempts have been made to describe at least the geometry of faulting analytically. Thus, JOBERT[2,3] assumed that the movement of the ground in the vicinity of a fault is (in plan) given by an analytic function (of complex variables). Most important is a study by ODÉ[4] who assumes that faults correspond to localized bands of plastic deformation. Since the equations of plasticity theory form a hyperbolic system (cf. Sec. 3.23) the characteristics of the velocity field may be lines of a velocity discontinuity and thus represent a fault.

There have been some attempts at using the concept of *dislocations* in connection with the fracture of materials. These studies will be discussed when we investigate the mechanism of earthquakes in Sec. 7.2. In fact, the phenomenon of an earthquake is probably one of slippage along a fault which occurs at the present time. It stands to reason that the faults now visible to geologists on the surface of the Earth were formed by the occurrence of earthquakes in the past.

[1] SANFORD, A. R.: Bull. Geol. Soc. Amer. **70**, 19 (1959).
[2] JOBERT, G.: Geofis. Pura Appl. **43**, 75 (1959).
[3] JOBERT, G.: Geofis. Pura Appl. **45**, 13 (1960).
[4] ODÉ, H.: Geol. Soc. Amer. Mémoir **79**, 293 (1960).

7.14. The Rôle of Pore Pressure. Some interesting investigations have been made of the origin of the large thrust-faulting sheets (nappes) that have been found in many places, which are, under any circumstances, very difficult to explain.

Thus, HUBBERT and RUBEY[1, 2] proposed that the pore pressure (referring to the water contained in the rocks) at depth may be so great as to effectively provide a lubricated path for these sheets to glide upon. It is well known, of course, that the cohesion of a material is greately reduced if the pore pressure reaches the vicinity of the overburden pressure. However, BIRCH[3] has pointed out that, if the shearing strength at the base of a thrust sheet is reduced by the presence of a high pore pressure, the shearing strength must also be reduced *within* the sheet so that, in fact, no "thrust" could be supported. Thus, what is in fact required is a high pore pressure in the layer *below* the thrust sheet rather than within it.

Nevertheless, the idea of the possible importance of the pore-water pressure in the rocks during faulting presents some intriguing possibilities[4].

7.15. Jointing. Finally, we shall make a few remarks on jointing. Although ANDERSON's theory gives a qualitative explanation of such features, attributing joints to tension fracture, some additional investigations of the subject have been made.

Thus, PRICE[5] noted that, apart from tension joints, shear joints may also occur in highly compressed rocks. However, true joints are always tension fractures.

Some interesting conditions for the origination of joints have been enumerated by various authors. Thus, JAEGER[6] calculated the stress system in cooling sills and showed how joints are created in the process. A similar effect has been shown to occur by HARLAND[7] in rocks after the removal of an ice cover.

7.2. Theory of Earthquakes

7.21. Requirements of a Theory of Earthquakes. The phenomenon of the occurrence of an earthquake has been described earlier (see Sec. 2.2). A valid explanation of this phenomenon has to account for

[1] HUBBERT, M. K., and W. W. RUBEY: Bull. Geol. Soc. Amer. **70**, 115 (1959).
[2] RUBEY, W. W., and M. K. HUBBERT: Bull. Geol. Soc. Amer. **70**, 167 (1959).
[3] BIRCH, F.: Bull. Geol. Soc. Amer. **72**, 1441 (1961).
[4] PLATT, L. B.: Amer. J. Sci. **260**, 107 (1962).
[5] PRICE, N. J.: Geol. Mag. **96**, 149 (1959).
[6] JAEGER, J. C.: In: Dolerite a Symposium, p. 77. Hobart: Geol. Dept. Univ. Tasmania 1958.
[7] HARLAND, W. B.: J. Glaciol. **3**, 8 (1957).

the various aspects of its characteristic features. First of all, one must find a mechanism which causes stresses to be built up in those layers of the Earth which are prone to contain earthquake foci; this problem can be disposed of by identifying the mechanism with that causing orogenesis, thus labelling earthquakes as geodynamic phenomena. Second, the particular cyclic pattern in which the energy is released in earthquake sequences must be explained; this leads to the strain rebound theory and the "field theory" of earthquakes. Third, one has to account for the characteristic way in which energy is radiated from the focus during an earthquake; this has prompted one to construct a variety of models for the focal mechanism which would suit the observed seismic phenomena. Finally, the actual process at the focus during the occurrence of an earthquake is of much interest. Needless to say, these are all very difficult problems and no complete explanation can be hoped for to-date.

7.22. Mechanisms of Stress Creation. Let us consider first the problem of stress creation. A suitable theory must produce the pattern of energy release in earthquake sequences established by BENIOFF (see Sec. 2.23). Accordingly, there seems to be a constant strain accumulation in the world wherein, however, the energy is not released at a corresponding steady rate, but rather in sudden bursts of short duration which occur at frequent intervals.

We have already referred to an interpretation of this phenomenon which may be called "strain rebound theory", originally proposed by BENIOFF (cf. Sec. 3.63). Accordingly, the strain in the world is assumed to be built up steadily by some basic orogenetic process (e.g. contraction). At a certain instant, the breaking strength of the rock is reached and a fracture (an earthquake) develops. As soon as this has happened, the resistance to further motion along the fracture surface (i.e. fault) is greatly decreased (the fault is "greased") so that subsequent adjustments (aftershocks) may occur at frequent intervals. The energy for the aftershocks is supplied by the strain rebound owing to a Kelvin effect (cf. Sec. 3.63) in the mantle of the Earth.

The above theory gives a good explanation for the phenomenological pattern of earthquake aftershock sequences. However, there seems to be a similar pattern of energy release apparent if the sequence of world earthquakes is investigated. As was pointed out in Sec. 2.23 (see also Fig. 37), time intervals of high seismic activity alternate with time intervals of low activity, the characteristic period being about 10 years. This is true for the sequence of world shallow earthquakes and seems to be also true for the earthquakes in any one seismic zone. It is hardly conceivable that this could be the outcome of the same Kelvin effect held responsible for aftershock sequences since the required relaxation times are quite different in the two cases. It seems therefore desirable to seek a mechanism which could produce such patterns of alternating activity.

An interesting attempt along these lines has been made by MATUZAWA[1] by the proposal of his "field theory" of earthquakes. An earthquake field is a part of the crust of the Earth which produces periodically swarms of earthquakes. Such a field is, according to MATUZAWA, a heat engine which functions intermittently. The driving force of this engine is the constantly supplied heat flow from the interior of the Earth where it is assumed that in seismic areas the heat flow is somewhat larger than in seismically inactive zones. This would cause the temperature in seismic areas to rise which in turn would produce stresses which eventually could cause an earthquake swarm as soon as the breaking strength of the rocks is exceeded. Once an earthquake swarm is started, it may itself be capable of dissipating the excess heat beneath it rather rapidly so that the process then could start all over again. This would produce the cyclic appearance of earthquake sequences.

MATUZAWA considered two thermal mechanisms which might be capable of producing stresses. The first is the obvious one of producing the stresses by thermal expansion. MATUZAWA found (with reasonable assumptions for the constants involved) that the seismic zones in the upper part of the mantle would have to be heated to a temperature by 100° C higher than the surrounding aseismic zones to reach stresses equal to the breaking strength of rocks. To expect this to occur every 10 years or so is certainly somewhat unreasonable. MATUZAWA therefore considered a second possibility, viz. the assumption that a solid-liquid phase transformation would occur at the bottom of the crust. Such phase transitions are governed by the well-known Clausius-Clapeyron equation

$$\Delta p = \frac{L}{T(v_1 - v_2)} \Delta T \tag{7.22-1}$$

where Δp is the increase in pressure necessary in order to maintain equilibrium if the temperature is raised by the amount ΔT (°K); L is the heat of melting (per gram) and v_1, v_2 are the specific volumes (per gram) of the liquid and solid phases, respectively. Using suitable constants ($v_1 = 0.385$, $v_2 = 0.346$, $T = 1500°$ K, $L = 360$ joule/gram) one obtains the result that a temperature increase of 5° K requires an increase of pressure of 3×10^8 dyne/cm² to maintain equilibrium. If the mantle is assumed to be in a critical equilibrium state, it follows that a greater pressure is required to maintain this equilibrium in a region

[1] MATUZAWA, T.: Bull. Earthquake Res. Inst. Tokyo **31**, 179, 249 (1953); **32**, 231, 341 (1954).

where the temperature is higher than in the surrounding region; for a $5°$ temperature increase, the required pressure increase is, as outlined above, 3×10^8 dyne/cm². Now, if a temperature increase of $5°$ is assumed as a reasonable one to occur in the region of an earthquake field, this means that, under the above assumption of a critical equilibrium state, the pressure beneath the crust in an earthquake area must be higher by the calculated amount as compared with the surrounding areas. MATUZAWA shows that such an increase in pressure is indeed capable of producing stresses in the crust which are of the order of the breaking strength of rocks. In order to do this he took as models of earthquake fields plates of various shapes (circular, elliptic) with fixed rim, subject to a pressure on one side[1]. In the case of a circular model, MATUZAWA chose the following constants and dimensions: thickness $h=25$ km, radius $a=2h$, modulus of elasticity $E=1.25 \times 10^{12}$ dyne/cm², POISSON's ratio $m=1/4$. The result was that the maximum shearing stress is of the same order of magnitude as the pressure increase (3×10^8 dynes/cm²) and hence of the order of the breaking strength of rocks. The corresponding volume increase ΔV, owing to the upbulging of the plate, turns out to be $\Delta V=5.1 \times 10^{16}$ cm³. It must be assumed, of course, that an equal volume ΔV below the plate is being filled with molten material during the build-up of stresses. As soon as the breaking occurs, the pressure below the considered part of the crust will collapse, heat will escape and great seismic activity will follow. This process may be assumed to repeat itself at regular intervals.

The possibility of the above mechanism hinges on whether enough heat can be supplied to cause the required differential heating. Taking the heat of melting as equal to 9.4×10^8 erg/cm³ yields that a total amount of heat of $H=9 \times 10^{27}$ erg is required to melt the above-calculated volume ΔV. This amount of heat must be conducted into and out of the area of the earthquake field during the time interval of one cycle (about 10 years). Thus, the total heat flow in an earthquake area consistent with the above mechanism would have to be

$$\frac{H}{\pi a^2} \times \frac{1}{10\,\text{years}} \cong 10^{-2}\,\text{cal/deg. sec. cm}^2. \tag{7.22-2}$$

This is by about the factor 10^4 higher than any normally observed heat flows.

The explanation of the required high heat flow, thus, constitutes one of the main difficulties in the field theory of earthquakes although

[1] These plates are assumed to be *elastic*. KASAHARA [KASAHARA, K.: Bull. Earthquake Res. Inst. **37**, 39 (1959)] amplified the above theory by assuming *plastic* plates.

some investigations regarding this point have recently been published by AKI[1]. In addition, LOMNITZ[2] has also investigated the thermodynamics of planets from a more general standpoint.

7.23. Models of Earthquake Foci. Models of earthquake foci have been referred to already in this book in Sec. 2.24. We shall investigate here somewhat more closely some of the assumptions adopted there.

The whole issue in the construction of models is to explain the pattern of displacements observed at seismological observatories in terms of a focal mechanism. The observed displacements are partly conditioned by the properties of the medium transmitting the waves; it is therefore first necessary to reduce them to the abstract case of seismic rays emanating in all directions from the focus into a homogeneous medium. This can be done by the device (also mentioned in Sec. 2.24) of introducing a focal sphere; i.e. a sphere of homogeneous material enclosing the focal region, the sphere being large compared with the focal region. In the focal sphere, all seismic rays are straight. As was outlined earlier, each seismic station corresponds to a point on the surface of the focal sphere; the equivalent displacement on that surface can be inferred from seismic observations. This can be done with the aid of travel-time curves and with the aid of results of wave-transmission theory which is beyond the scope of the present book. Then, the task is to construct such models of the focal mechanism so that the induced motions on the surface of the focal sphere tally with those which are inferred from actual observations.

The types of models that have been proposed for this purpose fall into three categories: (i) a tault with rigid motion of the two halves of the focal sphere; (ii) a point source (i.e. a suddenly applied single force, dipole etc.) at the center of the focal sphere, (iii) a dislocation at the center of the sphere. In the cases above, the medium is taken as perfectly elastic. Finally, we shall discuss some experimental investigations of these models and analyze the energy relationships.

Turning first to the model assuming a fault with rigid motion of the two halves, we note that the displacements can be read easily from the picture of the focal sphere in Fig. 38. Of particular interest is the position of the nodal lines of P, SH and SV (cf. Sec. 2.24 for the definition of these terms) for a given fault. These are lines on the surface of the focal sphere on which the P, SH or SV motion is zero. These nodal lines can be compared with the actually observed zero lines for the first impulse of the corresponding seismic phases.

---

[1] AKI, K.: J. Phys. Earth **4**, 53 (1956).

[2] LOMNITZ, C.: Geophys. J. Roy. Astr. Soc. **5**, 157 (1961).

With regard to the P component, an inspection of the focal sphere shows at once that firstly the trace of the fault plane and secondly the trace of the plane perpendicular to the motion vector (which has been termed "auxiliary" plane) separate regions of outward radial motion (○) from regions of inward radial motion (△). Therefore they represent the nodal lines for P. The situation corresponding to the same faulting as in Fig. 38 is outlined in Fig. 124. From the discussion given here it becomes obvious that, by P observations alone, one is not able to distinguish between fault plane and auxiliary plane.

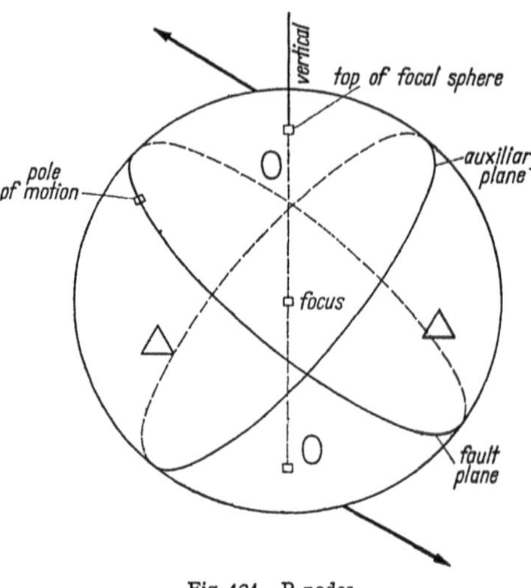

Fig. 124. P nodes

The nodes for SH can be calculated as follows. The sign of the SH motion is that of the horizontal component of the tangential displacement on the surface of the focal sphere, i.e. of that component which is normal to the plane through the focus and the vertical direction. It is obvious that this component is zero, firstly, on the fault plane, and secondly, on a plane containing the vertical direction, the focus, and the pole of the motion. The latter fact becomes apparent immediately upon inspection of the focal sphere in Fig. 38. Therefore, the nodal lines pertaining to SH upon the surface of the focal sphere are as shown in Fig. 125, where the position of the fault corresponds to that assumed in Fig. 38.

Finally, the sign of SV is identical to that of the component of the tangential displacement on the surface of the focal sphere which lies in the plane of the ray. The plane of the ray is that plane which contains the focus, the top of the focal sphere and that point S′ on the focal sphere which is under consideration. In order to determine the nodal lines of SV, one can note firstly, that the trace of the fault plane must be such a nodal line, and secondly, that one is faced with the task of determining the geometrical locus of all such points S where the SV component of the motion on the surface of the focal sphere is zero, i.e. where the tangential motion is orthogonal to the plane of the ray. This means that in the spherical triangle formed by the points Pole, S′ and Top on the focal sphere, the angle Pole-S′-Top must be a right one. This yields the following condition:

$$\cos\alpha = \tan b \cot c \qquad (7.23\text{–}1)$$

where α denotes the difference between the azimuth of S′ and the azimuth of the pole of the motion as seen from the top of the sphere, b is the declination of S′ and c the declination of the pole of the motion from the top of the sphere. The curve described by Eq. (7.23–1) is not a

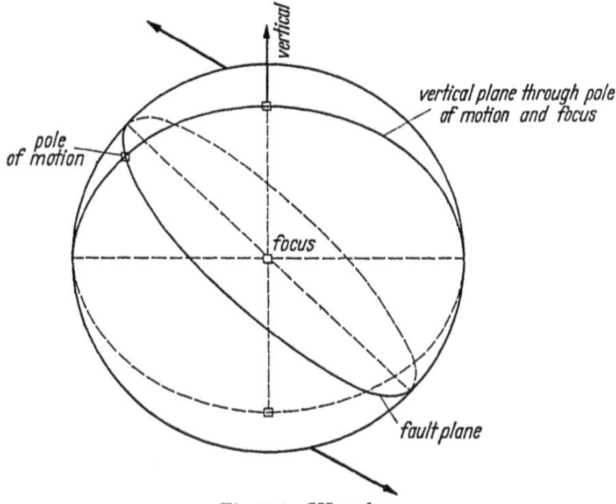

Fig. 125. SH nodes

simple one, it is not even plane. The stereographic projection of a family of such curves (for various values of c) is shown in Fig. 126.

It is quite obvious that the above model of an earthquake focus is a gross oversimplification of the physical facts. It is hardly believable that the motion at a distance from the focus is exactly that which would occur if the whole sphere were sliced and the halves would move rigidly. Nevertheless, it is interesting to note that the nodes observed in earthquakes correspond in the vast majority of cases to just those which have been postulated in the above theory. It thus appears that the model of "rigid faulting" at least describes the *signs* of the first impulses of the P, SH, and SV phases correctly.

The most obvious attempt to introduce a model which is physically more realistic is by assuming various types of multipole forces which are suddenly or periodically applied at the center of the focal sphere. The displacements at the surface of the sphere can then be calculated by methods of the theory of elasticity[1–3]. If this is done, it turns out

[1] MALINOVSKAYA, L. N.: Trudy Geofiz. In-ta Akad. Nauk SSSR., No. **22** (**149**) 143 (1954).

[2] NAKANO, H.: Seism. Bull. Centr. Met. Obs. Japan **1**, 92 (1923).

[3] GASSMANN, F.: Geofis. Pura Appl. **40**, 55 (1958).

that a dipole force with a moment, applied at the center of the focal sphere, produces exactly the same nodal lines as those obtained by the model of rigid faulting. Other types of forces have also been tried, and it turns out that various types of sources produce exactly the same

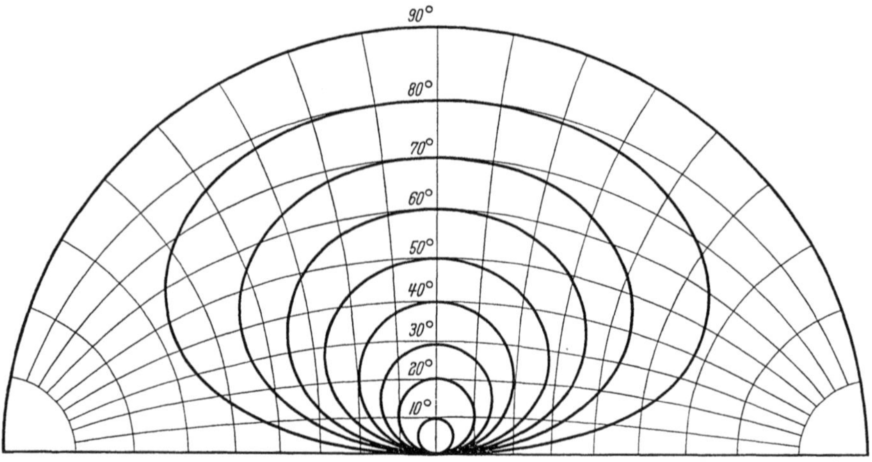

Fig. 126. SV nodes, in stereographic projection (after MALINOVSKAYA[1])

P-nodal lines (see Fig. 127). The attendant questions have been analyzed, for instance, by BYERLY and STAUDER[2]. Thus, a distinction between the various types of sources can be made only if either S-nodes or else

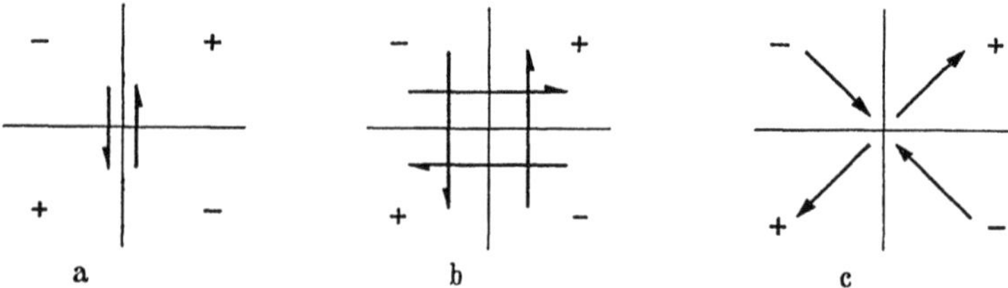

Fig. 127a–c. Identical quadrants of compression and dilatation produced (a) by a dipole force, (b) by a quadrupole force, (c) by a orthogonal regional compression and tension

amplitudes are taken into account. Attempts at such calculations and comparisons with observational data have been made by various people. Thus KNOPOFF and GILBERT[3] attempted to calculate radiation patterns, KEILIS-BOROK[4] compared some calculations with earthquake records

[1] MALINOVSKAYA, L. N.: Trudy Geofiz. In-ta Akad. Nauk SSSR., No. **22** **(149)** 143 (1954).

[2] BYERLY, P., and W. STAUDER: Earthq. Notes **29**, No. 3, 17 (1958).

[3] KNOPOFF, L., and F. GILBERT: Bull. Seism. Soc. Amer. **49**, 163 (1959).

[4] KEILIS-BOROK, V. I.: Izv. Akad. Nauk SSSR., Ser. Geofiz. **1957**, No. 4, 440 (1957). — Ann. Geofis., Roma **12**, 205 (1959) and references given there.

and found that a dipole source with a moment is the most likely type (at least in Russia), and HONDA and MASATUKA[1], in a similar investigation, found that quadrupole sources fit Japanese data best. It appears thus that the problem of the source type has not yet been finally resolved.

In addition to the above investigations, one might mention that KASAHARA[2] assumed an ellipsoidal instead of a spherical focal region so as to obtain an asymmetry.

An extension of the above models has been attempted by various authors by referring to the theory of *dislocations* in an elastic medium. The dislocation models advocated in this connection are shown in Fig. 128. A dislocation theory of earthquake faulting has been proposed particularly by SAITO[3], by VVEDENSKAYA[4] and by DROSTE and TESSEYRE[5]. STEKETEE[6] has analyzed such models rather closely mathematically and showed that a dislocation sheet can be built up from nuclei which correspond to those shown in Fig. 127 where the arrows are interpreted as displacements. He noted that the type (a) and (b) nuclei are dynamically significantly different: whilst the (b)-type is in equilibrium, the (a)-type is not; i.e., if the resulting force and torque are calculated for a part of the medium which contains the nucleus, these vanish for the (b)-nuclei but not for the (a)-nuclei. Thus, for an (a)-nucleus, at least in the static case, an external torque is required to maintain equilibrium (the force vanishes automatically). It should be noted, however, that the above remarks refer to *static* nuclei only; for dynamic nuclei, the conditions applying are still somewhat uncertain. STEKETEE's calculations have been further elaborated upon by CHINNERY[7].

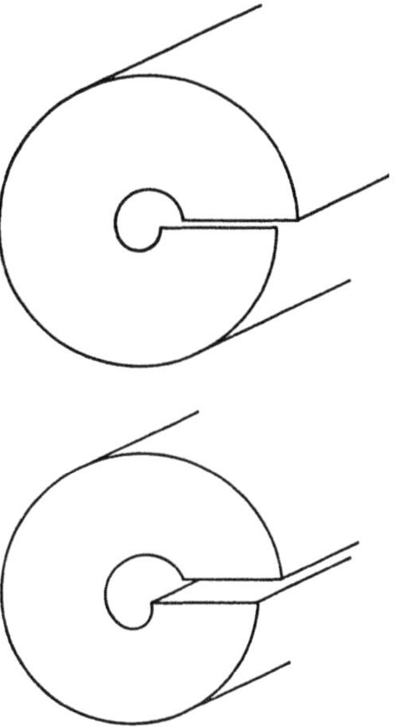
Fig. 128. Two dislocation models. After SAITO[3]

[1] HONDA, H., and A. MASATUKA: Sci. Rep. Tohoku Univ. (5) **8**, 30 (1952).
[2] KASAHARA, K.: Bull. Earthquake Res. Inst. **35**, 532 (1957); **36**, 21 (1958).
[3] SAITO, Y.: Geophys. Mag. **28**, No. 3, 329 (1958).
[4] VVEDENSKAYA, A. V.: Izv. Akad. Nauk SSSR., Ser. Geofiz. **1958**, 175 (1958); **1959**, 516 (1959); **1960**, 513 (1960); **1961**, 261 (1961).
[5] DROSTE, S., and R. TESSEYRE: Ann. Geofis., Roma **12**, 179 (1959). — Sci. Rep. Tohoku Univ., Ser. V, Geophys. **11**, No. 1, 55 (1959). — Bull. Seismol. Amer. **50**, 57 (1960); see also TESSEYRE, R.: Acta Geophys. Polon. **8**, No. 2, 107 (1960).
[6] STEKETEE, J. A.: Canad. J. Phys. **36**, 192, 1168 (1958).
[7] CHINNERY, M. A.: Bull. Seism. Soc. Amer. **51**, 355 (1961). — J. Geophys. Res. **65**, 3852 (1960).

It may be interesting to compare the above theory with model experiments. Such experiments have been performed by HEALY and PRESS[1-3] with the express purpose of elucidating the position of the P and S nodes in relation to the seismic source. Whereas for the P nodes the result was as expected, this was not the case for the S-nodes[3], if the classical theory of a dipole was taken as a basis of comparison. However, the validity of the models has admittedly not been established beyond doubt. In fact, some investigations based on observations of near earthquakes gave the result that something akin to a conical source should be assumed[4]. Incidentally, such conical models have also been studied theoretically[5-7].

Finally, we present some calculations of the size of the strained region V prior to an extreme earthquake. If S and μ be the strength and the rigidity of the medium, respectively, then BULLEN[8,9] showed that the following relationship is valid ·

$$12 q \mu E = S^2 V \tag{7.23-2}$$

where E is the energy released in seismic waves and qE the distortional strain energy in the focal region. The energy E can be obtained from the various magnitude relationships given in Sec. 2.23, and hence an estimate of V can be arrived at. For an earthquake of the maximum magnitude known ($M = 8.6$), BULLEN[9] obtained

$$V = 6 \times 10^9 \text{ cm}^3. \tag{7.23-3}$$

This is equivalent to a sphere of radius 25 km.

Analogous calculations have also been made by TESSEYRE[10].

7.24. The Friction at an Earthquake Fault. From the models of earthquake foci discussed above, it appears that the assumption of a simple fault has at least a great likelihood of being correct. In order to get a proper understanding of the phenomenon of an earthquake, it is necessary to have a detailed picture of the mechanism of slip along the fault.

---

[1] HEALY, J. H., and F. PRESS: Bull. Seism. Soc. Amer. **49**, 193 (1959).
[2] HEALY, J. H., and F. PRESS: Geophysics **25**, 987 (1960).
[3] PRESS, F.: Publ. Dom. Obs. Ottawa **20**, 271 (1958).
[4] MIKUMO, T.: Mem. Coll. Sci. Univ. Kyoto A **29**, No. 2, 221 (1959).
[5] INOUYE, W.: Bull. Earthquake Res. Inst. **14**, 582 (1936).
[6] TAKAGI, S.: Quart. J. Seism. **14**, No. 3, 1 (1950); **17**, No. 3, 53 (1953); **17**, No. 4, 1 (1953).
[7] HIRONO, T., and T. USAMI: Pap. Met. Geophys. **7**, 287 (1956).
[8] BULLEN, K. E.: Trans. Amer. Geophys. Un. **34**, 107 (1953).
[9] BULLEN, K. E.: Bull. Seism. Soc. Amer. **45**, 43 (1955).
[10] TESSEYRE, R. K.: Acta Geophys. Polon. **6**, 260 (1958).

This is of particular interest because the process of faulting in an earthquake is probably not much different from the process occurring in geological faulting.

At first glance it would appear that earthquakes are not even mechanically possible[1]. The surface-roughness of broken rocks causes, under ordinary (laboratory-) circumstances, the latter to have a coëfficient of friction of $f=2$. At 100 km depth the pressure is roughly

$$p=3\times10^{10}\ \text{dynes/cm}^2. \qquad (7.24\text{–}1)$$

This means that, with a coëfficient of friction of $f=2$, one needs a tangential stress of

$$\tau=f\,p=6\times10^{10}\ \text{dynes/cm}^2 \qquad (7.24\text{–}2)$$

to slide one rock face over another. However, this required tangential stress is about 10 times larger than the yield stress of rocks. Hence earthquakes appear to be impossible.

The only possible way to resolve the above dilemma is by assuming that the coëfficient of friction must, in effect, be much smaller than $f=2$ suggested by laboratory experiments. Indications that this is so, come from experiments by JAEGER[2] who found friction coëfficients of the order of 0.6 under pressure. Such values, however, are still too high for earthquakes to be possible. Further indications that the coëfficients of friction might be much lower at high pressures come also from the work of BOWDEN[3] on the microscopics of friction. Accordingly, under ordinary circumstances, the actual area of contact (not to be confused with the apparent area of contact) between two sliding surfaces is in fact very small, due to minute surface irregularities. During sliding, pressures across the actual area of contact are always very great; in fact so great as to cause plastic yielding and flow. If the pressure between the two sliding bodies is increased, the actual area of contact is simply increased proportionally which yields the customary linear law of dry friction implied in Eq. (7.24–2).

Extrapolating the above picture to very high pressures, it can be argued that one should expect that, at a certain stage, complete contact (i.e. a saturation point) would be reached between the sliding surfaces and hence that at that stage the linear law of dry friction should break down. It is easy to calculate the pressure beyond which the law of dry friction is *certain* to break down. BOWDEN'S (l.c., p. 31) data show that in the case of steel, the area of contact is a fraction of $1/(9.5\times10^3)$ of

---

[1] This has been noted, for instance, by OROWAN [OROWAN, E.: Geol. Soc. Amer. Mém. **79**, 323 (1960)].

[2] JAEGER, J. C.: Geofis. Pura Appl. **43**, 148 (1959).

[3] BOWDEN, F. P., and D. TABOR: The Friction and Lubrication of Solids. Oxford: Clarendon Press 1954.

the total area per pressure of 1 kg/cm². Thus, if the law of dry friction were valid to very high pressures, i.e. if the area of contact would remain proportional to the load, the saturation point would be reached at a pressure of 9.5×10^3 kg/cm² $=9.5\times10^9$ dynes/cm². Beyond this pressure, the mechanism of dry friction as envisaged by BOWDEN could certainly no longer be valid. This "saturation" pressure, however, is lower than the pressure at 100 km depth.

This would account for an apparently much lowered coëfficient of dry friction at depth. Furthermore, friction also causes much heat to be produced which, in turn, might help to soften the material adjacent to the fault surface, thus again lowering the resistance to sliding. The absence of pronounced chemical evidence regarding this point, however, shows that heating cannot be too great.

Unfortunately, no further quantitative corroborations of the above qualitative arguments are as yet available.

7.25. Fracture Theories of Earthquakes. We come now to the central problem of strain release during an earthquake. Since earthquakes have much in common with fracture processes, the logical thing to do is to attempt to go through the various theories of fracture and to see whether a suitable explanation of earthquakes can be obtained from them.

Earthquakes have always the appearance of sudden shocks, and therefore only those theories of fracture can be relevant which exhibit high-velocity crack propagation.

First of all, one would think of the earthquake mechanism as being one of brittle fracture. This would imply the validity of MOHR's fracture criterion, i.e. the postulate that the stresses producing the fractures are such that the fracture surface is inclined by about 30° toward the maximum principal pressure and contains the intermediate principal stress. Equally, turning to the microscopics of brittle fracture, one would assume the Griffith mechanism and a crack propagation velocity derived from Eq. (3.53–10). In this interpretation, an earthquake would correspond to a sudden extension at the margin of an old or to a creation of a new fracture, the sweep of the edge at high speed providing the shock; the energy for the fracture would be provided by the release of elastic energy in the strained region. This theory has been called "strain rebound theory" and seems to be due to REID[1].

Straightforward as such a theory might appear, it has some unsatisfactory aspects. Brittle fracture represents essentially the opening up of a crack, and it is very doubtful whether at the depths at which earthquakes occur, cleavage could occur at all. Furthermore, it is most doubtful whether the material can be assumed as brittle.

[1] REID, H. F.: Phys. Earth 6, 100 (1933).

The next possibility is therefore to think of some high-velocity ductile fracture. However, this also represents essentially the opening up of a crack with the extension of the crack at its edges causing the shock. This seems improbable.

It appears therefore indicated that one should look for a mechanism where the shock is produced by a sudden slippage along a pre-existing fault surface. The aspects of such a phenomenon are very much like that described in the intergranular type of fracture, although it is of course unlikely that it is crystal grains which provide the interlocking. In connection with the discussion of intergranular fracture, we have mentioned, however, that a similar type of stress release could be expected in any material that exists of two different structural elements.

Analytically, the intermittent sliding along a pre-existing fracture surface can be described by the mechanism of snapping dislocations[2] (for the definition of dislocations, see Secs. 3.22 and 3.54). The general idea of snapping dislocations has been explained in Sec. 3.54. However, it is now necessary to envisage physical conditions for the start and continuation of the snapping process along an earthquake fault.

It has already been stated that it is quite impossible to try to give a proper analytical description of a fracture process, and therefore it is also impossible to give an exact analysis of the snapping of dislocations. The best that can be hoped for, are therefore statistical considerations. In order to do this, HOUSNER[1] split the whole earthquake fault into slip areas A in which the slip is constant, i.e. areas which are active in any one earthquake. He then assumed that the expected number of slips having areas between A and $A+dA$ would be proportional to dA/A, i.e. he assumed a statistical frequency distribution f of slip areas A which may be written

$$f = C\frac{a_0}{A} \equiv C\frac{1}{x} \tag{7.25-1}$$

where a_0 is the minimum value of A. Normalization yields

$$C = \frac{1}{\log x_1 - \log x_0} \tag{7.25-2}$$

where x_1 is the largest and x_0 the smallest possible value of x. Hence the frequency distribution can be expressed as follows

$$f = \frac{1}{\log x_1 - \log x_0} \cdot \frac{1}{x}. \tag{7.25-3}$$

As a measure of an earthquake, HOUSNER took the average slip S occurring along the fault. This he connected by a logarithmic measure

[1] HOUSNER, G. W.: A Dislocation Theory of Earthquakes, 34 pp. Cal. Inst. Technol. Rep. N 6-onr-244. — Bull. Seism. Soc. Amer. **45**, 197 (1955).

with the Richter magnitude M which led to the following formula:

$$dM = dS/S. \qquad (7.25\text{-}4)$$

Integrating, HOUSNER obtained

$$S = S_0 \, e^M. \qquad (7.25\text{-}5)$$

If it be further assumed that all dislocations are geometrically similar, the slips must be proportional to the square root of the area A. Hence:

$$A = A_0 \, e^{2M} \qquad (7.25\text{-}6)$$

and finally with (7.25–1):

$$f = \text{const } e^{-2M}. \qquad (7.25\text{-}7)$$

The last Eq. (7.25–7) may be compared with the frequency distribution of earthquakes in any one area, M being taken as the Richter magnitude. According to HOUSNER, the agreement is good, at least for the earthquakes that occur in the Imperial Valley of California.

A similar test can be made with Eq. (7.25–6) regarding the areas affected by earthquakes of any one magnitude. From an analysis of observational data HOUSNER obtained

$$A_0 = 0.0012 \text{ sq. miles} = 3.1 \times 10^7 \text{ cm}^2. \qquad (7.25\text{-}8)$$

Similarly, he obtained for the constant S_0 in (7.25–5):

$$S_0 = 0.25 \text{ inches} = 0.64 \text{ cm}. \qquad (7.25\text{-}9)$$

These are values which appear to be reasonable. They permit one to calculate the average slip and area of slip for earthquakes of any one magnitude.

In spite of the apparent success of the above theory, there remain, in fact, very grave difficulties of a fundamental nature. These are connected with the observations made in Sec. 7.24; viz. it appears as impossible that any form of fracture could exist at the depths at which earthquakes occur. A very detailed analysis of these difficulties has been made by OROWAN[1] who came to the conclusion that the mechanism of seismic faulting must be based upon a *plastic* instability phenomenon such as gives rise to slip bands. This idea, which appears as most promising, however, has not yet been developed beyond qualitative arguments.

7.3. Analytical Theories of Folding

7.31. The Problem of Folding. A study of the mechanics of folding must yield an explanation of the very large contortions evident in the physiography of mountain ranges. It is obvious that this will have to

[1] OROWAN, E.: Geol. Soc. Amer. Mém. **79**, 323 (1960).

be achieved by a recourse to the rheology of the Earth. We have pointed out in Sec. 3.6 that one and the same material, although it may appear as very strong in tests involving only short time intervals, may have totally different rheological properties when it is subject to stresses of very long duration.

Therefore, the approach to the explanation of folding has been one of trying to apply each one of the basic rheological equations of state to the problem. First of all, one would think of folding as being simply due to elastic instability as exhibited by buckling. Although results suggestive of simple folds are obtained in this fashion, the main objections against such theories stem from the observation that the deformations in natural folding are certainly exceeding the elastic limit.

The next step, therefore, is to assume infinite flexibility to bending of the strata, but retaining most of the other concepts of elasticity theory. This does not lead to a very satisfactory explanation either.

The ultimate explanation of folding, therefore, can only be achieved by assigning to the material in question general rheological properties. Unfortunately, except for a few cases of plastic buckling, no such cases have ever been rigorously discussed. The reason for this is that any exact treatment of such questions involves extreme analytical difficulties.

7.32. Buckling. In order to demonstrate how elastic buckling could give rise to folds, let us consider the following idealized case. A rectangular piece of the crust of the Earth is represented by a thin plate of the same shape (but thought of as plane). A force F is acting in the original plane of this plate normal to one of its edges. Then, for an elastic material, the condition of equilibrium is

$$-F\,y(x)=M\,d^2y/dx^2 \qquad (7.32\text{–}1)$$

where F is the force, y the deflection of the plate from its original plane, x the co-ordinate in the direction of the acting force, and M a constant indicative of the resistance of the plate to bending depending on its elastic parameters. In Eq. (7.32–1) the left-hand side represents the moment of the froce exercised upon the plate at the point x, the right hand side is the resistance to bending of the plate, being assumed as proportional to the local curvature which is equal to d^2y/dx^2 for small deflections.

The solution of Eq. (7.32–1) is

$$y=A\sin\left(\sqrt{\frac{F}{M}}\,x\right)+B\cos\left(\sqrt{\frac{F}{M}}\,x\right) \qquad (7.32\text{–}2)$$

where A, B are constants of integration to be determined from the boundary conditions. If we assume for the latter, say,

$$y(0)=y(X)=0 \qquad (7.32\text{–}3)$$

it becomes at once obvious that, in general, no solutions of the assumed type (i.e. bulging) exist; that is one obtains

$$A = B = 0. \qquad (7.32\text{–}4)$$

The plate, if it be subjected to the indicated force, will simply contract a little under the load and will not bulge. However if F has a certain particular value given by (i.e. if it is equal to an eigenvalue of the system)

$$F = M \frac{n^2 \pi^2}{4X^2} \qquad (7.32\text{–}5)$$

with n denoting any even integer, then the solution becomes ($B=0$)

$$y = A \sin\left(\frac{n\pi}{2} \frac{x}{X}\right), \qquad (7.32\text{–}6)$$

and this satisfies the boundary conditions for *any* A. Thus, if F reaches an eigenvalue of the equation, the plate will buckle. The lowest eigenvalue of F produces a sinusoidal half-wave with arbitrarily great amplitude. For F above the first eigenvalue, the deformation becomes *unstable*. The first mode of buckling, being in the shape of a sinusoidal half-wave, has some resemblance with a fold. This resemblance has been used for an explanation of folding. A representative example of this kind has been calculated in Sec. 6.24 in connection with the contraction theory.

The above theory is obviously somewhat oversimplified. Calculations have therefore been made where the elastic plate was embedded in a viscous medium or floating upon it[1–4].

However, we have already mentioned that the assumption of a basically elastic behavior of the materials involved is hardly justified in view of the large permanent deformations that are evidenced in the crust of the Earth. It is therefore seen that buckling will at best give an indication where folds will start, with another mechanism taking over as soon as deformations become established. Accordingly, we shall return to the buckling hypothesis in connection with the discussion of *systems* of folds.

7.33. Theories Assuming Infinitely Flexible Strata. A semi-analytical solution of the folding problem assuming the mechanism of deformation to be that of infinitely flexible sheets, has been given by DE SITTER[5].

[1] SMOLUCHOWSKI, M.: Akad. Wiss. Krakau, math. Kl. **1909**, 3 (1909); **1910**, 727 (1910).

[2] GOLDSTEIN, S.: Proc. Cambridge Phil. Soc. **23**, 120 (1926).

[3] KIENOW, S.: Fortschr. Geol. Paläont. **14**, No. 46 (1942).

[4] BIOT, M. A.: Bull. Geol. Soc. Amer. **72**, 1595 (1961) and references given there.

[5] DE SITTER, L. U.: Proc. Kon. Ned. Akad. Wet. **52**, No. 5 (1939). See also DE SITTER, L. U.: Structural Geology. New York: McGraw-Hill Book Co. 1956.

According to DE SITTER, the mechanism of folding is the result of the following conditions: (a) during folding, the volume of the strata is conserved and (b) each infinitesimal layer undergoes only bending. Furthermore, DE SITTER assumed the "principle of concentric folding" which is expressed by the assumption that the surface of a folded layer is formed by three circles (cf. Fig. 129). DE SITTER thus arrived at the

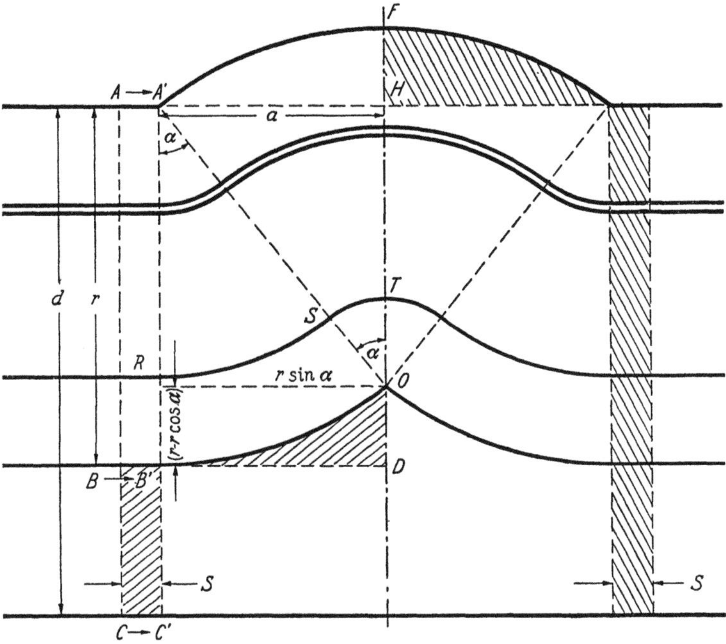

Fig. 129. DE SITTER's[1] model of folding

picture of folding illustrated in Fig. 129; in this Figure, unprimed letters refer to the situation before folding, whereas primed letters refer to the situation after folding. The folding has been caused by the compression by the amounts $2s$ of the original strata. It is easy to see that above the line through B, the two laws stipulated by DE SITTER are indeed satisfied, below that line this is, however, not the case. Thus, the equivalent of the material from the first shaded area must be transferred into that of the second one by plastic flow or some such phenomenon. Moreover, although DE SITTER is certainly satisfying his two assumptions above the line through B, it is quite obvious that this is not the *only* solution satisfying those principles. It may thus be observed that the solution of DE SITTER is not a real "explanation" of folds as the cause of the latter is not reduced to a field of forces, nor is any

[1] DE SITTER, L. U.: Proc. Kon. Ned. Akad. Wet. **52**, No. 5 (1939).

attempt made at a rationalization what forces could produce the particular type of bending assumed by DE SITTER. That the two laws are satisfied, is not sufficient to account for this, as they are only an expression of the conservation of area and matter.

It is obvious that DE SITTER's theory can be used to estimate the basal shearing plane (located at depth d beneath the surface in Fig. 129) from a knowledge of the volume of rocks pushed above the original level by the folding and the crustal shortening involved. DUSKA[1] has shown that d can also be estimated from the dip of the strata should the folds be eroded.

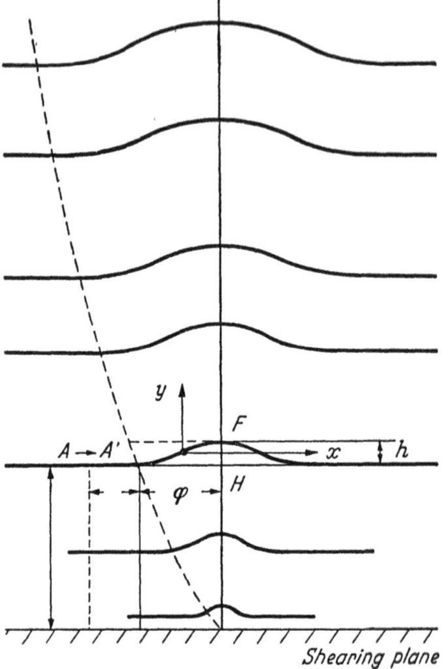

Fig. 130. TIEDEMANN's[2] model of folding

The theory of DE SITTER has been modified by TIEDEMANN[2]. The latter author replaced the circles which make up the form of a concentric fold, by sine-curves. As in DE SITTER's scheme, the two fundamental assumptions are adhered to. It is fairly easy to calculate the shapes of a series of sine-curves that make up the strata in a layer of the Earth, and one thus obtains a picture as shown in Fig. 130. It will be observed that the lower boundary of possible folding is now not a surface below which one has to assume plastic deformation or such like, but rather a "shearing plane" above which a displacement takes place, but below which everything remains fixed.

Referring to Fig. 130, one has for a point on the curve:

$$y = \tfrac{1}{2} h \sin \pi x/\varphi . \qquad (7.33\text{–}1)$$

From the geometry apparent in Fig. 130, one can form the following equations which are based upon the fundamental assumptions:

$$\text{area } A'FH = \tfrac{1}{2} \varphi h = s h, \qquad (7.33\text{–}2)$$

$$L \equiv \text{length } A' \text{ to } F = AF \equiv \varphi + s . \qquad (7.33\text{–}3)$$

[1] DUSKA, L.: J. Alberta Soc. Petrol. Geol. 9, 20 (1961).
[2] TIEDEMANN, A. W.: Geologie en Mijnbouw 3, 199 (1941).

However, the length of the sine curve can be calculated; one has:

$$dL = (1 + y'^2)^{\frac{1}{2}}\,dx = (1 + n^2)^{\frac{1}{2}}\left(1 - [n^2/(1 + n^2)]\sin^2 \pi x/\varphi\right)^{\frac{1}{2}} dx \qquad (7.33\text{–}4)$$

with $n^2 = \pi^2 h^2/4\varphi^2$, and hence

$$L = 2(1 + n^2)^{\frac{1}{2}} E(k) \cdot \varphi/\pi \qquad (7.33\text{–}5)$$

where $E(k)$ is a standard elliptic integral:

$$E(k) = \int_0^{\pi/2} (1 - k^2 \sin^2 x)^{\frac{1}{2}}\,dx \qquad (7.33\text{–}6)$$

with

$$k = n/(1 + n^2)^{\frac{1}{2}}. \qquad (7.33\text{–}7)$$

The integral $E(k)$ has been tabulated; using its values, one can calculate the values for h for various assumptions for s and d. It can be readily seen that the process of folding can be explained by a continuous movement, simply by adjusting the parameter n for the neighboring strata accordingly. One thus arrives at a series of folds as depicted in Fig. 130.

The theory of TIEDEMANN is open to the same criticisms as that of DE SITTER: The mechanism of folding is not uncovered as the sine-curve type of bending is nothing but an arbitrary shape satisfying two laws of continuity. Moreover, it is quite certain that the second law (which prohibits the areal extent of a stratum from being altered) is not fulfilled because of the almost certain preponderance of plastic extensions in the folding process.

7.34. General Rheology. The final section on folding should be devoted to the mathematics of fold formation assuming a general rheological equation of state of the material. Unfortunately, the theory of fold formation based on these assumptions is still unwritten and no report on it can therefore be given. Some calculations have been made of the behavior of a *viscous* plate embedded in a viscous medium[1], but no media of a more complicated nature seem to have been investigated. GOGUEL[2] made some qualitative investigations based on general energy dissipation. It is possible that the advent of high-speed computing devices will enable one to overcome the tremendous analytical difficulties involved so that some progress in this direction may be made in the years to come.

7.4. Model Experiments of Faults and Folds

7.41. Theory of Scale Models. Owing to the difficulty of describing the deformation of rocks apparent in faults and folds by analytical means, emphasis has been placed upon experimental investigations.

[1] BIOT, M.: Bull. Geol. Soc. Amer. **72**, 1595 (1961) and references given there.

[2] GOGUEL, J.: Introduction à l'étude mécanique des déformations de l'écorce terrestre. Paris: Mémoire du Ministère de la Production Industrielle et des Communications, Imprimerie Nationale. 1943.

The basis of such experimental investigations is the mechanical theory of scaling. The general principles of scaling have been known in physics for a long time, but it is to the credit of HUBBERT[1] to have pointed out their significance in connection with geodynamical problems.

The general principles of mechanical scaling are based upon the fact that the scaling factors between the model and nature must be chosen in such a manner so as to cause all the relevant dynamical equations to become identities if the scaling factors are inserted in place of the quantities themselves. This can be exemplified as follows. Assume a dynamical system whose behavior is completely described by NEWTON's law of motion

$$F = M\, d^2X/dT^2 \tag{7.41-1}$$

where F signifies the force, M the mass, X the displacement and T time. If we denote the corresponding scaling factors by f, m, x, t respectively and insert them into the equation of motion, we obtain the following scaling condition:

$$f = m\,x/t^2. \tag{7.41-2}$$

The relationship between the various scaling factors is therefore the same as that between the dimensions of the various quantities involved. Since, in a mechanical system, there are *three* independent dimensional units (usually chosen as *length*, *mass* and *time*) it follows that there are, in general, three independent scaling factors that one is able to choose at will. All other scaling factors are then prescribed.

This can be illustrated in a practical example. In the discussion of geodynamic phenomena, it is of importance to know by what material the rocks should be represented. Thus, let us assume that we want to make a model with a length reduction of $x = 5 \times 10^{-6}$ (i.e. 1 km is represented by 5 mm). For practical reasons, there are limits set to the reduction in density which may be assumed of the order of 1/2. A further quantity that is fixed is the ratio in gravity (the latter is an acceleration) which is unity since the gravitational attraction in the laboratory is the same as that in nature.

With the ratios of length, density and acceleration being given, all others will be prescribed. One finds easily for the mass reduction $m = 6.25 \times 10^{-17}$ and for the time reduction $t = 2.24 \times 10^{-3}$. In order to determine what material would be suitable to represent the rocks, one can now determine the required reduction in strength. The strength is expressed in terms of a stress, and one finds therefore for its ratio $s = 2.5 \times 10^{-6}$. If this be compared with the strength of granite of 2×10^9 dynes/cm², one finds that the model-material must have a

[1] HUBBERT, M. K.: Bull. Geol. Soc. Amer. **48**, 1459 (1937); **62**, 355 (1951). — Bull. Amer. Ass. Petrol. Geol. **29**, 1630 (1945).

strength of 5×10^3 dynes/cm². This represents a rather weak material; a cube of 3.3 cm to the side or larger would not stand up under its own weight. Thus it turns out that e.g. sand would be a good example.

After HUBBERT'S fundamental investigations, many people have worked on scaling relationships. Particularly GZOVSKII[1] has has amplified the above theory and has made many experiments, and BELOUSOV[2] has described investigations that were performed. Similarly, KÖSTER[3] and WUNDERLICH[4] have published papers on such investigations. In addition, GAKKEL[5] has shown that tectonic deformations may be modelled naturally in sea-ice.

It may be noted that in most geodynamic processes, one is faced with the condition that it is permissible to neglect inertia forces: the motions are usually so slow that one proceeds through a series of equilibrium states. Under such circumstances, it is possible to ignore the required scaling factors of time. One has only to insure that in the model the experiments are performed slowly enough to permit one to ignore inertia terms.

7.42. Faults. The theory of scaling outlined above shows that it is permissible to try to duplicate geological structures by means of experiments.

In order to duplicate *faulting*, it has been proven to be most convenient to take ganular materials such as sand, lead-shot etc. A series of beautiful experiments has been reported by NETTLETON and ELKINS[6] who achieved the duplication of many well-known types of faults, including cliffs.

Other experiments on faulting have been performed by CLOOS[7], HUBBERT[8], KÖSTER[9], SANDFORD[10], TANNER[11], and LEE and coworkers[12]. The last authors reported experiments with clay on shear fractures. In all these experiments, the general idea that geological faulting is nothing

---

[1] GZOVSKII, M. V.: Sovietsk. Geol. 1958, **4**, 53 (1958). — Publ. Bur. Centr. Séism. Int. A **20**, 383 (1959). — Int. Geol. Congr. 21st Sess., Norden, Dokl. Soviet. Geol. 17 (1960).

[2] BELOUSOV, V. V.: Vestn. Akad. Nauk SSSR. **28**, (9), 3 (1958). — Bull. Geol. Soc. Amer. **71**, 1235 (1960). — Sci. Amer. **204**, 96 (1961).

[3] KÖSTER, R.: Neues Jb. Geol. Paläont. Mh. 7/8, 289 (1957); 8/9, 337 (1958).

[4] WUNDERLICH, H. G.: Neues Jb. Geol. Paläont. Mh. 11, 477 (1957).

[5] GAKKEL', YA. YA.: Izv. Vsesoyuz. Geogr. Obshch-va **91**, No. 1, 27 (1959).

[6] NETTLETON, L. L., and T. A. ELKINS: Trans. Amer. Geophys. Un. **23**, 451.

[7] CLOOS, H.: Zbl. Min. Geol. & Pal. **1932**, B 273 (1932).

[8] HUBBERT, M. K.: Bull. Geol. Soc. Amer. **62**, 355 (1951).

[9] KÖSTER, R.: Proc. 21st Int. Congr. Geol. (Norden) **18**, 295 (1960).

[10] SANDFORD, A. R.: Bull. Geol. Soc. Amer. **70**, 19 (1959).

[11] TANNER, W. F.: J. Geol. **70**, 101 (1962).

[12] LEE, J. S., C. H. CHEN and M. T. LEE: Bull. Geol. Soc. China **28**, No. 1—2, 25 (1948).

but the expression of MOHR's ideas of fracture in rock strata, — as envisaged by ANDERSON, has been beautifully confirmed. A drawing of a typical model-experiment, showing both faulting and folding (cf. next Section) is shown in Fig. 131.

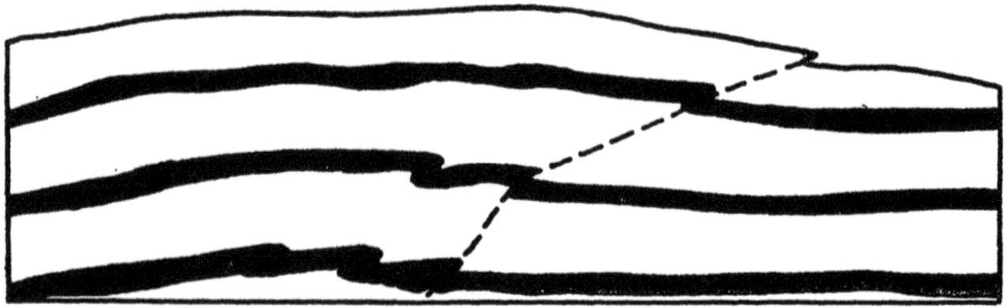

Fig. 131. Drawing of a model experiment showing faulting and folding. After HUBBERT[1]

7.43. Folds. The remarks of Sec. 7.41 on scaling are particularly important with regard to the explanation of folding. The fact that it has been shown that rock strata, in a small-scale model, must be represented by a very soft material, has the effect that it is no longer difficult to understand why the rock strata actually should have been contorted to the fabulous extent uncovered by field geology. In a material with a yield strength as low as that calculated in Sec. 7.41, plasticity and creep must have a major effect. This also makes it very doubtful whether elasticity (through buckling) could have had a major influence on folding. At best, buckling might give the folding process a start and perhaps determine the location of the final faults. It is thus not at all necessary to postulate that the contortions of the strata had to occur during some catastrophe at high temperature.

Experiments on folding have been performed by HUBBERT[1], BELOUSOV et al.[2], BUCHER[3], BHATTACHARJI[4], MCBIRNEY and BEST[5], KUENEN and DE SITTER[6], and others. The last-mentioned authors performed a particularly beautiful set of experiments, making use of various types of plastic materials. In every instance, results very suggestive of geological folding have been obtained.

The use of scale models to "explain" folding does not, in fact, provide an actual "explanation" of the process. The mechanism of producing the folds is not any better understood in the model than it is

[1] HUBBERT, M. K.: Bull. Geol. Soc. Amer. **62**, 355 (1951).

[2] BELOUSOV, V. V., E. I. CHERTKOVA i V. V. ÉZ: Byull. Mosk. Ob-va Ispyt. Priody, otd. geol. **30**, No 5 (1955).

[3] BUCHER, W. H.: Geotektonisches Symposium zu Ehren von H. STILLE, publ. by Dtsch. Geol. Ges., p. 396 (1956).

[4] BHATTACHARJI, S.: J. Geol. **66**, 625 (1958).

[5] MCBIRNEY, A. R., and M. G. BEST: Bull. Geol. Soc. Amer. **72**, 495 (1961).

[6] KUENEN, P. H., and L. U. DE SITTER: Leidsche Geol. Med. **10**, 217 (1938).

in nature. Nevertheless, the duplication of natural phenomena on a small scale shows that the evident geological effects of crustal shortening are nothing supernatural or catastrophic, but the reasonable outcome of a reasonable process. Owing to mathematical difficulties, the analytical calculation of the resulting folds from a given external stress field is probably still a long way off and model experiments will be the only way to treat the problem for some time to come.

7.5. Theory of Systems of Faults and Folds

7.51. The Problem. We finally have to provide a link between the mechanics of mountain-building discussed in Chap. VI and the mechanics of producing the small-scale elements of such mountains, i.e. faults and folds. This leads us to the discussion of *systems* of faults and folds. Such questions, of course, have already been touched upon in the chapter on orogenesis, particularly with regard to such theories of faults and folds that are of specific interest to only one type of hypothesis of mountain building. We shall give here a more complete review of all such theories of systems of folds and faults than has been done earlier, wherein of course proper references to cases that have already been treated elsewhere will be given.

A suitable classification of theories of systems of faults and folds seems to be suggested by the various types of rheological behavior of the Earth's crust which they surmise. These types reach from elastic behavior all the way to complete fluidity. We shall start with brittle fracture, then proceed to elasticity, plasticity, general "rheidity" and fluidity.

7.52. Fracture Systems. We have noted in Sec. 7.12 that faults and dykes often occur in the form of parallel systems. An explanation of this is at once suggested by assuming a state of uniform stress, wherein every single fault of the system would be caused by fracture as envisaged by ANDERSON (cf. Sec. 7.12).

The above argument can even be carried further. ANDERSON[1] notes that the occurrence of any fault in a uniform stress system will in general tend to restore the standard state. The pattern of a fault-system may therefore change after the development of some of the faults owing to the reaction of the latter onto the stress system.

ANDERSON calculated the change of stresses due to the development of a transcurrent fault by using the solution of INGLIS (cf. Sec. 3.21) of stress around an elliptic crack in a plate. Thus, he assumed that there exists a vertical transcurrent fault of length $2c$ with the co-ordinate x

[1] ANDERSON, E. M.: The Dynamics of Faulting and Dyke Formation with Applications to Britain. Edinburgh: Oliver and Boyd 1942. See p. 160 therein.

being taken along the strike and y normal to the strike of the fault in a horizontal plane. The "additional" stress system (i.e. the stresses "additional" to ANDERSON's standard state), which produced this fault must have had its principal axes inclined at 45° and 125° to the strike of the fault. Introducing elliptic co-ordinates α, β with

$$\left.\begin{aligned} x &= c\cosh\alpha\cos\beta, \\ y &= c\sinh\alpha\sin\beta \end{aligned}\right\} \tag{7.52–1}$$

permits one to express the additional stresses at infinity after the formation of the fault as follows:

$$\left.\begin{aligned} \tau_{\alpha\alpha} &= K\sin 2\beta, \\ \tau_{\alpha\beta} &= K\cos 2\beta, \\ \tau_{\beta\beta} &= -K\sin 2\beta \end{aligned}\right\} \tag{7.52–2}$$

where K is some constant related to the strength of the rock. After the fault has occurred, one has the further condition that all the stresses must vanish at the fault surface. INGLIS has given the solution for this case; one obtains from (3.21–33):

$$\left.\begin{aligned} \tau_{\alpha\alpha} &= K\sin 2\beta(\cosh 2\alpha - 1)\left(\frac{1}{F} - \frac{1}{F^2}\right), \\ \tau_{\alpha\beta} &= K\sinh 2\alpha\left(\frac{\cos 2\beta}{F} - \frac{1-\cos 2\beta}{F^2}\right), \\ \tau_{\beta\beta} &= -K\sin 2\beta\left(\frac{\cosh 2\alpha}{F} + \frac{1-\cos 2\beta}{F^2}\right) \end{aligned}\right\} \tag{7.52–3}$$

with

$$F = \cosh 2\alpha - \cos 2\beta. \tag{7.52–4}$$

In the stress system before the formation of the fault, the additional stresses are given by (7.52–2); the stress system after the formation of the faults is given by (7.52–3). The difference between the two stress systems is due to the creation of the fault.

The system of Eq. (7.52–3) shows that the fault causes a stress concentration near the tips of the original crack. Furthermore, the stress trajectories intersect the fault near its tip at roughly right angles. Thus, additional (transcurrent) faults caused by the stress concentration after the formation of the "main" fault, branch off from the latter at acute angles since, according to ANDERSON's theory, their strikes must bisect the stress trajectories. This explains the often-observed occurrence of "splay-faulting" in fault systems, i.e. of faults that branch off at an acute angle from the main faults in an otherwise more or less parallel system.

In a similar fashion, attempts have been made to explain the systems of joints discussed in Sec. 1.61. The lineaments have been thought as associated with the nodal lines on plates which are being flexed. Another possibility is that the joints might be caused by very old fractures in the basement which worked their way up to the surface owing to fatigue fracture under the periodic stresses caused by tidal forces.

Because of the analytical difficulties in investigating systems of faults and folds, many model experiments have been made (see Sec. 7.4).

7.53. Folding Systems Originated by Buckling. According to the discussion in Sec. 7.32 there is little justification for trying to explain the shape of single folds by elastic buckling. There is, however, the possibility of explaining the geometrical position of folds by buckling. In order to do this, it is convenient to follow an idea of DARWIN'S according to which use is made of the equations of elasticity theory and, after the deformation of the body is computed, the displacements are replaced by the velocities. In this fashion, one can attempt to obtain an explanation of the geometrical position of the folds in large systems of folding without having to attach much faith to the actual shapes of the folds as predicted by elasticity theory.

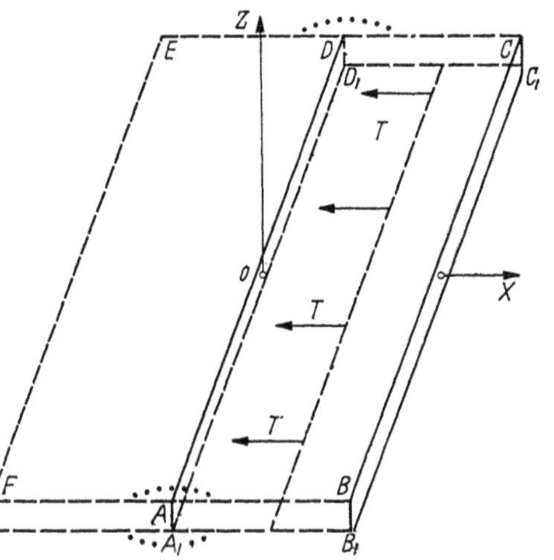

Fig. 132. Formation of a coastal mountain range. After JARDETZKY[1]

The above argument has been used by JARDETZKY[1] in an attempt to treat the problem of mountain chains. JARDETZKY considers four types of mountain ranges as fundamental which he calls *Precambrian type, coastal mountains, intercontinental ranges,* and *Himalaya type.* The different types of folds are obtained by assuming various thicknesses of the buckling layer and various modes of application of the compressive force. Thus, the *Precambrian type* of folding is obtained by assuming that a thin elastic layer is underlain by a plastic one which subjects the upper layer to tangential forces. This yields a multitude of small undulations which, allegedly, represent Precambrian mountain ranges. The *coastal mountains* are obtained by considering a strip which is dragged against another body (cf. Fig. 132). To simplify the problem, it is assumed that the deformation of the strip corresponds to plane strain. The *intercontinental ranges* are obtained by assuming that a strip of matter

[1] JARDETZKY, W. S.: Trans. Amer. Geophys. Un. **31**, 901 (1950).

(representing a geosyncline) is compressed by two shields or continents. The *Himalaya type* of folding, finally, is obtained by assuming that an elastic rectangular plate is fixed on two adjacent sides and that a force is applied diagonally to the free corner. This, according to JARDETZKY, produces a curved bulging in the plate.

Let us sketch JARDETZKY's calculation on the example of *coastal mountain ranges:* a strip of material is dragged against another, as represented in Fig. 132. The strip AC_1 is moderately thick: its thickness is denoted by $2h$ and its width by a. If, as indicated above, only plane strain in the plane Oxz is considered, the displacements u and w are solutions of the plane strain elasticity equations [from (3.12–20) and (3.21–5)]

$$\left.\begin{aligned} \operatorname{lap} u + \frac{\lambda+\mu}{\mu}\frac{\partial\Theta}{\partial x} &= 0, \\ \operatorname{lap} w + \frac{\lambda+\mu}{\mu}\frac{\partial\Theta}{\partial z} &= 0 \end{aligned}\right\} \tag{7.53–1}$$

where, as usual

$$\Theta = \frac{\partial u}{\partial x} + \frac{\partial w}{\partial z} \tag{7.53–2}$$

and λ, μ are LAMÉ's parameters. The sought-after solution can then be represented as follows

$$\left.\begin{aligned} u &= -\frac{\lambda+\mu}{2\mu}\,x\Theta + \Phi_1, \\ w &= -\frac{\lambda+\mu}{2\mu}\,z\Theta + \Phi_3 \end{aligned}\right\} \tag{7.53–3}$$

where $\Theta(x, z)$, $\Phi_1(x, z)$ and $\Phi_3(x, z)$ are solutions of the Laplace equation

$$\operatorname{lap}\Theta = 0, \quad \operatorname{lap}\Phi_1 = 0, \quad \operatorname{lap}\Phi_3 = 0. \tag{7.53–4}$$

By combining (7.53–2) and (7.53–3), JARDETZKY obtained the relation

$$\frac{\lambda+2\mu}{\mu}\Theta = -\frac{\lambda+\mu}{2\mu}\left[x\frac{\partial\Theta}{\partial x} + z\frac{\partial\Theta}{\partial z}\right] + \frac{\partial\Phi_1}{\partial x} + \frac{\partial\Phi_3}{\partial z}. \tag{7.53–5}$$

In order to complete the formulation of the problem, one has to introduce the stress-strain relations

$$\tau_{xx} = \lambda\Theta + 2\mu\frac{\partial u}{\partial x}; \quad \tau_{zz} = \lambda\Theta + 2\mu\frac{\partial w}{\partial z}; \quad \tau_{xz} = \mu\left(\frac{\partial u}{\partial z} + \frac{\partial w}{\partial x}\right). \tag{7.53–6}$$

Then, the problem is mainly one of finding suitable solutions of the Laplace equations (7.53–4) satisfying the correct boundary conditions. If one introduces harmonic polynomials P_n, Q_n

$$P_n + iQ_n = (x+iz)^n, \tag{7.53–7}$$

the general solution of the Laplace equation may be written as follows:

$$\Theta = a_0 + \sum_{n=1}^{\infty} (a_n P_n + b_n Q_n). \qquad (7.53\text{-}8)$$

Since all the Φ's, as well as Θ, can be expressed in terms of harmonic polynomials, one obtains the following expression for u and w

$$u = -\frac{\lambda+\mu}{2\mu} x \{a_0 + \sum (a_n P_n + b_n Q_n)\} + \alpha_0 + \sum (\alpha_n P_n + \beta_n Q_n), \qquad (7.53\text{-}9)$$

$$w = -\frac{\lambda+\mu}{2\mu} z \{a_0 + \sum (a_n P_n + b_n Q_n)\} + \gamma_0 + \sum (\gamma_n P_n + \delta_n Q_n). \qquad (7.53\text{-}10)$$

This set of equations can be solved by tackling the expansions term by term. After going through two pages of tedious algebra, JARDETZKY finally found for the quantity w which is the vertical component of the displacement which is alone of interest:

$$w = -\frac{\lambda a^2}{8\mu h(\lambda+\mu)} T_1 z - \frac{3\lambda+2\mu}{24 h \mu(\lambda+\mu)} T_1 z^3 + \frac{\lambda T_1}{8\mu h(\lambda+\mu)} x^2 z \qquad (7.53\text{-}11)$$

where T_1 is a constant indicative of the *strength* of the drag. At the surface of the Earth, this yields

$$w = \pm T_1(-K_0 + K_2 x^2) \qquad (7.53\text{-}12)$$

where K_2 and K_0 are both positive constants. The solution indicates that the strip thickens in one (or many) buckles, the first buckle being at $x=0$. This is the "coastal" mountain range.

7.54. Plastic Folding. The view that systems of faults and folds may be created as a result of large-scale plastic behavior of the Earth's crust has been taken by RUUD[1] and by GESZTI[2].

RUUD took the view that mountain-ranges would correspond to certain lines on a two-dimensional plate during plastic deformation. If small regions of weakness are assumed, it would turn out that plastic deformations would take place which are similar to those observed when a stamp is pressed into a soft steel plate (cf. Fig. 57). Thus, the prototype of a mountain range would be a circle. RUUD compared this with the craters on the Moon which he envisaged as being due to the same cause as mountains on the Earth. The last supposition, however, seems somewhat doubtful since it appears now fairly certain that the craters on the Moon were caused by meteorite impact and not by processes analogous to mountain building.

Less definite ideas are contained in GESZTI's article where it is simply shown that the volume of mountains can well be accounted for by assuming relatively small crustal shortening.

[1] RUUD, I.: Gerlands Beitr. Geophys. **52**, 123 (1938).

[2] GESZTI, J.: Gerlands Beitr. Geophys. **21**, 36 (1929).

A further application of the theory of plasticity to mountain building has been made by BIJLAARD[1]. The latter author assumed that the zones of orogenetic activity would correspond to the zones of local plastic deformation in his theory of the failure of thin steel plates under tension (cf. Sec. 3.23). The result of this idea is that the prototype of an orogenetic zone should be of the form of an "X" and not of that of curved arcs as would seem more appropriate from physiographic investigations. However, BIJLAARD's basic concept that orogenetic zones might correspond to some type of bands of rheological instability, certainly has some merit. The idea, however, to compare these bands with those occurring in steel plates under *tension*, would appear rather doubtful.

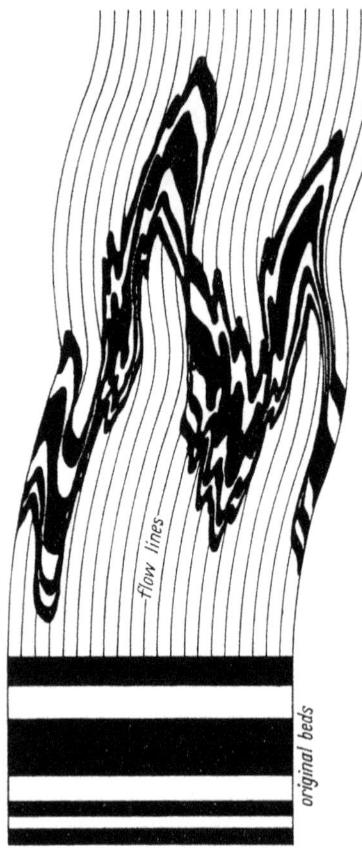

Fig. 133. Formation of folds owing to rheidity. After CAREY[2]

7.55. General Rheology. Systems of folds can also be created by an unspecified rheological behavior of the materials in question. By this we mean that the materials are capable of flow with well-defined flow lines, assuming that the stresses and their durations are of the right order of magnitude. The actual type of rheological behavior (plastic, Maxwell-, Bingham-type) need not be specified in detail. Such materials have simply been called[2] "rheid" and it stands to reason that, as long as flow *does* occur, it makes little difference what the actual equations of motion are.

From the above discussion it is clear that all the explanations based upon the assumption of "rheidity" of the strata, are rather qualitative. No detailed investigations into the stresses required to produce the various flow patterns seem to have been made.

[1] BIJLAARD, P. P.: Rap. Ass. Géod. U. G. G. I., Edimbourg, 1936. — Proc. Kon. Ned. Akad. Wet. **51**, No. 4 (1948). — Trans. Amer. Geophys. Un. **32**, 518 (1951).

[2] CAREY, S. W.: J. Geol. Soc. Australia **1**, 67 (1953).

The mechanism by which systems of folds are postulated to have been caused by the rheidity of the strata, is shown in Fig. 133. Accordingly, the unevenness of the Earth's surface is created by a difference in flow velocity along parallel flow lines that are more or less vertical.

7.56. Rift Systems. The above remarks are all essentially concerned with systems of faults and folds that are caused by basically compressional stress-fields such as must be assumed to be present in an area during the time when crustal shortening, and therewith mountain build-

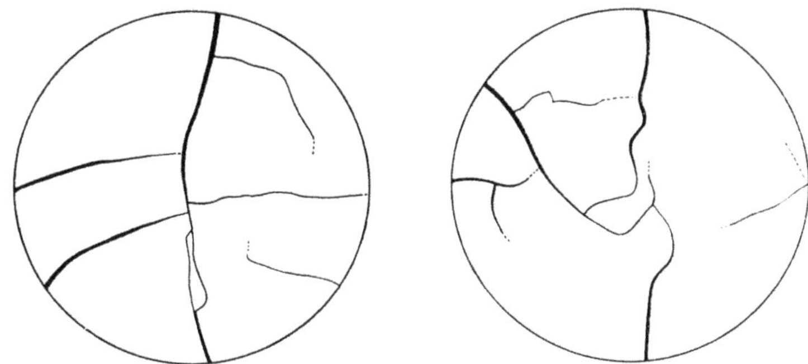

Fig. 134. Tension cracks on a shrinking spherical shell upon an unchanging interior. After BUCHER[1]

ing, is taking place. Entirely different effects occur in tensional stress systems. According to ANDERSON'S theory, it may be assumed that in such cases one obtains *rifts*. Since such rifts have been observed in the African Rift Valleys and in the mid-ocean ridges, this, in turn, is an indication that parts of the Earth's crust are subject to tension.

In accordance with the above remarks, there is a theory of the origin of mid-ocean ridges which postulates that[2] the latter constitute, in essence, giant chasms in the oceanic crust through which material from below was able to intrude to build up the ridges. Tear-faulting is still going on at the present time which causes e.g. the mid-Atlantic rift.

Again, because of the difficulties in treating rift systems analytically, model experiments have been resorted to (cf. Sec. 7.4). A particularly notable set of experiments has been performed by BUCHER[1] who produced tension cracks in a sphere by letting one spherical shell contract upon another of constant radius. The results of BUCHER'S experiments are shown in Fig. 134. BUCHER connected his results with the contraction hypothesis of orogenesis, but perhaps, one could also correlate them with mid-ocean rift systems.

---

[1] BUCHER, W. H.: Geotektonisches Symposium zu Ehren von H. STILLE, ed. by LOTZE, publ. by Dtsch. Geol. Ges., p. 396 (1956).

[2] See e.g. DIETRICH, G., u. K. KALLE: Allgemeine Meereskunde. Berlin: Bornträger 1957.

7.6. Evaluation of Theories of Faults and Folds

The various theories of faults and folds discussed in the present Chapter (VII) all have in common that they attempt to explain these features in terms of known behavior of matter under stress. In this instance, it must be said that there is general agreement regarding the origin of faults and folds: faults are fractures and folds are large continuous contortions of rock strata.

Differences of opinion occur only regarding comparatively minor details: what is the long-term rheological behavior of the (rock-) material undergoing fracturing and folding? This is not known. Accordingly, it is also not possible to obtain an exact idea as to how faults and folds develop as a function of time.

Nevertheless, in spite of these difficulties, the explanation of faults and folds is one instance in geodynamics where the features concerned have been shown to be the entirely reasonable result of entirely reasonable processes.

VIII. Dynamics of Some Special Features

8.1. Meteor Effects

8.11. Physical Principles. In this, the final Chapter of our treatise on geodynamics, we shall discuss the cause of some special features of the Earth's crust whose physiography has been outlined in Sec. 1.7.

We have shown in Sec. 1.71 that the Earth is pockmarked with craters that might conceivably be of meteoritic origin. Meteorites are small celestial objects that are found to strike the Earth's surface at infrequent intervals. We are therefore faced with the problem of explaining the mechanism of crater formation by impact and of estimating the size and speed of the objects causing the holes in the ground.

However, before discussing the attempts at explaining the physics of crater formation, it is of interest to note some important *correlations* with regard to the various parameters that describe the shape of the craters. This will be done in Sec. 8.12.

Then, we shall proceed to investigate the *formation* of the craters. For this purpose, two types of attempts have been made. In the first of these, it has been assumed that owing to the pressures created by the impact, rock loses its solid character and can be treated as a liquid. Thus, the problem is treated as a case of the impact of a liquid drop into a liquid medium. In the second type of attempts at explaining crater formation, it is assumed that the kinetic energy of the meteorite is equal to an equivalent amount of high explosive detonating instantaneously. One then makes estimates of the effect of high explosives and, thus, one arrives at conclusions about the size of the meteorite.

Finally, we shall make a few remarks about tektites which are thought by many to be of meteoric origin.

8.12. Crater-Correlations. A meteoritic origin has not only been claimed for some craters on the surface of the Earth, but also for many craters on the Moon[1–6]. The fact that many more craters are visible

---

[1] Baldwin, R. B.: The Face of the Moon. Chicago, Ill.: Univ. Chicago Press 1949.
[2] Gilvarry, J. J., and J. E. Hill: Publ. Astron. Soc. Pacific **68**, 223 (1956).
[3] Bülow, K. v.: Umschau **50**, No. 14, 430 (1959).
[4] Gilvarry, J. J.: Nature, Lond. **88**, 886 (1960).
[5] Salisbury, J. W.: An Introduction to the Moon. Geophys. Res. Dir., USAF. TN-60-456.
[6] LeRoy, L. W.: Bull. Geol. Soc. Amer. **72**, 591 (1961).

on the Moon than on the Earth has been attributed to the lack of detrition and sedimentation on our satellite. The frequency of impacts by meteorites upon the surface of the Earth and of the Moon might thus well differ only very little, what difference there is being caused by the protecting influence of the Earth's atmosphere. In connection with the above remarks, it should be noted, however, that KOZYREV[1] claims to

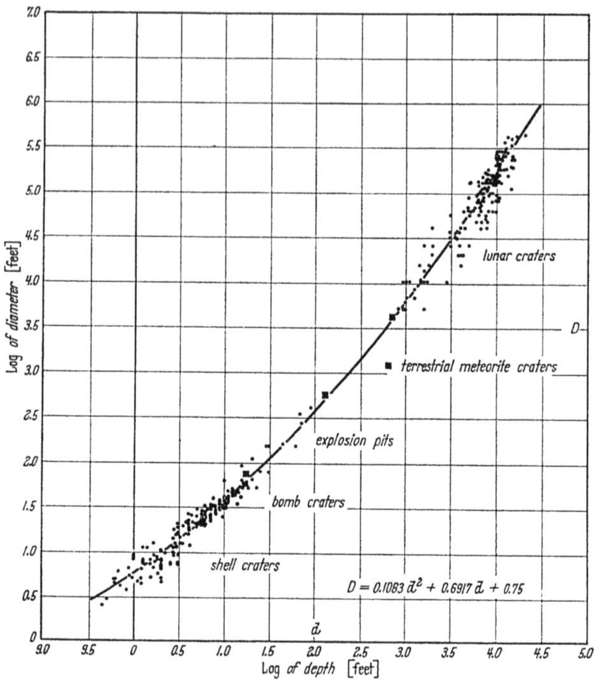

Fig. 135. BALDWIN'S[2] correlation between depth and diameter of craters

have seen volcanic activity on the Moon so that the meteoric origin of all lunar craters, though likely, is not absolutely ascertained.

The formation of meteoritic craters could be thought of as similar to the formation of explosion craters: a meteorite would strike the surface of the Earth or Moon at a speed of some 20 km/sec (the standard speed of meteorites), become vaporized instantly and thus create the effect of an exploding super-bomb. This comparison prompted BALDWIN[2] to expect that correlations between the various geometrical parameters of craters caused by explosions, of craters caused by meteorite impact on the Earth and of lunar craters could be established. Thus,

[1] KOZYREV, N. A.: Priroda **1959**, No. 3, 84 (1959).

[2] BALDWIN, R. B.: The Face of the Moon. Chicago, Ill.: Univ. Chicago Press 1949.

plotting the logarithm of the depths of all these craters against the logarithm of the diameter, he found that the corresponding points all fell very nearly on a continuous curve (see Fig. 135). The latter can be represented by the following equation:

$$D = 0.1083\,d^2 + 0.6917\,d + 0.75 \qquad (8.12\text{–}1)$$

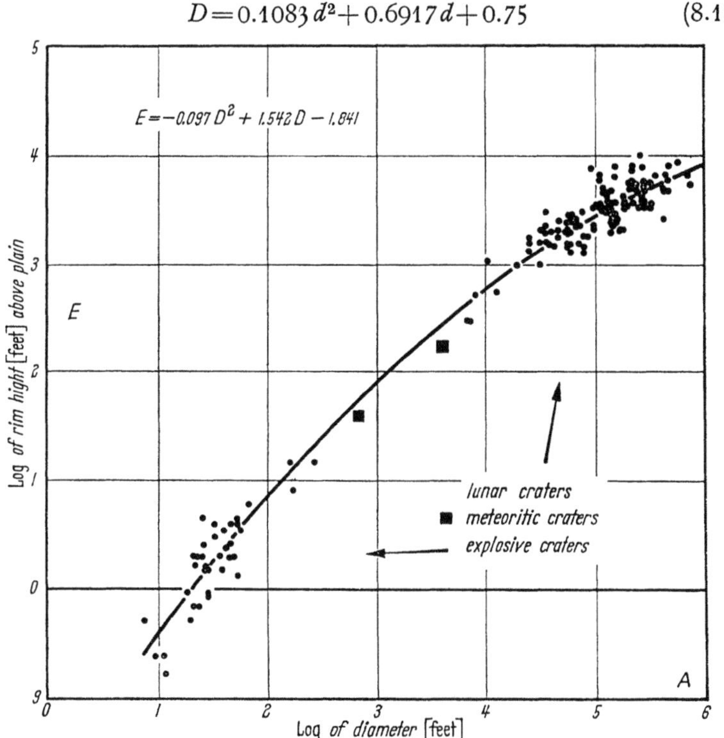

Fig. 136. BALDWIN's[1] correlation between height and diameter of craters

where D is the logarithm of the diameter in feet and d the logarithm of the depth in feet.

BALDWIN found a similar simple relationship between the diameter and rim height of all explosion pits and lunar craters (of a certain type), as demonstrated in Fig. 136.

$$E = -0.097\,D^2 + 1.542\,D - 1.841 \qquad (8.12\text{–}2)$$

where E is the logarithm of the rim height in feet and D again the logarithm of the diameter in feet.

The fact that such relationships as represented by Eqs. (8.42–1/2) exist is a strong indication that the assumption of a similar origin of the various types of craters might be correct.

---

[1] BALDWIN, R. B.: The Face of the Moon. Chicago, Ill.: Univ. Chicago Press 1949.

8.13. Liquid-Drop Model of Crater Formation. We shall proceed now with the discussion of the physics of crater formation. In the present Section we assume that the process can be explained by assuming it as equivalent to the impact of a liquid "projectile" upon a liquid. The chief exponent of this idea has been ÖPIK[1].

According to the above model, the minimum mass of the projectile can be estimated by noting that the mechanical work required to lift up the walls of the crater, throwing out the fragments and shattering the rocks must be furnished by the kinetic energy of the projectile. Since part of the available energy may also be dissipated in form of heat and seismic effects, letting the mechanical work necessary to form the crater be equal to the kinetic energy of the meteorite will lead to a minimum estimate of the latter's mass.

With the above reasoning, one obtains for the meteorite that caused the Arizona crater (cf. Sec. 1.71), a mass of 60000 (metric) tons. In order to arrive at this estimate it was assumed that the diameter $2r$ of the crater was 1200 meters and the depth of the base of solid rock $d=320$ m. If the angle of incidence of the meteorite was 20°, this yields a distance of penetration of $x_m=340$ m. With a density ϱ assumed as equal to 2.7 g cm^{-3}, this yields that the mass affected was

$$M=\pi r^2 \varrho\, d=10^9 \text{ tons}=10^{15} \text{ g}. \qquad (8.13\text{–}1)$$

Judging from the distance at which fragments were found, ÖPIK assumed that the mechanical work would be equivalent to lifting all the mass involved to a height of 1200 meters, or as equal to $\alpha=1.2\times10^8$ erg/g. He took the work of shattering the rock as equal to about 6×10^6 erg/g (from compressibility and crushing strength) which is quite negligible.

For a meteoritic velocity of $v=20$ km/sec ($=2\times10^6$ cm/sec) one can write the energy equation as follows:

$$E=\tfrac{1}{2}\, m v^2=\alpha M. \qquad (8.13\text{–}2)$$

Thus, ÖPIK obtained for the mass of the meteorite

$$m=2\,\frac{\alpha M}{v^2}=6\times10^{10} \text{ g}=60000 \text{ tons}. \qquad (8.13\text{–}3)$$

This would correspond to an iron sphere of 24 meters diameter.

ÖPIK then proceeded to calculate the penetration of the projectile into the ground by the liquid model. He assumed the resistance R to a body penetrating a fluid at high speed as given by

$$R=C\,\varrho\, v^2 \qquad (8.13\text{–}4)$$

---

[1] ÖPIK, E.: Publ. Obs. Astron. Univ. Tartu **28**, No. 6 (1936).

where C is some constant. If the projectile was originally of the form of a cylinder of radius r_0 and height $2r_0$, then during the impact it must have been flattened owing to its own "liquidity". After penetration to the distance x it will be compressed to, say, height $H(x)$ and radius $r(x)$ so that

$$r^2 H = 2r_0^3. \tag{8.13-5}$$

ÖPIK assumed that the velocity of the front surface be v', the velocity of the center of mass be v and, with a linear velocity gradient, the velocity of the rear surface would accordingly be $2v - v'$. The loss of momentum per unit time and cross section (i.e. the pressure) would then be equal to

$$p' = \tfrac{1}{2} \varrho v'^2 + \vartheta, \tag{8.13-6}$$

according to a well-known law of hydrodynamics. Here ϑ is the crushing strength which ordinarily can be neglected. If the velocity of sideways expansion is denoted by v'', one obtains

$$v'' = dr/dt. \tag{8.13-7}$$

Furthermore, one has

$$-2(v - v') = dH/dt, \tag{8.13-8}$$

and hence

$$v'' = \frac{r}{H}(v - v'). \tag{8.13-9}$$

The pressure on the lateral surface of the cylinder is in analogy with (8.13-6)

$$p'' = \tfrac{1}{2} \varrho v''^2 + \vartheta. \tag{8.13-10}$$

Furthermore, using another well-known hydrodynamical principle

$$\delta \frac{v''^2}{2} = p' - p'', \tag{8.13-11}$$

where δ is the density of the projectile, ÖPIK obtained

$$v'' = v' \sqrt{\frac{\varrho}{\varrho + \delta}}, \tag{8.13-12}$$

and finally

$$v' = \frac{v}{1 + \frac{H}{r}\sqrt{\frac{\varrho}{\varrho + \delta}}}. \tag{8.13-13}$$

NEWTON's equation of motion requires

$$\pi r^2 p' = -m \frac{dv}{dt} = -m \frac{d^2 x}{dt} \tag{8.13-14}$$

where m $(=2\pi r_0^3 \delta)$ is, as above, the mass of the projectile and x the depth of penetration. Thus

$$\frac{d^2x}{dt^2} = \frac{dv}{dt} = -\frac{1}{r_0}\left(\frac{1}{4}\frac{\varrho}{\delta}v'^2 + \frac{\vartheta}{2\delta}\right)\left(\frac{r}{r_0}\right)^2. \qquad (8.13\text{–}15)$$

The system of Eqs. (8.13–6/7/12/13/15) can be integrated numerically. ÖPIK's results are shown in Table 21.

Table 21. *Maximum penetration* (x_m) *and final radius* (r_m) *in units of initial radius* ($r_0=1$) *of a cylindrical projectile* ($H_0=2r_0$) *of initial velocity* v_0 (km/sec) *moving parallel to its axis* (*after* ÖPIK)

	Case a		Case b	Case c	Case d
	$v_0=60$	$v_0=20$	$v_0=60$	$v_0=60$	$v_0=60$
x_m	7.923	7.351	4.536	4.255	2.785
r_m	4.022	3.636	3.456	3.254	2.504
$2r_m/x_m$	1.02	0.99	1.52	1.54	1.80

Case a: iron projectile impact into stone $\delta/\varrho=3$, $\vartheta=2\times10^9$ cgs.
Case b: stone projectile impact into stone $\delta=\varrho$, $\vartheta=2\times10^9$ cgs.
Case c: iron projectile impact into iron $\delta=\varrho$, $\vartheta=2\times10^{10}$ cgs.
Case d: stone projectile impact into iron $\varrho/\delta=3$, $\vartheta=2\times10^{10}$ cgs.

The above results can be used to calculate the mass of a meteorite from the depth of its penetration. If one takes the observed penetration x_m in the case of the Barringer Crater in Arizona as equal to 340 meters, ÖPIK obtained with $\delta/\varrho=3$ and initial velocity $v=20$ km/sec, that the meteorite mass should have been equal to 4.8×10^6 tons. This is well above the minimum required from energy calculations (cf. 8.13–3) and is, in fact, unreasonably high.

8.14. Analogy with Explosion Craters. We shall discuss now the second possibility of attempting to explain the formation of meteoritic craters. This consists in drawing up an analogy with explosion craters. The procedure, thus, is as follows. First, an estimate is made of the amount of high explosive which would be required to produce a crater the size of that created by a certain meteorite. Second, the kinetic energy of the meteorite is set equal to the chemical energy contained in the explosive. Finally, the mass of the meteorite giving the correct energy is calculated. The above procedure has been proposed by WYLIE[1].

The analogy with explosion craters is very much in line with BALDWIN's attempt to establish correlations between meteorite craters and

[1] WYLIE, C. C.: Popular Astronomy **51**, 97 (1943).

explosion craters. The fact that such correlations could be found, is a strong indication that the two types of craters are due to similar causes.

By comparing, for instance, the size of the Barringer Crater in Arizona with explosion craters, WYLIE estimated that about 1.1×10^8 kg of nitroglycerin would be required to produce it. From the molecular weight and the heat of combustion, he calculated the energy of this amount of explosive to be equal to $E = 9 \times 10^{21}$ ergs. Setting this equal to the kinetic energy of the meteorite, one has

$$E = 9 \times 10^{21} \text{ ergs} = \tfrac{1}{2} m v^2. \tag{8.14–1}$$

With $v = 20$ km/sec this yields

$$m = 4.5 \times 10^9 \text{ g} = 4500 \text{ tons}. \tag{8.14–2}$$

This is considerably less than what had been estimated by means of the liquid drop model. In fact, it is much less than what had been estimated from the "work of excavation" necessary to produce the crater and thus may be an indication that the estimate is too low.

It is clear that the problem of the analytical description of explosion craters is a very complicated one and, thus, that the numbers obtained above are somewhat doubtful. Attempts at theories of crater formation by explosives have been published on many occasions in the literature[1–5]; this is not the place to review them all as this would almost require a monograph in itself. Nevertheless, the above discussion will give a good idea about the possibility of meteor crater formation.

8.15. Tektites. Finally, we shall make a few remarks regarding tektites. We have already noted that tektites occur in "swarms" in rather well-defined areas of the Earth. There is a possibility that they are of extra-terrestrial origin, although this is not yet entirely certain.

One of the most intriguing theories regarding the origin of tektites has been proposed by VARSAVSKI[6] who supposed that tektites come from the Moon, being ejected there when a meteorite strikes the lunar surface. He was able to show that the strange distribution of tektites shown in Fig. 20 could be connected with the possible missile trajectories between the Moon and the Earth.

---

[1] GILVARRY, J. J., and J. E. HILL: Astrophys. J. **124**, 610 (1956).
[2] HIRSCHFELDER, J. O., et C. F. CURTIS: Rev. Inst. Franç, Pétr. **13**, 586 (1958).
[3] RINEHART, J. S.: Quart. Colo. School Mines **55**, No. 4 (1960).
[4] HARRISON, M. A., and H. B. KELLER: Quart. Colo. School Mines **54**, No. 3, 333 (1959).
[5] See also the "Cratering Symposium" in J. Geophys. Res. **66**, No. 10 (October 1961).
[6] VARSAVSKI, C. M.: Geochim. et Cosmochim. Acta **14**, 292 (1958).

Other authors[1] have connected the origin of tektites with meteoric showers and other cosmic phenomena. The various attempts at explaining tektites are highly intriguing, but one will have to wait until more is known about these strange bodies in order to give a definitive statement as to their origin.

8.2. Boudinage

8.21. Experimental Approach. The next special feature to be discussed here is boudinage. In the chapter dealing with the physiographic description of boudinage (1.72) it has already been stated that, in order to produce this structure, it is necessary to have a competent layer wedged in between two incompetent ones[2-4]. It may then be assumed that an elongation of the system parallel to the layering would cause the incompetent rock to yield without rupture, whereas the competent layer would break so as to form the boudins. Numerous field observations seem to support this view.

A test of this hypothesis can be made by setting up a model-experiment simulating the incompetent rock with putty, and simulating the competent rock with various other substances such as modeling clay etc. A series of such experiments has been reported by RAMBERG[5]. He describes them as follows: "The competent materials were formed into evenly thick sheets from 2 to 5 mm thick. In each experiment, one competent sheet was placed between incompetent putty layers 1—2 cm thick. These layered cakes were then compressed between two stiff plates. In most runs, the cakes were allowed to expand in two dimensions. In other runs, the expansion was restricted to one dimension by performing the experiments in an oblong box. After the compression, which lasted a few minutes and was performed by hand pressure, the cakes were cut with a razor blade, and the cross section examined and photographed. In all cakes, the competent layers were ruptured and formed boudins, or necked-down to form pinch-and-swell structures. The most 'brittle' types of the competent layers formed relatively sharp-edged boudins, whereas the most plastic types of plasticene formed smooth, lenticular boudins and pinch-and-swell structures."

There is therefore little doubt that the general picture outlined above for the explanation of boudinage is essentially correct.

---

[1] See review by O'KEEFE, J. A., and B. E. SHUTE: Aerospace Engng. **20**, No. 7, 26 (1961); also by FRIEDMAN, I. *et al.*: Amer. J. Sci. **127**, 91 (1958).

[2] WEGMANN, C. E.: C. R. Soc. Géol. France **5** pt. 2, 477.

[3] CLOOS, E.: Trans. Amer. Geophys. Un. **28**, 626 (1947).

[4] Similar ideas have also been put forward by G. I. GUREVICH, Izv. Akad. Nauk SSSR., Ser. Geofiz. **1954**, 411 (1954).

[5] RAMBERG, H.: J. Geol. **63**, 512 (1955).

RAMBERG[1, 2] has also made an interesting modification of the above argument. In the cases under investigation above, it was assumed that the competent rock layer is essentially at right angles to the compression. If it is parallel to the compression, then one would expect ptygmatic folds (sinusoidal folds) to result. If it is cross-cutting, mixed features result (Fig. 137).

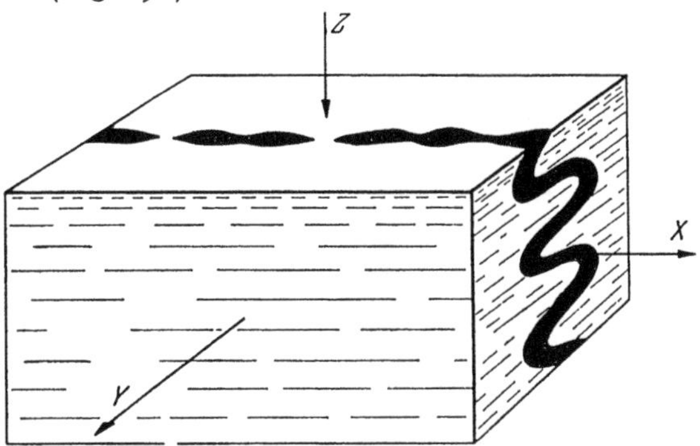

Fig. 137. Simultaneous development of ptygmatic folds and pinch-and-swell structures in a cross-cutting layer (z is direction of maximum compression, x of maximum extension). After RAMBERG[1]

8.22. Theoretical Approach. In order to substantiate the explanation of boudinage suggested by experimental analysis, it will be necessary to investigate theoretically the dynamics of such models as were discussed above. This has also been done by RAMBERG[3]. It can be achieved easily if (i) the deformation of the incompetent rock layers is treated as viscous, incompressible flow and (ii) if the competent layer is assumed as rigid and incompressible before rupture, the latter occurring at a critical tensile stress.

For the convenience of the calculation, RAMBERG introduced a Cartesian co-ordinate system x, y, z; here z is assumed to be perpendicular and x, y parallel to the layering of the beds. The origin of the co-ordinate system is assumed in the center of the competent bed, the latter is wedged in between two parallel incompetent beds. For the sake of simplicity, deformation in the x-direction only was taken into account; i.e. it is assumed that there is a constraint preventing the material from flowing in the y-direction.

If now the three layers are compressed uniformly in the z-direction, the assumed incompressibility effects that a certain volume V of incompetent rock is forced to flow outward from the center of the system. At a distance x from the center, the volume-flow in the incompetent

[1] RAMBERG, H.: Norsk Geol. Tids. **39**, 99 (1959).
[2] RAMBERG, H.: Amer. J. Sci. **258**, 36 (1960).
[3] RAMBERG, H.: J. Geol. **63**, 512 (1955).

layers is

$$\frac{\partial V}{\partial t} = y' x \frac{\partial z}{\partial t} \qquad (8.22\text{–}1)$$

where y' is the width of the layer and $\partial z/\partial t$ is the rate of compression in each layer.

The rate of volume flow is connected with the pressure gradient $\partial p/\partial x$ by the following equation[1]

$$\frac{\partial V}{\partial t} = -\frac{z^3 y'}{12\eta} \frac{\partial p}{\partial x} \qquad (8.22\text{–}2)$$

where z is the thickness of the flowing layer and η the viscosity. Combining (8.22–1) and (8.22–2) yields

$$\frac{\partial z}{\partial t} x = -\frac{z^3}{12\eta} \frac{\partial p}{\partial x}. \qquad (8.22\text{–}3)$$

For a constant rate of compression, this can be integrated and one obtains

$$p_x = p_0 - \frac{6\eta\, \partial z/\partial t}{z^3} x^2. \qquad (8.22\text{–}4)$$

Of particular interest is the drag force which creates tension in the competent layer. The shearing stress τ at the distance x is

$$\tau = \frac{z}{2} \frac{\partial p}{\partial x}. \qquad (8.22\text{–}5)$$

The total force is then found by integrating this from the end of the feature, say L, to x. Finally, in order to calculate the tensile stress σ in the competent layer, it must be recalled that the shearing stress calculated in Eq. (8.22–5) acts on both of its sides. One thus obtains:

$$\sigma = \frac{z}{T} \int_L^x \frac{\partial p}{\partial x} dx = \frac{6\eta}{T z^2} \frac{\partial z}{\partial t} (L^2 - x^2), \qquad (8.22\text{–}6)$$

where T is the thickness of the competent layer. In particular, the tensile stress at the center of a structure of length $2L$ is

$$\sigma_0 = \frac{6\eta}{T z^2} \frac{\partial z}{\partial t} L^2 \qquad (8.22\text{–}7)$$

which can also be written

$$\sigma_0 = \frac{z}{T} (p_0 - p_L). \qquad (8.22\text{–}8)$$

This shows that the tensile stress in the competent layer increases with the square of the length of the structure. The largest length possible is therefore that for which σ_0 is equal to the critical tensile stress of the

[1] See LAMB, H.: Hydrodynamics, 6th ed. New York: Dover Publ. Co. 1945. Equation (4) on page 582.

competent layer at which the latter ruptures. This, automatically, gives rise to boudinage-structures, as the competent layer must break in such intervals as correspond to the maximum length compatible with its strength to tensional forces.

8.23. Tectonic Lenses. Phenomena related to boudins have been found which were given a corresponding explanation by SORSKII[1]. This concerns the transformation of a continuous layer of rock into a lenticular thread (tectonic lenses) which may be observed in regions of violent orogenetic diastrophisms among deformed Archean rocks. SORSKIĬ suggests that if a plastic mass is compressed by high vertical pressures, the compressed stratum flows in a lateral direction. This would give rise to the tectonic lenses. It is in this instance the *incompetent* rock which is supposed to be collected into a string of disconnected lenses; the less-yielding rock above and below would simply close up in between the lenses.

In order to substantiate the above theory, one would have to investigate the behavior of a thin plastic layer in between two elastic plates, under pressure. No such calculations, however, appear to have been made.

8.3. Domes

8.31. Principles of a Theory of Domes. The striking circular features discussed in Sec. 1.73 have held the interest of geologists for a long time. Drilling and other direct procedures have established that such features are domes and the question has arisen as to the physics of their origin. After many unsatisfactory conjectures, it is now pretty well accepted that dome-formation is a case of plastic intrusion of a less dense layer into a denser overburden under the action of gravity. The originator of this idea was ARRHENIUS[2] who reasoned that the intruding masses (usually salt) being less dense than the overburden, would be in an unstable state owing to this condition. They would thus tend to rise independently of any tectonic forces. The theory has latter been developed particularly by NETTLETON[3] and DOBRIN[4].

To test the above theory, it seems appropriate to represent the mechanism of (salt) dome formation by constructing (theoretically and experimentally) models in which the intruding layer as well as the overburden are represented by layers of liquids of appropriate viscosity,

[1] SORSKIĬ, A. A.: Dokl. Akad. Nauk SSSR. **72**, 937 (1950).

[2] ARRHENIUS, S.: Med. K. Vetenskabsakad. Nobelinst. **2**, No. 20 (1912). — Geol. Rdsch. **3** (1912).

[3] NETTLETON, L. L.: Bull. Amer. Ass. Petrol. Geol. **18**, 1175 (1934).

[4] DOBRIN, M. B.: Trans. Amer. Geophys. Un. **22**, 528 (1941).

density etc. Analytical attempts to calculate the intrusion of one layer into the other have been made, but it is obvious that any attempt at an exact calculation of the hydrodynamical phenomena would be beset with tremendous difficulties. The chief emphasis in the study of dome formation has therefore been on experiments.

We shall discuss these attempts below.

8.32. Analytical Attempts. Turning first to analytical attempts at elucidating dome-formation, we note an investigation by DOBRIN[1] in which the following assumptions were made: (i) the model-dome is being formed in the center of a large cylindrical box, (ii) the dome is cylindrical and has a flat top, and (iii) the dome is considered as a *solid* of variable height pushing its way upward through a *viscous* liquid of greater density. These assumptions certainly oversimplify the problem to a great extent. It appears, however, that they should nevertheless lead to a valid indication of the physical processes involved.

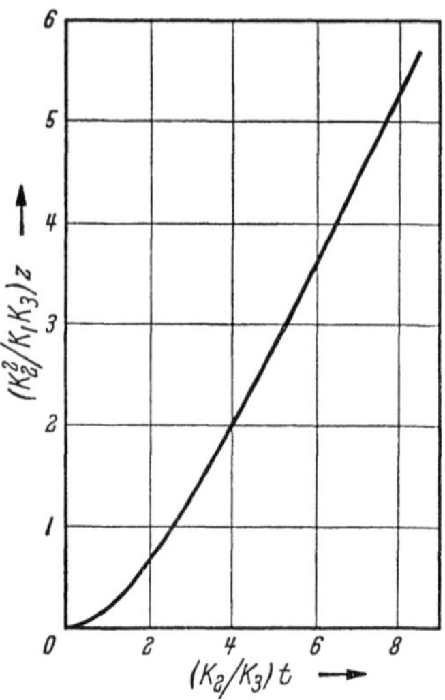

Fig. 138. DOBRIN's[1] solution for the rise of a dome

For his analysis, DOBRIN defined the following symbols: z is the height of the top of the dome above the surface of the layer from which it originates; t is the time of the beginning of the intrusive process, v is the velocity of intrusion $(=dz/dt)$; R is the radius of the dome; w is a characteristic distance expressing proportionality between the velocity and the velocity-gradient, ϱ_1 is the density of the fluid, ϱ_2 is the density of the dome (with $\varrho_D=\varrho_1-\varrho_2$), η is the viscosity of the liquid and ψ is the Newtonian form-resistance coëfficient.

When the dome is at height z, there will be three forces acting which must be in equilibrium at all times. They are (i) the buoyant force F_B given by

$$F_B=\varrho_D\,\pi R^2 g z \tag{8.32-1}$$

(ii) the viscous drag F_V on the side of the cylinder

$$F_V=-\frac{2\pi}{w}R\eta z\frac{dz}{dt} \tag{8.32-2}$$

(iii) the turbulent resistance F_F at the front of the cylinder

$$F_F=-\frac{1}{2}\psi\pi R^2\varrho_1\left(\frac{dz}{dt}\right)^2. \tag{8.32-3}$$

[1] DOBRIN, M. B.: Trans. Amer. Geophys. Un. **22**, 528 (1941).

Using the abbreviations

$$K_1 = \varrho_D g R, \quad K_2 = 2\eta/w, \quad K_3 = \psi R \varrho_1/2, \tag{8.32–4}$$

DOBRIN obtained for the equilibrium condition the following expression

$$\left(\frac{dz}{dt}\right)^2 + \left(\frac{K_2}{K_3}\right) z \frac{dz}{dt} - \left(\frac{K_1}{K_3}\right) z = 0. \tag{8.32–5}$$

The solution of this differential equations is[1]

$$t = \frac{K_3}{K_2} \operatorname{lognat} \frac{1}{1 - (K_2/K_1)\,v} - \frac{K_2}{K_1} \frac{v}{1 - (K_2/K_1)\,v} \tag{8.32–6}$$

with

$$v = \frac{K_2}{2K_3} \left\{ -z \pm \sqrt{z^2 + \frac{4K_1 K_3}{K_2^2} z} \right\}. \tag{8.32–7}$$

This solution is shown in Fig. 138. It may be noted that, if z becomes very large, the expression for v tends towards

$$v \cong \frac{K_1}{K_2} = \frac{Rw}{2} \frac{\varrho_D g}{\eta}, \tag{8.32–8}$$

which shows that the curve in Fig. 138 becomes a straight line for large z.

8.33. Model Studies of Domes. Because of the tremendous analytical difficulties in treating the problem of domes accurately, recourse has been taken to model studies. An excellent summary of such studies has been provided by TRAVIS and MCDOWELL[2]. Accordingly, it is to the credit of NETTLETON[3] to have originated much of the experimental work; others followed suit. DOBRIN[1] has compared the experimental formation of model salt domes with his analytical theory.

In making model experiments leading to domes, cognisance has to be taken of the dynamical theory of scaling (see Sec. 7.41). The conditions for the scaling of salt domes have been determined by HUBBERT in his general discussion of scaling in geology (see Sec. 7.41). He found that the ratio η of viscosities must satisfy the following relationship

$$\eta = \varrho x t, \tag{8.33–1}$$

if ϱ signifies the density-ratio, x the length-ratio and t the time-ratio. In addition, the usual similarity conditions have to be fulfilled except, of course, that the considerations regarding the strength of the materials do not come into play (viscous fluids as are here under consideration have no finite yield strength).

[1] DOBRIN, M. B.: Trans Amer. Geophys. Un. **22**, 528 (1941).

[2] TRAVIS, J. P., and A. N. MCDOWELL: Bull. Amer. Ass. Petrol. Geol. **39**, 2384 (1955).

[3] NETTLETON, L. L.: Bull. Amer. Ass. Petrol. Geol. **18**, 1175 (1934); also Bull. Amer. Ass. Petrol. Geol. **27**, 51 (1943); **39**, 2373 (1955).

Observing the above conditions, various investigators have made experiments with suitable liquids. They were indeed able to simulate dome-formation. The shape of the domes is what one would expect it to be, the rate of rise is reasonably fast so as to correspond to the formation of a dome in the time interval available, say, since the Eocene epoch. The rise-*versus*-time curve of a particular experiment (performed by DOBRIN) is shown in Fig. 139.

If one compares the empirical curve of Fig. 139 with the theoretical one of Fig. 138, one observes immediately many points of similarity. The general form of the beginning is convex downward in both cases, and both curves approach a straight line for large dome-heights.

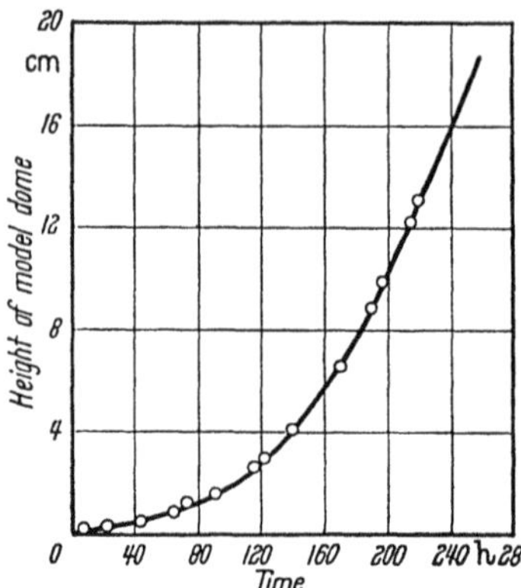

Fig. 139. Rise characteristics of an asphalt model dome. After DOBRIN[2]

In view of the above, it must be held that ARRHENIUS' idea of explaining the formation of domes by the assumption of plastic intrusion is substantially correct. A different view, however, has been taken by GZOVSKIĬ[1] who was able to obtain dome-like structures by pressing a stamp from the bottom into an *elastic* overburden. The upper surface of the overburden assumes indeed a dome-like structure; the stress trajectories inside the dome can be traced by means of photo-elasticity. In view of the general rheological properties of the Earth (particularly because of its low yield strength) it seems, however, that there is little likelihood that domes were actually formed in the manner envisaged by GZOVSKIĬ.

8.4. Volcanism

8.41. The Shape of Volcanoes. The peculiar cone-like structures that represent volcanoes suggest upon a very first inspection that they are simply piles of ash and other materials ejected from the Earth. The steepness of their slopes would be determined by that angle at which a mound of volcanic material could support itself. The conical form of volcanoes would simply result from the fact that the critical slope angle must be reached everywhere.

The stability of mounds of various materials has been discussed in Sec. 3.64 where the Terzaghi equation (3.64-2) has been given. This equation enables us to deduce a slope angle β from the height H of

[1] GZOVSKIĬ, M. V.: Izv. Akad. Nauk SSSR., Ser. Geofiz. **1954**, 527.

[2] DOBRIN, M. B.: Trans. Amer. Geophys. Un. **22**, 528 (1941).

volcanoes, the density ϱ of volcanic material and from its yield strength ϑ. However, the yield strength may vary in wide limits so that it is presumably always possible to adjust it in such a manner as to produce the desired slope angle. It would thus appear as a more honest procedure to assume the slope angle β as given in the first place and to calculate therefrom the required yield strength. It can be tested, then, whether the latter has a reasonable order of magnitude. If so, the shape of volcanoes has been "explained".

With $H=800$ m, $\varrho=1$ g cm^{-3}, $\beta=30°$ (and hence $N=6.5$) one obtains from the Terzaghi equation

$$\vartheta \sim 1.2\times 10^7 \text{ cgs}. \tag{8.41-1}$$

This is by about a factor of 100 less than the yield stress obtained in the case of mountains which consist of granite. This appears as reasonable in view of the difference between granite and the ash-materials of which volcanoes are composed.

The above argument explains the cone-like appearance of volcanoes. It is noted, however, that the cones are not always straight, but sometimes somewhat concave. It must be assumed that the latter feature is due to exogenetic agents as discussed by the writer in his treatise on geomorphology[1].

8.42. Volcanic Heat and Orogenesis. The eruption of a volcano is certainly an event which is very impressive to a human observer. It is therefore rather amazing that one can show that volcanism as such plays only a very insignificant rôle in geodynamics.

We have already mentioned that the *heat* released during volcanic eruptions is quite a small fraction of the total heat flow from the interior of the Earth into space (cf. Sec. 2.61). A corroboration of this statement will be given at the end of the present section. In addition, it is also possible to show that the *volume of lava* ejected in any one geological period is small in comparison with the volume of mountains thrust up during the corresponding orogeneses.

To illustrate this point, VERHOOGEN[2] assumed that 30 outpourings of lava (certainly an over-estimate in the light of geological findings) occurred since the beginning of the Cambrian epoch, each of which may be of the order of 10^{21} cm^3,—corresponding to a plateau-type outpouring of 10^6 km^2 in area and 1 km in thickness. This leads to a volume produced which is equal to 3×10^{22} cm^3. On the other hand, we have calculated in Sec. 6.12 that the volume which is upthrust in one orogenetic cycle, is approximately equal to 32×10^6 km$^3 \cong 3\times 10^{22}$ cm^3. This shows

[1] SCHEIDEGGER, A. E.: Theoretical Geomorphology. Berlin-Göttingen-Heidelberg: Springer (Englewood Cliffs: Prentice-Hall) 1961.

[2] VERHOOGEN, J.: Amer. J. Sci. **244**, 745 (1946).

that the volume of *all* the lava produced since the end of the Precambrian is equal to the volume upthrust in *one* orogenetic cycle. Since there were several orogenetic cycles (at least 2, possibly more) since the end of the Precambrian, this shows that volcanism can play only a minor part in orogenesis.

The above estimate of the volume of lava produced enables one to calculate the heat lost to the Earth by the outpouring of that lava. Assuming that the heat lost by the lava owing to cooling and crystallization is equal to 400 cal/g (following VERHOOGEN[1]), the total heat lost in this fashion would turn out to be 4×10^{25} cal. During the same period the heat lost due to the ordinary heat flow through the Earth's surface (cf. Sec. 2.61) was 8.2×10^{28} cal. The last value was obtained by assuming the mean surface heat flow as equal to 1.2×10^{-6} cal cm^{-2} sec^{-1}; VERHOOGEN obtained only 7.5×10^{28} cal as total heat lost because he assumed the mean heat flow as equal to 1.1×10^{-6} cal cm^{-2} sec^{-1}. The estimate of the total heat flow is, in any case, low, because it must be assumed that the heat flow was higher during early geological epochs than it is at present owing to the continual decay of radioactive and hence heat-creating material. The above estimates show decisively that the heat produced by volcanism is entirely insignificant. Similar conclusions were also arrived at in an extensive study by YOKOYAMA[2] who showed that the fraction of total heat energy liberated in volcanism is of the same order of magnitude as that liberated in seismic effects. Some people have thought, therefore, that volcanism may be *caused* by earthquakes (or faulting processes), but this is certainly doubtful.

It follows from the above discussion that volcanism is really an insignificant phenomenon in the evolution of the Earth's surface. The energetics of a volcano is no problem,—owing to the small amount of extra heat (over and above the ordinary heat flow) that is required. The only problem that remains to be solved is that of finding an actual mechanism which would produce all the impressive puff and smoke.

8.43. Mechanism. Unfortunately, no definte answer exists to the question what the actual mechanism of a volcano is. GRATON[3] has given a review of the problem but the latter has a negative character. Of all the processes considered, each one meets with some serious objection.

A very old explanation of volcanic activity is the assumption that rising gases may act as heating agents. However, GRATON invalidates this theory by drawing attention to the fact that expanding gases are refrigerants, not heating agents.

[1] VERHOOGEN, J.: Amer. J. Sci. **244**, 745 (1946).

[2] YOKOYAMA, I.: Bull. Earthquake Res. Inst. **34**, 185 (1956); **35**, 75, 99 (1957).

[3] GRATON, L. C.: Amer. J. Sci. **243** A, 135 (1945).

Other possibilities that have been considered are various chemical reactions, but none of those investigated seem to fill the bill. There is, of course, always the possibility that further reactions might be postulated: VERHOOGEN[1], in a discussion of GRATON's paper, states "the fact that no reactions are known which could provide much energy at the surface does not imply that other reactions do not occur with important thermal effects at some depth".

Another hypothesis, that of postulating some "internal heat" to cause volcanoes, is so vague that it is even difficult to state.

Another interesting theory of volcanism has been advanced by RITTMANN[2]. Accordingly, it is postulated that the viscosity of the material below the Earth's crust is highly pressure-dependent. The viscosity is supposed to be very high (10^{22} cgs) at high pressures, viz. at such pressures that subsist in the undisturbed state at the depth in question. As soon as the pressure is lowered, e.g. by the opening-up of a fissure due to orogenetic activity, the viscosity drops sharply (to 10^0—10^5 cgs) and the material can flow freely as "lava", producing the eruption of a volcano. Interesting as RITTMANN's theory is, it does not seem entirely certain whether substances exist that would exhibit the postulated rheological properties.

A final attempt at elucidating the mechanism of volcanoes has been made by VERHOOGEN[1]. In it he postulates that the temperature at a certain level in the Earth (being located slightly below the crust) is not constant, but subject to small fluctuations in time and in space. This would have the effect that there exists a finite chance for the temperature in any one spot to be appreciably higher than in its vicinity. Consequently, the masses affected might become molten and cause a volcano to come into existence. The chief difficulty with such a statistical theory is that the distribution of volcanoes on the Earth's surface is not entirely random. The concentration of volcanoes in orogenetic belts is too marked so as not to suggest a connection between volcanism and orogenesis. Within the belts there is in fact a certain random arrangement of volcanoes so that there is the possibility of a statistical effect being present within those belts. However, it is difficult to see how such an effect could alone be responsible for volcanic activity.

As stated above, the problem of explaining the mechanism of volcanic eruptions has therefore obviously not yet been solved.

8.44. Heat Flow and Volcanic Intrusions. Some interesting calculations have been made on the heat flow that might be associated with

---

[1] VERHOOGEN, J.: Amer. J. Sci. **244**, 753 (1946).

[2] RITTMANN, A.: Bull. Volcanol. (2) **19**, 85 (1958). Also: RITTMANN, A.: Vulkane und ihre Tätigkeit. Stuttgart: Ferdinand Enke 1960.

magmatic intrusions. Workers active in this field have been particularly REILLY[1] and RIKITAKE[2].

Thus, let a sphere (representing the magmatic mass) of radius a be embedded at the depth ζ (this is the depth of the center of the sphere) in the Earth (see Fig. 140). Originally, the temperature in the sphere is T_0, the temperature in the Earth being assumed as zero. If it be surmised that the heat conductivity within the sphere and without is the same (=0.01 cgs.), then the temperature distribution for a particular case as calculated by RIKITAKE is shown in Fig. 141.

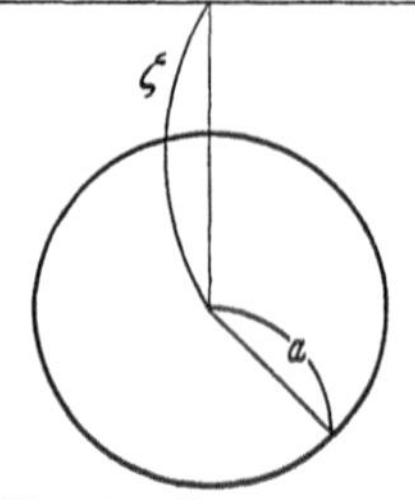

Fig. 140. Geometry of a magma intrusion. After RIKITAKE[2]

Of particular interest is the effect upon the geothermal gradient at the Earth's surface which might be caused by the presence of a magmatic mass. RIKITAKE has calculated this effect, in dependence of ζ (for $a = 2s$ km). The result is shown in Fig. 142 from which it is evident that, unless the magma is very close to the surface, no appreciable effect is observed at the surface so that it is impossible to detect the presence of the intrusion from heat flow measurements.

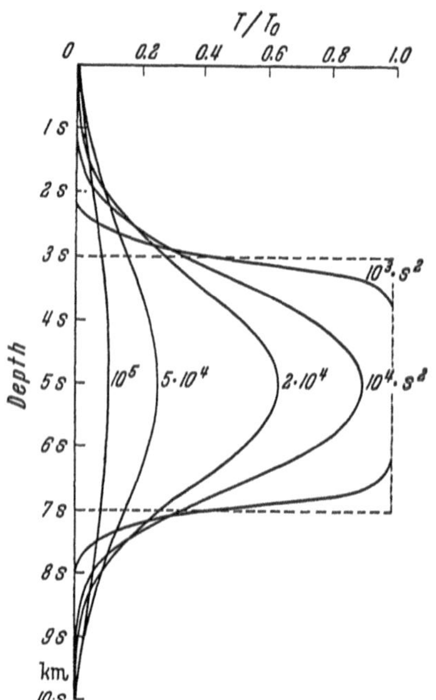

Fig. 141. Temperature distribution on a vertical line through the center of a magma intrusion ($a = 2s$ km, $\zeta = 5s$ km) as a function of time (years). After RIKITAKE[2]. Note that s is an arbitrary factor

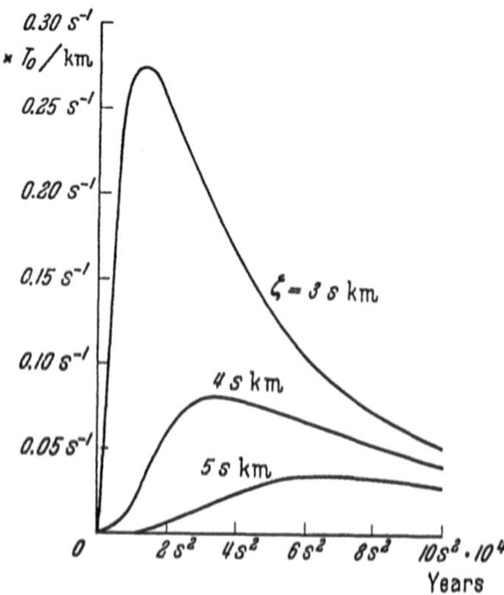

Fig. 142. Geothermal gradient changes due to the presence of a magma intrusion (after RIKITAKE[2])

[1] REILLY, W. I.: New Zeal. J. Geol. Geophys. 1, 364 (1958).
[2] RIKITAKE, T.: Bull. Earthquake Res. Inst. 37, 233 (1959).

8.5. Postglacial Uplift

8.51. General Remarks. In the final Sections of this book, we turn our attention to the uplift of land in the vicinity of Fennoscandia (cf. Sec. 1.75). For an explanation of this phenomenon, the theory of isostasy has been greatly favored. Accordingly, one explains the present uplift of land by assuming that Fennoscandia would have sunk after having been covered with a heavy, extensive load of ice which must have been present there during the Pleistocene ice age. Now, after the melting of much of the ice has been accomplished, it would be rising again to preserve isostatic equilibrium corresponding to its present loading condition.

The idea of isostatic rising has been accepted as logical by many scientists, notably BARRELL[1], DALY[2], SAURAMO[3], NISKANEN[4] and many others[5-8]. Signs of rising occur in other areas where a similar explanation could be advocated: parts of North America, Scotland, Iceland, Spitzbergen, Novaya Zemlya, South New Zealand, Antarctica and others,—which would make the above concept of load recovery appear as reasonable.

However, there are certainly large regions on the Earth which are not isostatically compensated and which, in spite of this, do not show any signs of rising or subsidence. India with large negative gravity anomalies is the most notable example. This would render doubtful any theory which postulates isostatic adjustment of moderately unbalanced regions. Furthermore, the observations of the rising of the coast on Hudson's Bay in North America (near Churchill) has been severely questioned[9]. LYUSTIKH[10], scrutinizing geologic evidence, maintains that the Fennoscandian shield has been rising even before the last ice age and that therefore some phenomenon other than unloading with ice should be its cause. On the other hand, HEAPS[11] has shown that failure *must* occur if an ice cap of even only a moderate thickness ($\frac{3}{4}$ km) is formed upon the Earth's surface. This has been demonstrated by assuming that the Earth's crust is elastic and has a yield strength

[1] BARRELL, J.: Amer. J. Sci. **40**, 13 (1915).
[2] DALY, R. A.: Bull. Geol. Soc. Amer. **31**, 303 (1920).
[3] SAURAMO, M.: Fennia **66**, No. 2, 3 (1939).
[4] NISKANEN, E.: Publ. Int. Isostat. Inst. No. **6** (1939).
[5] VENING MEINESZ, F. A.: Proc. Kon. Ned. Akad. Wet. **57**, 142 (1954).
[6] BURGERS, J. M., and B. J. COLETTE: Proc. Kon. Ned. Akad. Wet. B **61**, 221 (1958).
[7] USHAKOV, S. A., i G. LAZAREV: Dokl. Akad. Nauk SSSR. **129**, 785 (1959).
[8] SLICHTER, L. B., and M. CAPUTO: J. Geophys. Res. **65**, 4151 (1960).
[9] JOHNSTON, W. A.: Amer. J. Sci. **237**, 94 (1929).
[10] LYUSTIKH, E. N.: Izv. Akad. Nauk SSSR., Ser. Geofiz. **1956**, 360.
[11] HEAPS, H. S.: Trans. Roy. Soc. Canada **47**, Sec. 4, 17 (1953).

equal to that of granite, and that it is floating upon a slightly denser (by 1 g/cm³) substratum. Assumption of plastic yielding instead of fracturing would increase this effect. There is no doubt, therefore, that the assumption of an ice cap causing downpunching, and hence recovery after the removal of the ice, is, to say the least, reasonable.

8.52. The Haskell Theory. If we assume that the idea of isostatic adjustment is the correct explanation of the Fennoscandian uplift, then it is possible to arrive at an estimate of the viscosity consistent with the observational data, simply by discussing the hydrodynamics of the problem. The most rigorous discussion of this question has been given by HASKELL[1], whose analysis we shall sketch here.

The equations of motion of a viscous fluid (viscosity η, density ϱ) in a gravitational field are [cf. Eq. (3.32–3)]

$$\eta \operatorname{lap} \boldsymbol{V} = \operatorname{grad} p + \varrho \boldsymbol{g}, \tag{8.52–1}$$

$$\operatorname{div} \boldsymbol{V} = 0 \tag{8.52–2}$$

where $\boldsymbol{g}$ is the gravity vector, $\boldsymbol{V}$ the local velocity vector; inertia terms have been neglected. Transforming to cylindrical co-ordinates (r, z, Φ; z is downwards), assuming radial symmetry and setting

$$\bar{p} = p - \varrho g z, \tag{8.52–3}$$

one obtains

$$\frac{1}{r}\frac{\partial}{\partial r}\left(\frac{\partial V_r}{\partial r}\right) - \frac{V_r}{r^2} + \frac{\partial^2 V_r}{\partial z^2} = \frac{1}{\eta}\frac{\partial \bar{p}}{\partial r}, \tag{8.52–4a}$$

$$\frac{1}{r}\frac{\partial}{\partial r}\left(r\frac{\partial V_z}{\partial r}\right) + \frac{\partial^2 V_z}{\partial r^2} = \frac{1}{\eta}\frac{\partial \bar{p}}{\partial z}, \tag{8.52–4b}$$

$$\frac{1}{r}\frac{\partial}{\partial r}(r V_r) + \frac{\partial V_z}{\partial z} = 0. \tag{8.52–4c}$$

The stress components of interest are

$$\tau_{zz} = -(\bar{p} + \varrho g z) + 2\eta\, \partial V_z/\partial z, \tag{8.52–5a}$$

$$\tau_{rz} = \eta(\partial V_r/\partial z + \partial V_z/\partial r). \tag{8.52–5b}$$

The boundary conditions require that on the surface τ_{rz} shall be zero and τ_{zz} shall be equal to the applied stress, and at infinity the stresses and velocities shall vanish. Let the equation of the surface be

$$z = \zeta(r, t) \tag{8.52–6}$$

and take as the undisturbed surface (i.e. that before there was any ice)

$$z = 0. \tag{8.52–7}$$

[1] HASKELL, N. A.: Physics 6, 265 (1935).

If we assume that ζ remains small in comparison with other distances entering, we may replace the value $\partial V_z/\partial z$ at $z=\zeta$ by its value at $z=0$;— and similarly with all the other quantities involved, except $\varrho g z$. Since we are interested in the case of load recovery after the ice has melted, the external applied pressure is zero and the boundary conditions become:

$$\bar{p}(r,0,t) + \varrho g \zeta(r,t) - 2\eta \frac{\partial V_z}{\partial z}(r,0,t) = 0, \qquad (8.52\text{–}8\text{a})$$

$$\left[\frac{\partial V_r}{\partial z} + \frac{\partial V_z}{\partial r}\right]_{z=0} = 0, \qquad (8.52\text{–}8\text{b})$$

$$\frac{\partial \zeta}{\partial t} = V_z(r,0,t). \qquad (8.52\text{–}8\text{c})$$

Setting $V_r = R_1(r)\, Z_1(z)$, $V_z = R_2(r)\, Z_2(z)$, $\bar{p} = R_3(r)\, Z_3(z)$ in Eq. (8.52–4) and separating the variables, one ends up with the following differential equations:

$$d/r\,dr\,(r\,dR_1/dr) - R_1/r^2 + \lambda^2 R_1 = 0, \qquad (8.52\text{–}9\text{a})$$

$$d/r\,dr\,(r\,dR_2/dr) + \lambda^2 R_2 = 0, \qquad (8.52\text{–}9\text{b})$$

$$R_3 = \text{const}\, R_2. \qquad (8.52\text{–}9\text{c})$$

The solutions of these equations are Bessel functions, λ being eigenvalue of the system. The solutions are

$$R_1 = J_1(\lambda r); \quad R_2 = R_3 = J_0(\lambda r), \qquad (8.52\text{–}10)$$

where the constant factor has been included in the Z factor. The equations for the latter are

$$d^2 Z_1/dz^2 - \lambda^2 Z_1 = -\lambda Z_3/\eta, \qquad (8.52\text{–}11\text{a})$$

$$d^2 Z_2/dz^2 - \lambda^2 Z_2 = dZ_3/\eta\, dz, \qquad (8.52\text{–}11\text{b})$$

$$\lambda Z_1 + dZ_2/dz = 0. \qquad (8.52\text{–}11\text{c})$$

Eliminating Z_1 and Z_3, one obtains the following fourth-order equation for Z_2

$$d^4 Z_2/dz^4 - 2\lambda^2\, d^2 Z_2/dz^2 + \lambda^4 Z_2 = 0. \qquad (8.52\text{–}12)$$

The solutions of this equation are $\exp(\pm\lambda z)$, $z\exp(\pm\lambda z)$ of which only those with negative exponents are appropriate to the present problem. The solutions of the set (8.52–11) are then

$$Z_2 = \exp(-\lambda z)\,(A + Bz), \qquad (8.52\text{–}13\text{a})$$

$$Z_1 = \exp(-\lambda z)\,(A - B/\lambda + Bz), \qquad (8.52\text{–}13\text{b})$$

$$Z_3 = 2\eta B \exp(-\lambda z). \qquad (8.52\text{–}13\text{c})$$

In order to satisfy the boundary condition, one has $B=\lambda A$. One must now satisfy (8.52–9) with functions of the form

$$V_r = z\int_0^\infty A(\lambda)\exp(-\lambda z)\, J_1(\lambda r)\, d\lambda, \qquad (8.52\text{–}14a)$$

$$V_z = \int_0^\infty A(\lambda)\exp(-\lambda z)(1+\lambda z)\, d\lambda\, J_0(\lambda r), \qquad (8.52\text{–}14b)$$

$$\bar{p} = 2\eta\int_0^\infty A(\lambda)\exp(-\lambda z)\, J_0(\lambda r)\,\lambda\, d\lambda. \qquad (8.52\text{–}14c)$$

From (8.52–14b) one has $dV_z/dz=0$ at $z=0$, hence the first boundary condition becomes:

$$2\eta\int_0^\infty A(\lambda)\, J_0(\lambda r)\,\lambda\, d\lambda + \varrho g\zeta = 0. \qquad (8.52\text{–}15)$$

The quantity A will evidently have to be a function of time in order to satisfy this equation, hence we may differentiate with respect to t and use the third boundary condition. This yields

$$\int_0^\infty J_0(\lambda r)\left\{2\eta\frac{\partial A}{\partial t} + \varrho g\frac{A}{\lambda}\right\}\lambda\, d\lambda = 0. \qquad (8.52\text{–}16)$$

This yields the following differential equation for A

$$2\eta\frac{\partial A}{\partial t} + \varrho g\frac{A}{\lambda} = 0. \qquad (8.52\text{–}17)$$

The solution is

$$A = K(\lambda)\exp\left(-\frac{\varrho g t}{2\eta\lambda}\right) \qquad (8.52\text{–}18)$$

where $K(\lambda)$ must be determined from the initial conditions, *viz.* either from the initial velocity or from the initial configuration of the surface. From (8.52–14b) one obtains

$$V_z(r,0,0) = \int_0^\infty K(\lambda)\, J_0(\lambda r)\, d\lambda. \qquad (8.52\text{–}19)$$

Hence by inversion

$$K(\lambda) = \lambda\int_0^\infty V_z(r,0,0)\, J_0(\lambda r)\, r\, dr. \qquad (8.52\text{–}20)$$

One also has

$$\left.\begin{aligned} \zeta &= \zeta(r,0) + \int_0^t V_z(r,0,t)\, dt \\ &= \zeta(r,0) + \frac{2\eta}{\varrho g}\int_0^\infty K(\lambda)\left[1-\exp\left(-\frac{\varrho g t}{2\eta\lambda}\right)\right] J_0(\lambda r)\,\lambda\, d\lambda. \end{aligned}\right\} \qquad (8.52\text{–}21)$$

As t becomes infinite, ζ must approach zero, therefore

$$\zeta(r,0) = -\frac{2\eta}{\varrho g}\int_0^\infty K(\lambda)\, J_0(\lambda r)\, \lambda\, d\lambda, \tag{8.52–22}$$

or, upon inversion

$$K(\lambda) = -\frac{\varrho g}{2\eta}\int_0^\infty \zeta(r,0)\, J_0(\lambda r)\, r\, dr. \tag{8.52–23}$$

Thus, the subsequent motion is completely determined if one knows either $\zeta(r, 0)$ or $V_z(r, 0, 0)$.

In order to choose an initial condition, it is reasonable to suppose that $-V_z(r, 0, 0)$ will have a maximum at the center and will decrease outward in a way that can be represented with sufficient accuracy by an exponential function:

$$V_z(r,0,0) = -a\exp(-b^2 r^2). \tag{8.52–24}$$

By substituting this into (8.52–20), one has

$$\left.\begin{aligned} K(\lambda) &= -a\lambda\int_0^\infty \exp(-b^2 r^2)\, J_0(\lambda r)\, r\, dr \\ &= -(a\lambda/2b^2)\exp(-\lambda^2/4b^2). \end{aligned}\right\} \tag{8.52–25}$$

By using (8.52–21/22) one obtains:

$$\zeta = \frac{a\eta}{\varrho g b^2}\int_0^\infty \exp\left(-\frac{\lambda^2}{4b^2} - \frac{\varrho g t}{2\eta\lambda}\right) J_0(\lambda r)\, \lambda^2\, d\lambda. \tag{8.52–26}$$

At $r=0$, $t=0$ we have

$$\zeta(0,0) = \frac{a\eta}{\varrho g b^2}\int_0^\infty \exp\left(-\frac{\lambda^2}{4b^2}\right)\lambda^2\, d\lambda = 2\sqrt{\pi}\,\frac{\eta a b}{\varrho g}. \tag{8.52–27}$$

The viscosity is thus given by

$$\eta = \varrho g \zeta(0,0)/[2\sqrt{\pi}\, a b]. \tag{8.52–28}$$

In order to calculate a viscosity from this equation, all that remains to be done is to make a reasonable estimate of the quantities a, b, ζ. A probable assumption is[1]:

$$a/\zeta = k \tag{8.52–29}$$

with $1/k$ being the duration-time of the isostatic adjustment. Setting

$$1/k = 10000 \text{ years} = 2.8\times 10^{11}\,\text{sec}, \tag{8.52–30}$$

$$1/b = 750\text{ km} = 7.5\times 10^{7}\,\text{cm} \tag{8.52–31}$$

[1] NISKANEN, E.: Ann. Acad. Sci. Fenn., Ser. A, **53**, 10 (1939).

yields with $\varrho = 3$

$$\eta = 2 \times 10^{22} \text{ cgs}. \qquad (8.52\text{–}32)$$

This is the value for the viscosity listed as coming from the uplift of Fennoscandia when the rheology of the Earth was under discussion.

8.53. Postglacial Uplift Interpreted as a Kelvin Effect. It is also possible to interpret the uplift of Fennoscandia as a Kelvin effect[1], but the physical picture is then somewhat different from that underlying the discussion of Sec. 8.52: The driving force is now the expression of elastic afterworking (and not of buoyancy) after the ice has melted. The relaxation constant in a Kelvin body is equal to

$$T = \eta_K / \mu_K \qquad (8.53\text{–}1)$$

[this follows immediately upon integrating (3.43–1) upon setting $\tau = 0$]. If the relaxation constant is set equal to 10000 years, one has

$$\eta_K / \mu_K = 2.8 \times 10^{11} \text{ cgs}. \qquad (8.53\text{–}2)$$

Strictly speaking, no value for the rigidity can be given, as there is no way of determining it for the "long" time range here under discussion. Nevertheless, if the "short"-time rigidity is substituted, it is interesting to note that one ends up with a similar "formal" value of viscosity as that determined for the Maxwell-type behavior discussed earlier.

Inspecting the two types of possible interpretation of the recovery of areas upon the Earth after the disappearance of an additional load, one is faced with either having to accept that, in two time ranges, the Earth has the *same* type of rheological behavior (Kelvin-type), but with different time constants, or else different types of behavior ("Kelvin" in the "intermediate" range and "Maxwell" in the long range) with correspondingly non-related constants. The latter possibility would appear as more reasonable, especially since it is known that many bodies show increasing tendency towards Maxwell-behavior in long time-intervals.

8.6. Conclusion

Looking back over the way we have come in our discussion of geodynamics, we note that throughout, we were faced with the condition that no coherent theory is known from which all the details would follow. In contrast to most physical theories where a fundamental equation is postulated whose consequences more or less agree and thereby "explain" the facts of nature which they concern, no such fundamental equation exists in geodynamics. The only approach to the problem is

[1] SCHEIDEGGER, A. E.: Canad. J. Phys. **35**, 383 (1957).

therefore by induction. This has the effect that the causes of small-scale surface phenomena, such as faults and folds, are much better understood than the causes of fundamental processes such as orogenesis.

The only way open to the investigator is thus to make various guesses as to the possible cause of geologic phenomena, and to test their reasonableness on the consequences which they entail. In spite of over a hundred years of research in this fashion, this procedure has not yet led to entirely satisfactory results. One can only hope that some day it will be possible to obtain some *really* pertinent information regarding the rheological state of the pertinent layers of the Earth. This then would automatically sift what is reasonable from what is unreasonable and eliminate those speculations that border on the realm of the supernatural.

Author Index[1]

[1] Slavonic names have been transliterated according to the Cambridge system; this is the system used, for instance, by *Physics Abstracts*.

Subject Index